P9-DFS-968

14 95 A

Third Edition

GENERAL STATISTICS

ADDISON-WESLEY
PUBLISHING COMPANY

Reading, Massachusetts
Menlo Park, California
London • Amsterdam
Don Mills, Ontario • Sydney

Audrey Haber
University of California,
Los Angeles

Richard P. Runyon
C. W. Post College
of Long Island University

Third Edition

GENERAL STATISTICS

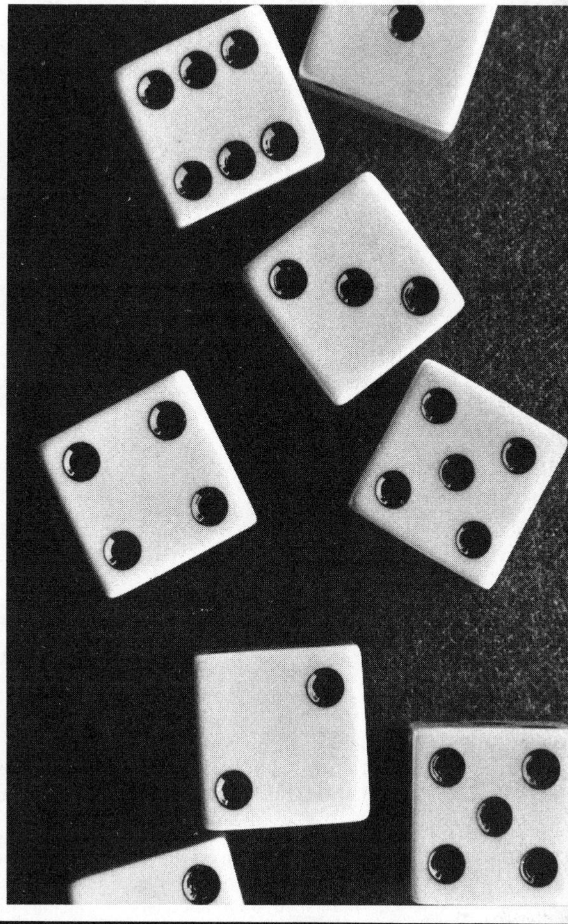

This book is in the
ADDISON-WESLEY SERIES IN STATISTICS

Copyright © 1977, 1973, 1969, by Addison-Wesley Publishing Company, Inc. Philippines copyright 1977, 1973, 1969, by Addison-Wesley Publishing Company, Inc.

All rights reserved. No part of this publication may be reproduced, stored in a retrieval system, or transmitted, in any form or by any means, electronic, mechanical, photocopying, recording, or otherwise, without the prior written permission of the publisher. Printed in the United States of America. Published simultaneously in Canada. Library of Congress Catalog Card No. 76-23985.

ISBN 0-201-02729-1
BCDEFGHIJK-MA-7987

Preface to Third Edition

Thanks to the response of many users of *General Statistics,* we received the encouragement to undertake a third edition of this text.

Specifically, we have attempted to respond to four broad suggestions:

1. We have provided a greater variety of examples and exercises drawn directly from the literature. The student is thereby given the opportunity to carry out all the steps in the decision-making process from the handling and processing of raw data to the drawing of appropriate inferences from these data. To illustrate, in an age that appears destined for continued concern about energy and resource management, we have introduced actual data obtained from the Environmental Protection Agency. These data involve the estimated miles-per-gallon of 1976 model cars in various weight classes and with varying amounts of engine displacement. Students are able to organize the data so that they can compare the percentile ranks of various model cars, find variables that correlate with the miles-per-gallon data, and then draw inferences concerning factors related to automobile mileage performance.

2. The chapter on probability has been almost completely rewritten so that it is an integral part of the text and provides a meaningful liaison between descriptive and inferential statistics. Rather than illustrating probability theory with tired and largely irrelevant examples involving selections from a deck of cards or tosses of a die, we have zeroed in on the business of inferential statistics as it relates to the many disciplines that employ statistical methods. Sampling experiments with which the student is completely familiar by the time he reaches the chapter on probability are employed to introduce the basic concept of probability theory.

3. We have adopted single definitions of the variance and standard deviation of a sample and have applied these definitions consistently throughout the text.

4. While increasing both the number and scope of the end-of-chapter exercises, we have vastly expanded the Answers to Selected Exercises in the end matter of the text. Indeed, answers to most of the exercises can be found.

We are, of course, indebted to many people for assistance in the preparation of this third edition. We would particularly like to express our deepest appreciation to users of the text who have taken the time and effort to make constructive criticisms. Finally, our deepest appreciation to Nancy Bromley for her cheerful patience and perseverance in typing the final manuscript.

Los Angeles, California A. H.
Tucson, Arizona R. P. R.

Preface to Second Edition

One of the most pleasant aspects of writing a textbook is that it brings the authors into contact with so many colleagues throughout the world. We were surprised at the number of people who took time out of busy schedules to offer comments and constructive criticisms of our first edition. We are most appreciative of their thoughtful comments and have given them serious consideration when we prepared this second edition. While we have introduced a number of changes in this edition, we have attempted to retain many of the features that were so well received in the previous edition.

In rewriting the text to include new material, we have made many improvements in the presentation of topics previously covered. We have both consolidated and expanded. We have attempted to present certain topics in a more concise and economical fashion. In other instances, however, we have elaborated on our discussion in an attempt to help clarify some of the more difficult concepts. We have added illustrative material, including several cartoons, a change which is consonant with our view that statistics is an interesting and exciting field which touches on many vital aspects of life.

It is obviously impossible to cite the names of all the people who have directly or indirectly contributed to the present edition. We have attempted to acknowledge their contributions through personal correspondence. Also, we should like to express our appreciation to Lois Runyon and Jerry Jassenoff for their patience and understanding during the preparation of this manuscript. Finally, we wish to express our appreciation to Miss Pamela Zeller, a graduate student at C. W. Post College, who prepared many of the exercises accompanying each chapter, and to Linda Moody for her invaluable assistance in preparing the index.

Los Angeles, California A. H.
Greenvale, L.I., New York R. P. R.
October 1972

Preface to First Edition

Many of us have grown in the tradition which assumes that statistics courses must be dull and uninteresting gauntlets that test the stamina, tenacity, and frustration tolerance of the student.

We take the view that statistics can be an interesting and exciting field which touches on many vital aspects of life. Accordingly, we have attempted to produce a book which is both readable and instructive. Particular efforts have been made to include many real-life examples and problems from newspapers and almanacs to demonstrate that statistics is not a concatenation of abstract and esoteric formulas which have little relevance to practical applications. Included in the text are many side discussions which were selected with these considerations in mind. For example, many interesting, humorous, and relevant excerpts have been borrowed from Daryl Huff's excellent book *How To Lie With Statistics*.

In an attempt to reflect the latest statistical advances and technological changes, a deliberate effort has been made to eliminate all materials which, from our experience, are rarely employed on the contemporary scene. Thus coded score methods, which were vital in the days of large sample statistics and before the advent of the high-speed calculators, will not be found in the text. While some instructors, in a moment of acute nostalgia, may lament the loss of this relic from the "horse and buggy" days of statistics, we feel that instruction in the coded score methods is wasteful of time, complicates the life of the harried student who comes to believe that knowledge of formulas is the essence of statistics, and advances no fundamental insights into the nature of statistical analysis. The use of the correlational charts has suffered a similar death for the same reason. To critics, we may only point out that we have never, in our research, had occasion to employ coded score methods or the correlational chart, and we have never met anyone who has!

On the other hand, we have not hesitated to introduce new statistical techniques which we feel represent an advance over prior methods. Thus the Sandler *A*-statistic, which is algebraically equivalent to the Student *t*-ratio with correlated samples, has been introduced because it drastically reduces the computational procedures required to arrive at a statistical decision.

Furthermore, we have attempted to preserve the distinction between population parameters and sample statistics in the testing of hypotheses. Since our approach is consistent throughout the text, we hope to eliminate some of the confusion that sometimes arises in this context.

A word about the organization of the text. Many recent statistics textbooks have relegated descriptive statistics to a place of secondary importance. While it is not our contention that descriptive statistical techniques represent anything beyond the fundamentals of statistical analysis, we do feel that a mastery of these techniques is prerequisite to the understanding and application of the concepts and procedures involved in inferential statistics. We have attempted, through the text, to demonstrate the continuity of these two branches of statistics.

Secondly, the statistical tables in Appendix IV have been carefully prepared to minimize the student's difficulties in his use of them. For example, most tables are preceded by a brief description of the procedures involved in their application. In addition, wherever appropriate, critical values for rejecting the null hypothesis are shown in terms of one- and two-tailed values at various levels of significance. Finally, some tables have been reduced in complexity, making it possible to locate the relevant information in a shorter period of time with less chance of error.

The exercises at the end of each chapter are an extremely important and integral part of the text since, in addition to illustrating fundamental relationships, they require the student to formulate many significant statistical concepts himself.

Finally, the book has been organized so that the first fifteen chapters constitute, in our opinion, a thorough introduction to the fundamentals of descriptive and inferential statistics. For the instructor desiring more advanced statistical procedures, we have included, in the last three chapters of the text, such topics as analysis of variance and several of the more widely employed nonparametric tests of significance.

In brief, then, we have attempted to produce a textbook for a one-semester introductory course in statistics for students representing diverse fields and interests. Our aim has been to provide a background in statistical techniques which are applicable to a wide variety of fields.

It is our hope that the student will gain an appreciation of the usefulness of the statistical method in his professional field, that he will have a good understanding of the assumptions and logic underlying the application of the statistical tools, that he will be able to select the appropriate statistical technique and perform the necessary computations, and, finally, that he will know how to interpret and understand the results of his efforts.

We are grateful to many people who have contributed to this book. We are deeply indebted to Millicent Cowit and Ruth DeMarco for their painstaking efforts in typing the manuscript.

Finally, we wish to express our gratitude to the many authors and publishers who have permitted us to adapt or reproduce material originally published by them. We have cited each source wherever it appears. We are indebted to the Literary Executor of the late Sir Ronald A. Fisher, F. R. S., Cambridge, to Dr. Frank Yates, F. R. S., Rothamsted, and to Messrs. Oliver and Boyd, Ltd., Edinburgh, for permission to reprint and adapt tables from their book, *Statistical Tables for Biological, Agricultural, and Medical Research*.

Greenvale, L.I., N.Y. A. H.
January 1969 R. P. R.

Contents

SECTION I—DESCRIPTIVE STATISTICS

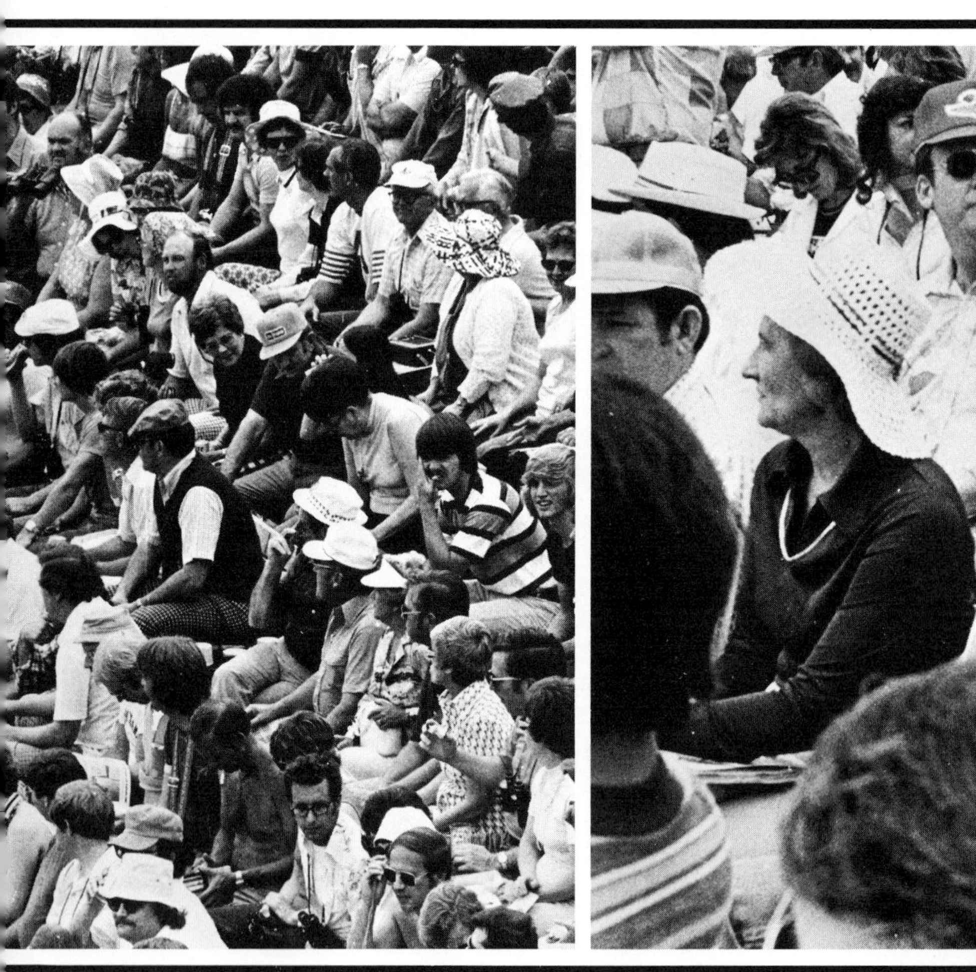

Section I

Descriptive Statistics

1 The Definition of Statistical Analysis

BOX 1.1

Each year The Environmental Protection Agency publishes a gas mileage guide for new car buyers based on tests conducted under strict laboratory conditions. Each car is subjected to exactly the same test, permitting comparisons among various makes and models. Presented below are the mileage data for 124 of the 1976 model cars and trucks.

Most potential car buyers utilize these data only for purposes of checking the mileage capabilities of vehicles in which they are interested. However, there is a wealth of information to be derived from these data. For example, we may ask such questions as: What is the average city and/or highway mileage for all cars listed? Where does a specific car stand in relation to all other cars? Is there any relationship between the car's weight and its mileage capability? Would an economy-minded buyer be well advised to consider a manual transmission over an automatic? In the chapters that follow we will refer back to these data and show how answers can be obtained to these and other questions.

				Mileage	
Manufacturer	Model	Automatic/ manual	Weight	City	Highway
American Motors	Gremlin	M	2831	16	26
	Pacer	M	3180	16	26
	Hornet	M	2975	16	26
	Hornet Wagon	M	3106	16	26
	Matador	A	3643	11	15
	Matador Wagon	A	4069	11	15
Audi	Fox	M	2086	24	36
	Fox Station Wagon	M	2250	24	36
	100	M	2613	18	26

Austin	M G Midget	M	1827	23	30
Buick	Skylark	A	3425	16	22
	Skylark	M	3004	17	26
	Century Regal	A	3837	16	22
	LeSabre	A	4456	12	18
	Estate Wagon	A	5279	11	15
	Electra	A	4880	12	18
	Riviera	A	4677	11	16
Cadillac	Seville	A	4232	13	19
	Cadillac	A	5025	11	15
	Eldorado	A	5085	11	15
Chevrolet	Chevette	M	1924	22	33
	Vega	M	2443	16	29
	Vega Kammback	M	2534	19	30
	Monza	M	2688	19	30
	Nova	A	3188	15	21
	Camaro	A	3421	15	21
	Chevelle	A	3650	15	21
	Malibu Wagon	A	4238	11	16
	Chevrolet	A	4175	11	16
	Chevrolet Wagon	A	4912	10	15
	Monte Carlo	A	3709	13	18
	Corvette	A	3445	13	16
Chrysler	Cordoba	A	3990	12	16
	Chrysler	A	4100	11	17
	Chrysler Wagon	A	5100	10	14
Datsun	B-210	M	2250	29	41
	710	M	2500	22	33
	710 Wagon	M	2600	21	30
	610	M	2500	21	30
	610 Wagon	M	2600	21	30
Dodge	Dart	M	3115	16	23
	Aspen	A	3250	13	17
	Aspen Wagon	A	3650	13	17
	Coronet/Charger	A	3965	12	16
	Coronet Wagon	A	4455	11	17
	Monaco	A	4420	11	17
	Monaco Wagon	A	5035	10	14
Fiat	128	M	1950	22	35
	128 Wagon	M	2020	21	31
	131 Mirifiori	M	2460	19	30
	131 Estate Wagon	M	2510	17	30
	124 Sport	M	2370	19	31
	XI/9	M	2085	21	31

Lincoln/ Mercury	Comet	M	3067	16	23
	Monarch	M	3417	15	20
	Montego	A	4263	12	17
	Montego Wagon	A	4327	12	17
	Cougar	A	4212	12	17
	Mercury	A	4637	11	16
	Mercury Wagon	A	4888	11	16
	Lincoln Continental	A	5248	11	15
	Continental Mark IV	A	5264	11	15
Oldsmobile	Omega	A	3341	15	21
	Starfire	M	2985	17	26
	Cutlass	A	3751	15	21
	Cutlass Wagon	A	4449	12	16
	Delta 88	A	4383	12	16
	Custom Cruiser Wagon	A	5111	11	16
	Olds 98	A	4657	10	15
	Toronado	A	4781	10	15
Peugot	504 Diesel	M	3000	27	35
	504 Diesel Wagon	M	3230	27	35
Plymouth	Valiant/Duster	M	3142	16	23
	Volare	A	3285	13	17
	Volare Wagon	A	3650	13	17
	Fury	A	3830	12	16
	Fury Wagon	A	4284	11	17
	Gran Fury	A	4405	11	17
	Gran Fury Wagon	A	5000	10	14
Pontiac	Astre	M	2439	19	30
	Astre Safari Wagon	M	2552	19	30
	Sunbird	M	2662	19	30
	Ventura	A	3362	15	21
	Firebird	A	3389	15	21
	Lemans	A	3660	15	21
	Lemans Safari Wagon	A	4407	13	17
	Pontiac	A	4266	13	17
	Pontiac Safari Wagon	A	4961	11	16
	Gran Prix	A	4038	13	18
Porsche	911S	M	2557	16	23
	912E	M	2557	19	32
Rolls Royce	Silver Shadow	A	4700	10	13
	Corniche	A	4850	10	13
Subaru	Subaru	M	1985	24	34
	Subaru Wagon	M	2145	22	32
Toyota	Corolla	M	2225	20	35
	Corolla Wagon	M	2325	20	35
	Corona	M	2576	19	32
	Corona Wagon	M	2645	19	32

	Celica	M	2545	19	32
	Corona Mark II	M	2845	16	22
	Corona Mark II Wagon	M	2905	16	22
Triumph	Spitfire	M	1828	23	30
Volkswagen	Beetle	M	2110	26	38
	Dasher	M	2040	24	36
	Dasher Wagon	M	2481	24	36
Volvo	240	M	3000	18	27
	245 Wagon	M	3180	18	28
	260	M	3330	15	25
	265 Wagon	M	3370	15	25
Chevrolet	Pickup	A	6400	12	16
	Van	A	6400	12	16
	El Camino	A	3857	15	21
Datsun	Pickup	M	2380	20	31
Ford	Pickup	M	3587	16	23
	Van (Econoline Club Wagon)	M	3873	15	23
	Bronco	A	3565	12	16
	Ranchero	A	4095	12	17
GMC	Pickup	A	3645	12	16
	Van	A	3594	12	16
	Sprint	A	3800	15	21
Jeep	Jeep	M	2945	13	18
Toyota	Hilux	M	3864	19	29
Volkswagen	Bus	M	3043	18	28

1.1 INTRODUCTION

If we were to ask the "man on the street" what statistics means to him, we would, in all likelihood, obtain some answers such as, "Statistics is 'hocus pocus' with numbers. By manipulating these numbers according to certain secret and well-guarded rules, we can prove anything we have a mind to." Or, "Statistics is the refuge of the uninformed. When we can't prove our point through the use of sound reasoning, we fall back upon statistical 'mumbo-jumbo' to confuse and demoralize our opponents." Or "Statistics is merely a collection of facts. Statisticians concern themselves with such vital issues as the number of bath tubs in the State of Kentucky in 1929, the number of men who grow mustaches to irritate their spouses, and the number of wives who retaliate by growing beards."

It is true that all these activities, and more, are widely attributed to the field of statistics. It is *not* true, however, that statisticians engage in them. What, then, is statistics all about? Although it would be virtually impossible to obtain a general consensus on the definition of statistics, it is possible to make a distinction between two definitions of statistics.

1. Statistics is commonly regarded as a *collection* of numerical facts which are expressed in terms of summarizing statements and which have been collected either from several observations or from other numerical data. From this perspective, statistics constitutes a collection of statements such as, "The average I.Q. of eighth-grade children is . . . ," or "Seven out of ten people prefer Brand X to Brand Y," or "The New York Yankees hit 25 home runs over a two week span during . . ."

2. Statistics may also be regarded as a *method* of dealing with data. This definition stresses the view that statistics is a tool concerned with the collection, organization, and analysis of numerical facts or observations.

The second definition constitutes the subject matter of this text.

A distinction may be made between the two functions of the statistical method: descriptive statistical techniques and inferential or inductive statistical techniques.

The major concern of **descriptive statistics** is to present information in a convenient, usable, and understandable form. **Inferential statistics**, on the other hand, is concerned with generalizing this information, or, more specifically, with making inferences about populations which are based upon samples taken from the populations.

In describing the functions of statistics, certain terms have already appeared with which you may or may not be familiar. Before elaborating on the differences between descriptive and **inductive statistics**, it is important to learn the meaning of certain terms which will be employed repeatedly throughout the text.

1.2 DEFINITIONS OF COMMON TERMS USED IN STATISTICS

Variable: a characteristic or phenomenon which may take on different values. Thus weight, I.Q., and sex are **variables** since they will take on different values when different individuals are observed. A variable is contrasted with a constant, the value of which never changes, for example, pi.

Data: numbers or measurements which are collected as a result of observations. They may be head counts (frequency data), as a number of individuals stating a preference for the Republican presidential candidate, or they may be scores, as on a psychological or educational test. Frequency data are also referred to as enumerative or categorical data.

Population or universe: a complete set of individuals, objects, or measurements having some common observable characteristic. Thus, all American citizens of voting age constitute a **population (universe)**.

Parameter: any characteristic of a population which is measurable, e.g., the proportion of registered Democrats among Americans of voting age. In this text we shall follow the practice of employing Greek letters (e.g. μ, σ) to represent population **parameters**.

Sample: a subset of a population or universe.

Statistic: a number resulting from the manipulation of raw data according to certain specified procedures. Commonly, we use a statistic which is calculated from a sample in order to estimate the population parameter, e.g., a sample of Americans of voting age is employed to estimate the proportion of Democrats in the entire population of voters. We shall employ italic letters (e.g. $\bar{X}$, s) to represent sample statistics. More will be said about the fascinating problem of sampling later in the text.

Example: Imagine an industrial firm engaged in the production of hardware for the space industry. Among its products are machine screws which must be maintained within fine tolerances with respect to width. As part of its quality control procedures, a number of screws are selected from the daily output and carefully measured. These screws constitute the **sample**. The variable is the width of the screw. The **data** consist of the measurements of all screws collected in the sample. When the data are manipulated according to certain rules to yield certain summary statements, such as the "average" width of the screws, the resulting numerical value constitutes a statistic. The population to which we are interested in generalizing is the entire daily output of the plant. The "average" width of all the screws produced in a day constitutes a parameter (see Fig. 1.1). Note that it is highly unlikely that the parameter will ever be known, for to determine it would require the measurement of every machine screw produced during the day. Since this is usually unfeasible for economic and other reasons, it is rare that an exhaustive study of populations is undertaken. Consequently, parameters are rarely known; but, as we shall see, they are commonly estimated from sample statistics (see Fig. 1.2).

Let us return to the two functions of statistical analysis for a closer look.

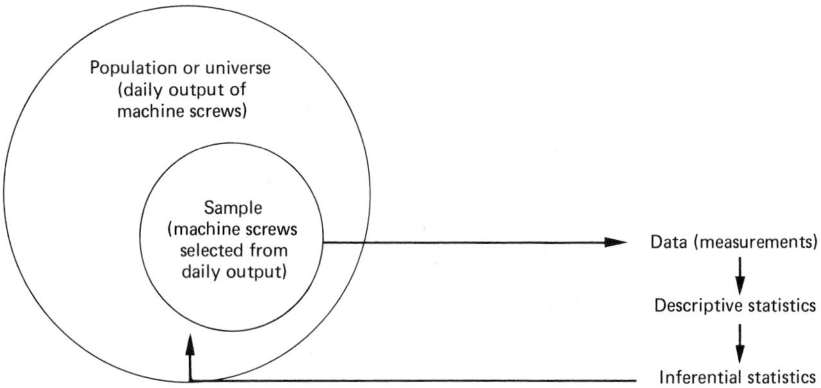

Fig. 1.1 A *sample* is selected from some *population* or *universe*. *Data* are collected and summarized, employing *descriptive statistics*. In *inferential statistics*, we attempt to estimate one or more *population parameters* (e.g., the "average" width of the screws in the daily output).

© 1954 United Feature Syndicate, Inc.

Fig. 1.2 What parameter is Lucy trying to estimate from what statistic?

1.3 DESCRIPTIVE STATISTICS

When an investigator conducts a study, he or she characteristically collects a great deal of numerical information or data about the problem at hand. The data may take a variety of forms: frequency data (head counts of voters preferring various political candidates), or scale data (the weights of the contents of a popular breakfast cereal, or the I.Q. scores of a group of college students). In their original form, as collected, these data are usually a confusing hodge-podge of scores, frequency counts, etc. In performing the descriptive function, the statistician formulates rules and procedures for presentation of the data in a more usable and meaningful form. Thus the statistician states rules by which data may be represented graphically. He also formulates rules for calculating various statistics from masses of raw data.

Let us imagine that a guidance counsellor administered a number of measuring instruments (e.g., intelligence tests, personality inventories, aptitude tests) to a group of high school students. What are some things which the counsellor may do with the resulting measurements or scores to fulfill the descriptive functions? He or she may:

1. Rearrange the scores and group them in various ways in order to be able to see at a glance an overall picture of the data (Chapter 3, "Frequency Distributions").

2. Construct tables, graphs, and figures to permit visualization of the results (Chapter 4, "Graphing Techniques").

3. Convert raw scores to other types of scores which are more useful for specific purposes. Thus, the counsellor may convert these scores into either percentile ranks, standard scores, or grades. Other types of conversions will also be described in the text (Chapter 5, "Percentiles", and Chapter 8, "The Standard Deviation and The Normal Distribution").

4. Calculate averages, to learn something about the typical performances of the subjects (Chapter 6, "Measures of Central Tendency").

5. Employing the average as a reference point, describe the dispersion of scores about this central point. Statistics which quantify this dispersion are known as measures of variability or measures of dispersion (Chapter 7, "Measures of dispersion").

6. Obtain a relationship between two different measuring instruments. The statistic for describing the extent of the relationship is referred to as a *correlation coefficient*. Such coefficients are extremely useful to the guidance counsellor. For example, it may be desirable to determine the relationship between intelligence and classroom grades, personality measures and aptitudes, or interests and personality measures. Once these relationships are established, scores obtained from one measuring instrument may be employed to predict performance on another (Chapter 9, "Correlation," and Chapter 10, "Regression and Prediction").

1.4 INFERENTIAL STATISTICS

The investigator's task is not nearly over when the descriptive function is completed. To the contrary, this is often nearer to the beginning than to the end of the task. The reason for this is obvious when we consider that the purpose of the research is often to explore hypotheses of a general nature rather than to simply compare limited samples.

Let us imagine that you are a pharmacologist who is interested in determining the effects of a given drug upon the performance of a task involving psychomotor coordination. Consequently, you set up a study involving two conditions, experimental and control. You administer the drug to the experimental subjects at specified time periods before they undertake the criterion task. To rule out "placebo effects," you administer a pill containing inert ingredients to the control subjects. After all subjects have been tested, you perform your descriptive function. You find that "on the average" the experimental subjects did not perform as well as the controls. In other words, the arithmetic mean of the experimental group was lower than that of the control group. You then ask the question, "Can we conclude that the drug produced the difference between the two groups?" Or, more generally, "Can we assert that the drug has an adverse effect upon the criterion task under investigation?" To answer these questions, it is not sufficient to rely solely upon descriptive statistics.

"After all," you reason, "even if the drug had *no effect*, it is highly improbable that the two group means would have been *identical*. *Some* difference would have been observed." The operation of uncontrolled variables (sometimes referred to rather imprecisely as "chance factors") is certain to produce some disparity between the group means. The critical question, from the point of view of inferential statistics, becomes: Is the difference great enough to rule out uncontrolled variation in the experiment as a sufficient explanation? Stated another way, if we were to repeat the experiment, would we be able to predict with confidence that the same differences (i.e., one mean greater than another) would systematically occur?

As soon as we raise these questions, we move into the fascinating area of statistical analysis which is known as inductive or inferential statistics. As you will see, much of the present text is devoted to procedures which the researcher employs to arrive at conclusions extending beyond the sample statistics themselves.

1.5 LYING WITH STATISTICS

A common misconception held by laypersons is that statistics is merely a rather sophisticated method for fabricating lies or falsifying our descriptions of reality. The authors do not deny that some unscrupulous individuals employ statistics for just such purposes. However, such uses of statistics are anathema to the person who is dedicated to the establishment of truth. From time to time, references will be made to various techniques which are used for lying with statistics. However, the purpose is not to instruct you in these techniques but to make you aware of the various misuses of statistical analyses so that you do not inadvertently "tell a lie," and so that you may be aware when others do (Fig. 1.3).

Fig. 1.3 How to lie with statistics—Charley Brown style.

1.6 A WORD TO THE STUDENT

The study of statistics need not and should not become a series of progressive exercises in calculated tedium. If it is approached with the proper frame of mind, statistics can be one of the most exciting fields of study; it has applications in virtually all areas of human endeavor and cuts across countless fields of study. H. G. Wells, the 19th century prophet, remarked, "Statistical thinking will one day be as necessary for efficient citizenship as the ability to read and write." Keep this thought constantly in mind. The course will be much more interesting and profitable to you if you develop the habit of "thinking statistically." Constantly attempt to apply statistical concepts to all daily activities, no matter how routine. When you are stopped at an intersection which you cross frequently, note the time the traffic light remains red. Obtain some estimate of the length of the green cycle. If it is red three minutes and green two, you would expect that the chances are three in five that it will be red when you reach the intersection. Start collecting data. Do you find that it is red 60% of the time as expected? If not, why not? Perhaps you have unconsciously made some driving adjustments in order to change the statistical probabilities.

When you see statistical information being exhibited, develop a healthy attitude of skepticism. Ask pertinent questions. When a national magazine sends a shapely reporter to ten different diet doctors and receives an unneeded prescription from each, do not jump to the conclusion that "diet doctors" are frauds. Do not say, "After all, ten out of ten is a rather high proportion" and dismiss further inquiry at this point. Ask how the reporter obtained her sample. Was it at random or is it possible that the doctors were selected on the basis of prior information indicating they were rather careless in their professional practices? Question constantly, but reserve judgment until you have the answers (Fig. 1.4).

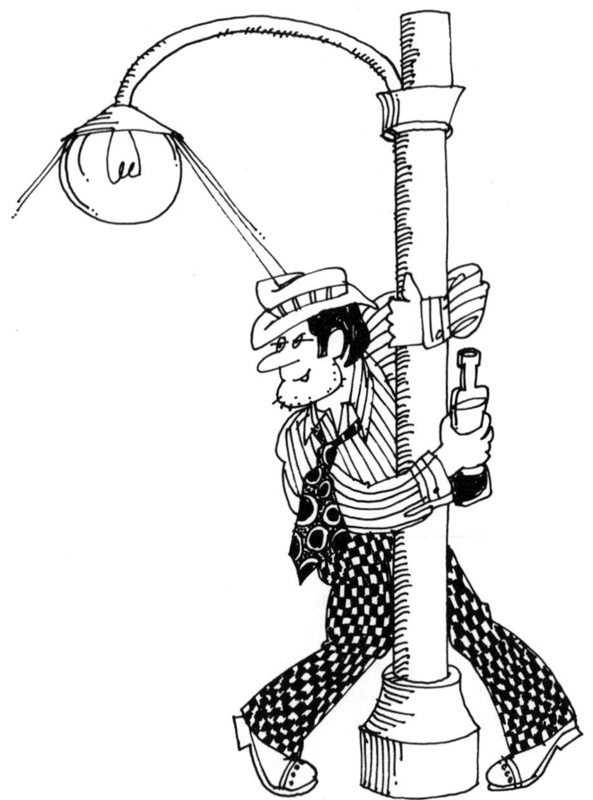

Fig. 1.4 It has been said that many people use statistics in much the same way that an inebriate uses a lamp pole: more for support than for illumination.

Watch commercials on television; read newspaper advertisements. When the pitchman claims, "Dodoes are more effective," ask, "More effective than what? What is the evidence?"

If you make statistical thinking an everyday habit, not only will you find that the study of statistics becomes more interesting, but the world you live in will appear different and, perhaps, more interesting.

CHAPTER SUMMARY

In this chapter, we have distinguished between two definitions of statistical analysis, one stressing statistics as a collection of numerical facts and the other emphasizing statistics as a method or tool concerned with the collection, organization, and analysis of numerical facts. The second definition constitutes the subject matter of this text.

A distinction is made between two functions of the statistical method, descriptive and inferential statistical analyses. The former is concerned with the organization and presentation of data in a convenient, usable, and communicable form. The latter is addressed to the problem of making broader generalizations or inferences from sample data to populations.

A number of terms commonly employed in statistical analysis were defined.

Finally, it was pointed out that statistics is frequently employed for purposes of "telling lies." Such practices are inimical to the goal of establishing a factual basis for our conclusions and statistically-based decisions. However, you should be aware of the techniques for telling statistical lies so that you do not inadvertently "tell one" yourself or fail to recognize one when someone else does.

New terms or concepts that have been introduced in a chapter will be listed at the end of each chapter. Some of these terms will be more precisely defined in other chapters and consequently may appear again.

Terms to Remember:

Statistical method
Descriptive statistics
Inferential or inductive statistics
Variable
Data
Population or universe
Parameter
Sample

EXERCISES

1.1 Which of the following most likely constitutes a *statistic*, a *parameter*, *data*, *inference from data*?

a) A sample of 250 wage earners in Ressag City yielded a *per capita* income of $5650.
b) The proportion of boys in a business arithmetic class is 0.58.
c) Her "statistics" are 36, 25, 35.
d) My tuition payment this year was $2860.
e) The number of people viewing Monday night's television special was 23,500,000.
f) The birth rate in the United States decreased by 5% over the previous month.
g) The birth rate in Perry Township decreased by 3% over the previous month.

1.2 In your own words, describe what you understand the study of statistics to be.

1.3 A friend who is taking a course in accounting asks, "How does statistics differ from accounting? Both work with numbers and both are used in the field of business." What is your answer?

1.4 In the example cited in Section 1.2, let us imagine that, on a given day, the entire output of machine screws was measured and the results were summarized as follows: "The arithmetic average of the widths of all the screws was 0.23 mm." May we assume that we have established "truth" with respect to the widths of the screws for that particular day? What about the measurement problem?

1.5 While listening to the radio or viewing television, note the number of times that statistical data are cited during the commercials. How detailed are the citations? Is there possibly some "lying with statistics"?

1.6 Bring in newspaper examples citing recent survey or poll results. In how many articles is the method of sampling mentioned? Do the articles reveal where the financial support for the surveys came from? Why is this information important? Why is it so commonly not revealed?

1.7 List four populations. *Do not* include any that are defined by geographical boundaries.

1.8 Describe how you would select a random sample of:
   a) 50 registered Democrats in Phoenix, Arizona.
   b) 25 stocks from all those listed on the New York Stock Exchange.
   c) 30 coffee pots from the population of those that are produced in a factory in Billings, Montana.
   d) 20 college students from all those enrolled at Scott University.

1.9 Many populations studied in experimental situations are hypothetical in nature. Give some examples of hypothetical populations that may be dealt with in research.

1.10 Describe how you would select samples from the following populations:
   a) Ages of entering freshmen at Felix University.
   b) Years of education completed by parents of entering freshmen at Felix University.
   c) Amount of money borrowed by undergraduate students for the next school year.
   d) Child-rearing practices among parents of delinquent children.
   e) Productivity levels with new automation techniques.
   f) I.Q. scores (as measured by the Stanford Binet) of all males between the ages of 14 and 17.

1.11 Compare the dictionary definition of *statistics* with the one given in this text.

1.12 Is statistics merely an extension of common sense? Compare the answer you give now with your answer after you complete the course.

1.13 Indicate whether each of the following represents a variable or a constant:
   a) Number of days in the month of August.
   b) Number of shares traded on the New York Stock Exchange on January 5, 1977.
   c) Age of freshmen entering college.
   d) Time it takes to complete an assignment.
   e) Age at which an individual is first eligible to vote.
   f) Scores obtained on a 100-item multiple choice examination.
   g) Maximum score possible on a 100-item multiple choice examination.
   h) Amount of money spent on books per year by a group of 10 students.
   i) Number of workers using private auto as transportation to work.

1.14 Which of the following situations most likely involve the use of descriptive statistics, and which involve inferential statistics?

a) A baseball club statistician wishes to determine the current batting averages of the team members.

b) An instructor employs different teaching methods with each of his two sections of Speech. He then compares the final examination scores obtained by the students in these two sections in order to determine which method is more successful.

c) The Surgeon General studies the relationship between cigarette smoking and heart disease.

d) An economist records population growth within a particular area.

e) The Committee on Smog Prevention studies the relationship between smog level and the number of automobiles in various areas.

f) The committee just mentioned analyzes the effect of decreasing automobile traffic on smog levels.

g) A psychologist studies the effect of new automation techniques on work production.

h) An educator studies the effect of personalized instruction on final examination grades.

i) An insurance company analyzes the effect of no-fault insurance on premium rates.

j) A student studies the distribution of scores on the Graduate Record Examination in order to determine the percentage of people scoring below him.

1.15 Suppose you were interested in comparing the number of three different albums purchased in your town during the last month. You identified four stores that carried the albums and found that store I had sold 4 of album X, 3 of album Y and 1 of album Z; store II had sold 8X, 4Y, and 3Z; store III had sold 8X, 6Y, and 1Z; and store IV had sold 10Z, 7Y, and 0Z. How might you arrange these data to facilitate the communication of information? Identify the population and data. Does the information yield a statistic or inference from data?

1.16 Suppose you were interested in comparing the number of these three albums purchased in the United States during the last month. Would it be necessary to contact every merchant? Identify the population. Would this example show a statistic or inference from data?

1.17 An investigator studied the amount of time it took a person to perform a task on an assembly line. The person completed 420 such tasks a day. Because the investigator could not spend all day measuring the time, she sampled 45 time periods and found the times for completion to be:

1.1	1.0	0.6	1.1	0.9	1.1	0.8	0.9	1.2
1.0	1.5	0.9	1.4	1.0	0.9	1.1	1.0	1.0
0.8	0.9	1.2	0.7	0.6	1.2	0.9	0.8	0.7
1.0	1.2	1.0	1.0	1.1	1.4	0.7	1.1	0.9
0.8	1.1	1.0	1.3	0.5	0.8	1.3	1.3	1.1

Identify the population, variable, data, and sample. Specify how the investigator might arrange the measurements in an orderly and meaningful manner.

Basic Mathematical Concepts

BOX 2.1

A common form of data collection involves the "head-count" technique, i.e., recording the number of individuals, objects, or observations that fall into various categories. To make sense of these data, the head counts are frequently converted to proportions or percentages. However, under certain circumstances, these percentages can be deceptive and misleading.

The following is an excerpt from *Winning with Statistics** which illustrates the problem of the Paradoxical Percentages.

Let me show you some hypothetical data in which the proportion of females accepted in each job category is actually higher than the proportion of males accepted in each of these categories. Nevertheless, when the proportions of males and females accepted for the various positions are combined over all of the categories, it is found, as if by magic, that the females have a much lower overall proportion of acceptances. So if you were to look at the overall figures you'd be led inexorably to the conclusion that there is discrimination against females. However, if you looked at the same evidence category by category you'd be led to precisely the opposite conclusion, namely, that there is discrimination against males. Let's look at these data in the table below.

Take a careful look at the table. It has been purposely constructed so that in every single category, the percentage of women accepted is double the percentage of men accepted. Nevertheless, when all of the applicants among the males are combined, and the number is divided into the number of males accepted, and the same thing is done with the female category, you find that in fact the overall percentage of males accepted (23%) is higher than the overall percentage of females accepted (15%). How is it possible to obtain this sort of numerical sleight-of-hand? Very simple. If you look at the table again you will find that most of the female applications, 600, were for positions in Grades K through 8. Conversely, very few of the male applicants competed for those positions. It turns out that the number of openings for grades K through 8 is the smallest. What you have then is a situation in which an overwhelming majority of the women were applying for "difficult-to-get" positions. On the other hand, most of the men were applying for positions that were more plentiful. When you combine all of these categories in which there

* Runyon, R.P., *Winning with Statistics*. Addison-Wesley, Reading, Mass., 1977.

Hypothetical data showing the number of male and female applicants for various teaching categories, the number accepted, and the percent accepted.

Job category: Teacher	Male No. appl.	Male no. accp.	Male % accp.	Female No. appl.	Female no. accp.	Female % accp.
Grades 13–14 (2-yr coll)	150	30	20	40	16	40
Grades 11–12	200	70	35	50	35	70
Grades 9–10	100	15	15	50	15	30
Grades K–8	50	2	4	600	48	8
Total	500	117	23	740	114	15

are different numbers of applicants for different positions, the lower percentage of individuals of females in the K–8 group is devastating to the female teachers. Why? Better than 80% of the female teachers applied for the very positions where the number of openings was extremely low in relation to the number of applicants.

There is a moral here. Better to apply for a position where few are called but many (relatively speaking) are chosen than to seek entrance where many are called but few are chosen.

2.1 INTRODUCTION

"I'm not much good in math. How can I possibly pass statistics?" The authors have heard these words pass through the lips of countless undergraduate students. For many, this is probably a concern which legitimately stems from prior discouraging experiences with mathematics. A brief glance through the pages of this text may only serve to exacerbate this anxiety, since many of the formulas appear quite imposing to the novice and may seem impossible to master. Therefore it is most important to set the record straight right at the beginning of the course.

You do not have to be a mathematical genius to master the statistical principles enumerated in this text. The amount of mathematical sophistication necessary for a firm grasp of the fundamentals of statistics is often exaggerated. As a matter of actual fact, statistics requires a good deal of arithmetic computation, sound logic, and a willingness to stay with a point until it is mastered. To paraphrase Carlyle, success in statistics is an infinite capacity for taking pains. Beyond these modest requirements, little is needed but the mastery of several algebraic and arithmetic procedures which most students learned early in their high school careers. In this chapter, we review the grammar of mathematical notation, discuss several types of numerical scales, and adopt certain conventions for the rounding of numbers.

For the student who wishes to brush up on basic mathematics, Appendix I contains a review of all the math necessary to master this text.

2.2 THE GRAMMAR OF MATHEMATICAL NOTATION

Throughout the textbook, we shall be learning new mathematical symbols. For the most part, we shall define these symbols when they first appear. However, there are three notations which shall appear with such great regularity that their separate treatment at this time is justified. These notations are $\sum$ (pronounced sigma), X, and N. However, while defining these symbols and showing their use, let's also review the grammar of mathematical notation.

It is not surprising to learn that many students become so involved in the forest of mathematical symbols, formulas, and operations, that they fail to realize that mathematics has a form of grammar which closely parallels the spoken language. Thus, mathematics has its nouns, adjectives, verbs, and adverbs.

Mathematical nouns. In mathematics, we commonly use symbols to stand for quantities. The notation we shall employ most commonly in statistics to represent quantity (or a score) is X, although we shall occasionally employ Y. In addition, X and Y are employed to identify variables; for example, if weight and height were two variables in a study, X might be used to represent weight and Y to represent height. Another frequently used "noun" is the symbol N which represents the number of scores or quantities with which we are dealing. Thus if we have ten quantities,

$$N = 10.$$

Mathematical adjectives. When we want to modify a mathematical noun, to identify it more precisely, we commonly employ subscripts. Thus if we have a series of scores or quantities, we may represent them as X_1, X_2, X_3, X_4, etc. We shall also frequently encounter X_i, in which the subscript may take on any value that we desire.

Mathematical verbs. Notations which direct the reader to do something have the same characteristics as verbs in the spoken language. One of the most important "verbs" is the symbol already alluded to as $\sum$. This notation directs us to sum all quantities or scores following the symbol. Thus,

$$\sum(X_1, X_2, X_3, X_4, X_5)$$

indicates that we should add together all these quantities from X_1 through X_5. Other "verbs" we shall encounter frequently are $\sqrt{\ }$, directing us to find the square root, and exponents (X^a), which tell us to raise a quantity to the indicated power. In mathematics, mathematical verbs are commonly referred to as operators.

Mathematical adverbs. These are notations which, as in spoken language, modify the verbs. We shall frequently find that the summation signs are modified by adverbial notations. Let us imagine that we want to indicate that the following quantities are to be added:

$$X_1 + X_2 + X_3 + X_4 + X_5 + \cdots + X_N.$$

Symbolically, we would represent these operations as follows:

$$\sum_{i=1}^{N} X_i.$$

The notations above and below the summation sign indicate that i takes on the successive values from 1, 2, 3, 4, 5 up to N. Stated verbally, the notation reads: We should sum all quantities of X starting with $i = 1$ (that is, X_1) and proceeding through to $i = N$ (that is, X_N).

Sometimes this form of notation may direct us to add only selected quantities; thus

$$\sum_{i=2}^{5} X_i = X_2 + X_3 + X_4 + X_5.$$

2.3 SUMMATION RULES

The summation sign is one of the most frequently occurring operators in statistics. Let us summarize a few of the rules governing the use of the summation sign.

Imagine a sample in which $N = 3$ and $X_1 = 3$, $X_2 = 4$, and $X_3 = 6$. The sum of the three values of the variable may be shown by

$$\begin{aligned}\sum_{i=1}^{N} X_i &= X_1 + X_2 + X_3 \\ &= 3 + 4 + 6.\end{aligned}$$

Let a be a constant. To show the sum of the values of a variable when a constant has been added to each,

$$\begin{aligned}\sum_{i=1}^{N} (X_i + a) &= (3 + a) + (4 + a) + (6 + a) \\ &= 3 + 4 + 6 + (a + a + a) \\ &= 13 + 3a.\end{aligned}$$

Thus

$$\sum_{i=1}^{N} (X_i + a) = \sum_{i=1}^{N} X_i + Na.$$

Generalization: *The sum of the values of a variable plus a constant is equal to the sum of the values of the variable plus N times that constant.*

To show the sum of the values of a variable when a constant has been subtracted from each,

$$\begin{aligned}\sum_{i=1}^{N} (X_i - a) &= (3 - a) + (4 - a) + (6 - a) \\ &= 3 + 4 + 6 - (a + a + a) \\ &= 13 - 3a.\end{aligned}$$

Thus

$$\sum_{i=1}^{N} (X_i - a) = \sum_{i=1}^{N} X_i - Na.$$

Generalization: *The sum of the values of a variable when a constant has been subtracted from each is equal to the sum of the values of the variable minus N times the constant.*

Example: $\sum_{i=1}^{N} (X_i - \bar{X}) = \sum_{i=1}^{N} X_i - N\bar{X}.$

2.4 TYPES OF NUMBERS

Cultural anthropologists, psychologists, and sociologists have repeatedly called attention to humankind's tendency to explore and understand the world that is remote from their primary experiences, long before they have investigated that which is closest to them. Thus, while people were probing distant stars and describing with striking accuracy their apparent movements and their interrelationships, they virtually ignored the very substance which gave them life: air which they each inhale and exhale over four hundred million times a year. In the authors' experience, a similar pattern exists in relation to students' experiences with numbers and their concepts of them. In our very quantitatively oriented western civilization, the child employs and manipulates numbers long before being expected to calculate the batting averages of the latest baseball hero. Nevertheless, ask a child to define a number, or to describe the ways in which numbers are employed, and you will likely be met with expressions of consternation and bewilderment. "I have never thought about it before," will frequently be the reply. After a few minutes of soul searching and deliberation, he will probably reply something to the effect that numbers are symbols which denote amounts of things which can be added, subtracted, multiplied, and divided. These are all familiar arithmetic concepts, but do they exhaust all possible uses of numbers? At the risk of reducing our student to utter confusion, you may ask: "Is the symbol 7 on a baseball player's uniform such a number? What about your home address? Channel 2 on your television set? Do these numbers indicate amounts of things? Can they reasonably be added, subtracted, multiplied, or divided? Can you multiply the number on any football player's back by any other number and obtain a meaningful value?" A careful analysis of our use of numbers in everyday life reveals a very interesting fact: most of the numbers we employ do not have the arithmetical properties we usually ascribe to them. For this reason, we prefer to differentiate between two terms, "numbers" and "numerals." Numerals refer to symbols such as Y, 10, IX. Numbers are specific types of numerals which bear fixed relationships to other numerals. Thus two numerals such as 4 and 5 are numbers if, and only if, they can be meaningfully added, multiplied, subtracted, and divided. From this point on the terms "number" and "numeral" will be differentiated on this basis. The list of such numerals is large. A few examples in addition to those enumerated above are: the serial number on a home appliance, a Zipcode number, a telephone number, a home address, an automobile registration number, and the numbers on a book in the library.

The important point is that numerals are used in a variety of ways to achieve many different ends. Much of the time, these ends do not include the representation of an amount or a quantity. In fact, there are three fundamentally different ways in which numerals are used.

1. To name (**nominal numerals**)
2. To represent position in a series (**ordinal numerals**)
3. To represent quantity (**cardinal numbers**)

2.5 TYPES OF SCALES

The fundamental requirements of observation and measurement are acknowledged by all the physical and social sciences as well as by any modern-day corporation interested in improving its competitive position. The things that we observe are often referred to as **variables** or **variates**. For example, if we are studying the price of stocks on the New York exchange, our variable is price. Any particular observation is called the **value of the variable**, or a score.

2.5.1 Nominal Scales

It is probable that when the majority of people think about measurement, they conjure up mental images of wild-eyed men in white suits manipulating costly and incredibly complex instruments in order to obtain precise measures of the variable that they are studying. Actually, however, not all measurements are this precise or this quantitative. If we were to study the sex of the offspring of female rats which had been subjected to atomic radiation during pregnancy, sex would be the variable that we would observe. There are only two possible values of this variable: male and female (barring an unforeseen mutation which produced a third sex!). Our data would consist of the number of observations in each of these two classes. Note that we do not think of this variable as representing an ordered series of values, such as height, weight, speed, etc. An organism which is female does not have any more of the variable sex, than one which is male, in spite of what Hollywood tries to tell us.

When is a woman a woman? Assignment of individuals or objects to classes is not always as clear-cut as it might first appear, since the properties defining the class are not always universally agreed upon. The recent controversy over the sex of females in international athletic competition is a case in point. Since all female athletes are required to submit to a series of sex tests prior to international competition, a number of renowned "female" athletes from behind the "iron curtain" have disappeared from the international scene. One is led to suspect that they might not have passed the physical, for one reason or another. The most fascinating and controversial case involves the great Polish athlete Ewa Klobkowska who passed the physical examination but was later disqualified when the study of her chromosomes revealed the presence of "masculine" *Y*-chromosomes. Is "she" female or male? An interesting outgrowth of the controversy has been the demand by some female athletes—perhaps with tongue-in-cheek—that the women doctors charged with examining them submit themselves to a prior physical examination to verify their "true" sex.

Observations of unordered variables constitute a very low level of measurement and are referred to as a **nominal scale** of measurement. We may assign numerical values to represent the various classes in a nominal scale but these numbers have no quantitative properties. They serve to identify the class.

The data employed with nominal scales consist of frequency counts or tabulations of the number of occurrences in each class of the variable under study. In the aforementioned radiation study, our frequency counts of male and female progeny would comprise our data. Such data are often referred to interchangeably as **frequency data**, **enumerative data**, **attribute data**, or **categorical data**. The only mathematical relationships germane to nominal scales are those of equivalence $(=)$ or of nonequivalence $(\neq)$.

2.5.2 Ordinal Scales

When we move into the next higher level of measurement, we encounter variables in which the classes *do* represent an ordered series of relationships. Thus the classes in **ordinal scales** are not only different from one another (the characteristic defining nominal scales) but they stand in some kind of *relation* to one another. More specifically, the relationships are expressed in terms of the algebra of inequalities: a is less than b $(a < b)$ or a is greater than b $(a > b)$. Thus the relationships encountered are: greater, faster, more intelligent, more mature, more prestigious, more disturbed, etc. The numerals employed in connection with ordinal scales are nonquantitative. They indicate only position in an ordered series and not "how much" of a difference exists between successive positions on the scale.

Examples of ordinal scaling include: rank ordering of baseball players according to their "value to the team," rank ordering of laboratory rats according to their "speed" in learning to run a maze, rank ordering of potential candidates for political office according to their "popularity" with people, and rank ordering of officer candidates in terms of their "leadership" qualities. Note that the ranks are assigned according to the ordering of individuals within the class. Thus, the most popular candidate may receive the rank of 1, the next popular may receive the rank of 2, and so on, down to the least popular candidate. It does not, in fact, make any difference whether or not we give the most popular candidate the highest numerical rank or the lowest, *so long as we are consistent in placing the individuals accurately with respect to their relative position in the ordered series.* By popular usage, however, the lower numerical ranks (1st, 2nd, 3rd) are usually assigned to those "highest" on the scale. Thus the winning horse receives the rank of "first" in a horse race; the pennant winner is "first" in its respective league; the rat requiring fewest trials to run a maze is "first" in its running performance. The fact that we are not completely consistent in our ranking procedures is illustrated by such popular expressions as "first-class idiot" and "first-class scoundrel."

Few people can match the ingenuity of horse-racing fans in devising "systems" to predict the running behavior of their favorite quadrupeds. Many of these systems involve no more than wild speculations and reveal great depths of superstitious behavior. Others, however, represent genuine efforts to understand the variables which affect the outcome of a race and to take them into account prior to placing bets.

One of the favorite systems among many fans is to bet "post position" in races held on circular tracks. The post position of a horse represents its ordinal position relative to the rail on the inside of the track. Thus first post position is occupied by the horse closest to the rail, whereas the eighth post position, in an eight-horse race, is farthest from the inside rail. The rationale for this system is based on the fact that, other things being equal, horses closest to the rail at the outset of a race will have to negotiate a shorter total distance in getting to the finish line than those in less favorable positions.

Is there any support for this "system"? Table 2.1 presents an analysis of the number and proportion of wins, according to post position, of one full month of racing at a particular circular track.

Table 2.1

Number of wins at eight post positions on a circular track*

	Post position								
	1	2	3	4	5	6	7	8	Total
Number of wins	29	19	18	25	17	10	15	11	144
Proportion of wins	0.20	0.13	0.12	0.17	0.12	0.07	0.10	0.08	

* From S. Siegel, *Non-Parametric Statistics*. New York: McGraw-Hill, 1956. Adapted with permission.

From the descriptive statistics, it would appear that there is some relationship between the post position and the likelihood of winning. However, the results above represent the outcome of a sample consisting of 144 observations taken from a single track during a single month. To draw a conclusion from these data would require us to make an inference from the sample to the population. Although we are not yet prepared to do this, we might anticipate certain questions that will be raised later in the course. What is the population from which the sample was drawn? May it be considered a representative sample? If it were to be repeated at the same track at another time, is it likely that the same general results would be observed? What if the study were conducted at another circular track? Would these results be likely to apply?

2.5.3 Interval and Ratio Scales

Finally, the highest level of measurement in science is achieved with scales employing cardinal numbers (**interval** and **ratio scales**). The numerical values associated with these scales are truly quantitative and therefore permit the use of arithmetic operations such as adding, subtracting, multiplying, and dividing. In interval and ratio scales equal differences between points on any part of the scale are equal. Thus the difference between 4 feet and 2 feet is the same as the difference between 9231 and 9229 feet.

In ordinal scaling, as you will recall, we could not claim that the difference between the first and second horses in a race was the same as the difference between the second and third horses.

There are two types of scales based upon cardinal numbers: interval and ratio. The only difference between the two scales stems from the fact that the interval scale employs an arbitrary zero point, whereas the ratio scale employs a true zero point. Consequently, only the ratio scale permits us to make statements concerning the ratios of numbers in the scale; e.g., 4 feet are to 2 feet as 2 feet are to 1 foot. A good example of the difference between an interval and a ratio scale is height as measured from a table top (interval) vs. height as measured from the floor. The difference between these scales can be further clarified by examining a well-known interval scale, e.g., the centigrade scale of temperature. Incidentally, the Fahrenheit scale of temperature is also an interval scale.

The zero point on the centigrade scale does not represent the complete absence of heat. In fact, it is merely the point at which water freezes at sea level and it has therefore an arbitrary zero point. Actually, the true zero point is known as absolute zero which is approximately $-273°$ centigrade. Now, if we were to say that 40°C is to 20°C as 20°C is to 10°C, it would appear that we were making a correct statement. Actually, we are completely wrong since 40°C really represents $273° + 40°$ of heat; 20°C represents $273° + 20°$ of heat; and 10°C represents $273° + 10°$ of heat. The ratio 313:293 as 293:283 clearly does not hold. These facts may be better appreciated graphically. In Fig. 2.1, we have represented all three temperature readings as distances from the true zero point which is -273°C. From this graph it is seen that the distance from -273°C to 40°C is not twice as long as the distance from -273°C to 20°C. Thus, 40°C is not twice as warm as 20°C, and the ratio 40°C is to 20°C as 20°C is to 10°C does not hold.

Apart from the difference in the nature of the zero point, interval and ratio scales have the same properties and will be treated alike throughout the text.

It should be clear that one of the most sought-after goals of the behavioral scientist is to achieve measurements which are at least interval in nature. Unfortunately, behavioral science has met with little success along these lines. Indeed, there are certain epistemological considerations which make the authors doubt that we will ever achieve interval scaling for many of the types of dimensions which behavioral scientists measure. In several instances where interval scaling is claimed, it is by virtue of certain assumptions which the claimant is willing to make. For example, some specialists in scaling procedures will assume that "equal-appearing intervals are equal" as the basis for claiming cardinality. The approximation of such scales to cardinal measurement is, of course, only as good as the validity of the assumption.

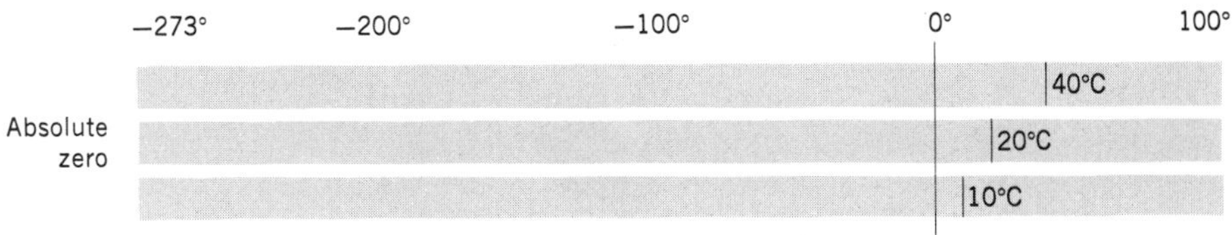

Fig. 2.1 Relationships of various points on a centigrade scale to absolute zero.

2.6 CONTINUOUS AND DISCONTINUOUS SCALES

Let us imagine that you are given the problem of trying to determine the number of children per American family. Your scale of measurement would start with zero (no children) and would proceed, by *increments of one*, to perhaps fifteen or twenty. Note that, in moving from one value on the scale to the next, we proceed by *whole numbers* rather than by fractional amounts. Thus a family has either 0, 1, 2, or more children. In spite of the statistical abstraction that the American family averages two and a fraction children, the authors do not know a single couple that has actually achieved this marvelous state of family planning.

Such scales, in which the variable can take on a finite number of values, are referred to as **discontinuous** or **discrete scales**, and they have equality of **counting units** as their basic characteristic. Thus, if we are studying the number of children in a family, each child is equal with respect to providing one counting unit. Such scales involve cardinality insofar as they permit arithmetic operations such as adding, subtracting, multiplying, and dividing. Thus we can say that a family with four children has twice as many children as one with two children. Observations of discrete variables are always exact as long as the counting procedures are accurate. Examples of discontinuous variables are: the number of hats sold in a department store during the month of January, the number of white blood cells counted in one square centimeter, the number of alpha particles observed in a second, etc.

You should not assume from the above discussion that discrete scales necessarily involve *only* whole numbers. A spinner, such as the one shown in Fig. 2.2, would be discrete, yet the seven values of the variable proceed by half units. The important point, in this example, is that the variable cannot take on values *between* 0 and 0.5, or 0.5 and 1, etc.

In contrast, a scale in which the variable may take on an unlimited number of intermediate values is referred to as a **continuous scale**. For example, let us take the same range of values which is illustrated in the spinner above (0 to 3). If it were possible for this variable to take on such values as 1.75, 2.304, etc., then we would be dealing with a continuous scale of measurement.

It is important to note that, although our measurement of discrete variables is always exact, our measurement of continuous variables is always approximate. If

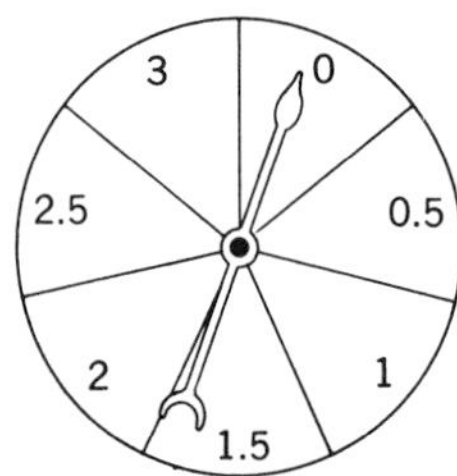

Fig. 2.2 A spinner illustrating a discrete scale in which the variable changes by half units.

we are measuring the height of American males, for example, any particular measurement is inexact because it is always possible to imagine a measuring stick which would provide greater accuracy. Thus, if we reported the height of a man to be 68 inches, we would mean 68 inches give or take one half an inch. If our scale is accurate to the nearest tenth, we can always imagine another scale providing greater accuracy, say, to the nearest hundredth or thousandth of an inch. The basic characteristic of continuous scales, then, is equality of **measuring units**. Thus if we are measuring in inches, one inch is always the same throughout the scale. Examples of continuous variables are length, velocity, time, weight, etc.

2.6.1 Continuous Variables, Errors of Measurement, and "True Limits" of Numbers

In our preceding discussion, we pointed out that continuously distributed variables can take on an unlimited number of intermediate values. Therefore, we can never specify the exact value for any particular measurement, since it is possible that a more sensitive measuring instrument can increase the accuracy of our measurements a little more. For this reason, we stated that numerical values of continuously distributed variables are always approximate. However, it is possible to specify the limits within which the true value falls; e.g., the **true limits of a value** of a continuous variable are equal to that number plus or minus one half of the unit of measurement. Let us look at a few examples. You have a bathroom scale, which is calibrated in terms of pounds. When you step on the scale the pointer will usually be a little above or below a pound marker. However, you report your weight to the nearest pound. Thus, if the pointer were approximately three quarters of the distance between 212 pounds and 213 pounds, you would report your weight as 213 pounds. It would be understood that the "true" limit of your weight, assuming an accurate scale, falls between 212.5 pounds and 213.5 pounds. If, on the other hand, you are measuring the weight of whales, you would probably have a fairly gross unit of measurement, say 100 pounds. Thus, if you reported the weight of a whale at 32,000 pounds, you would mean that the whale weighed between 31,950 pounds and 32,050 pounds. If the scale were calibrated in terms of 1000 pounds, the true limits of the whale's weight would be between 31,500 pounds and 32,500 pounds.

2.7 ROUNDING

Let us imagine that we have obtained some data which, in the course of conducting our statistical analysis, require that we divide one number into another. There will be innumerable occasions in this course when we shall be required to perform this arithmetic operation. In most cases, the answer will be a value which extends to an endless number of decimal places. For example, if we were to express the fraction $\frac{1}{3}$ in decimal form, the result would be 0.33333+. It is obvious that we cannot extend this series of numbers *ad infinitum*. We must terminate at some point and assign a value to the last number in the series which best reflects the remainder. When we do this, two types of problems will arise:

1. To how many decimal places do we carry the final answer?
2. How do we decide on the last number in the series?

The answer to the first question is usually given in terms of the number of significant figures. However, there are many good reasons for not following the mathematical stricture to the letter. For simplicity and convenience, we shall adopt the following policy with respect to **rounding**:

After every operation, carry to three and round to two more places than were in the original data.

Thus, if the original data were in whole-numbered units, we would carry our answer to the third decimal place and round to the second decimal. If in tenths, we would round to the third decimal, and so forth.

Once we have decided the number of places to carry our final figures, we are still left with the problem of representing the last digit. Fortunately, the rule governing the determination of the last digit is perfectly simple and explicit. If the remainder beyond that digit is greater than 5, increase that digit to the next higher number. If the remainder beyond that digit is less than 5, allow that digit to remain as it is. Let's look at a few illustrations. In each case, we shall round to the second decimal place:

6.546	becomes	6.55,
6.543	becomes	6.54,
1.967	becomes	1.97,
1.534	becomes	1.53.

You may ask, "In the above illustrations, what happens if the digit at the third decimal place is 5?"

You should first determine whether or not the digit is exactly 5. If it is 5 plus the slightest remainder, the above rule holds and you must add one to the digit at the second decimal place. If it is almost, but not quite 5, the digit at the second decimal place remains the same. If it is *exactly 5, with no remainder*, then an arbitrary convention which is accepted universally by mathematicians applies: Round the digit at the second decimal place to the *nearest even* number. If this digit is already even, then it is not changed. If it is odd, then *add* one to this digit to make it even. Let's look at several illustrations in which we round to the second decimal place:

6.545001	becomes	6.55.	Why?
6.545000	becomes	6.54.	Why?
1.9652	becomes	1.97.	Why?
0.00500	becomes	0.00.	Why?
0.01500	becomes	0.02.	Why?
16.89501	becomes	16.90.	Why?

2.8 RATIOS, FREQUENCIES, PROPORTIONS, AND PERCENTAGES

Of all the statistics in everyday use, perhaps the most misunderstood and misused involves the representation of ratios, proportions, and percentages. Concomitantly, but not incidentally, these statistics also provide the most fertile grounds for misleading and outright fraudulent statements. It is possible for a congressman, who needs to impress "the people back home" of his vital interest in consumer affairs, to charge that a given drug company is making a 300% profit on its sales to retail outlets while the drug company, looking at precisely the same statistical facts, may reply with righteous indignation that its profit is only 75%. How could this come about? Let us see.

Let us assume that the cost to manufacture a given drug is \$2 per gross. In turn, a gross is sold to the retailer at \$8. The ratio of profit (\$8 − \$2 = \$6) to manufacturing cost is 6:2 or 3:1. Stating this ratio as a percentage (multiplying by 100) the resulting percentage is 300%. It would appear that the congressman is correct. But the drug company replies, "The selling price is \$8 and our profit is \$6. The ratio of profit to selling price is 6:8 or 0.75. Stated as a percentage, our profit is 75% of the selling price. When considering the cost of research and development and all the inherent financial risks, the profit is not excessive."

Which statement better describes the facts? Actually, both statements are correct. The confusion stems from the fact that two different values (\$2 and \$8) have been employed as the base, or denominator, in arriving at the final percentage figures. There is nothing wrong with either procedure *as long as it is made perfectly clear* which base has been employed in the initial calculations. Knowing the base, one can freely move from one to the other and there need be little confusion. Thus if we know that the congressman has used the production cost as the base, we may employ elementary algebra to translate his statement to one employing selling price as a base: Let

$$x = \text{production cost},$$
$$y = \text{selling price}.$$

The following equation represents the percentage of profit, employing production cost as a base:

$$\frac{y - x}{x} \times 100 = \text{percent of profit (employing production cost as a base)}$$

Using the above values and substituting in the equation, we find that

$$\frac{y - x}{x} \times 100 = \frac{8 - 2}{2} \times 100 = 300\%.$$

The formula for calculating percentage of profit employing selling price as a base is

$$\frac{y - x}{y} \times 100 = \text{percent of profit (related to selling price)}.$$

Substituting into the above formula, we obtain

$$\frac{y - x}{y} \times 100 = \frac{6}{8} \times 100 = 75\%.$$

The following excerpt from *Winning with Statistics** illustrates "stealing by the base."

Any baseball enthusiast will understand immediately what is meant when the announcer proclaims, "Joe Morgan just stole second base." What the baseball fan may not realize is that there are many ways, besides playing baseball, of committing larceny with bases. I just happened to have in my collection of tapes a conversation between the redoubtable Sam Stickfinger and his financial consultant, Esab Rewol. Esab explains how a markup of 100% can become a 200% reduction during a sale in which the price of an item is cut 25%. All that is required is a little flexibility in selecting the base from which the calculations are made.

Sam: "We pay 50 cents for the item and sell it for $1.00. I thought of advertising that our markup is only 100%."
Esab: "That's suicide, Sam. Nobody will buy an item when they know the markup is 100%. But your markup is really only 50%."
Sam: "How do you figure that?"
Esab: "The item costs the customer $1.00, right?"
Sam: "Right."
Esab: "The profit is 50 cents, right?"
Sam: "Right."
Esab: "Well, 50 over 100 is 50 percent. It's just a matter of which base you use, 50 or 100. But you never really sell the item for $1.00, do you?"
Sam: "Oh no, it's always on sale. We take 25 cents off and advertise a 25% reduction in price."
Esab: "Bad, Sam, bad. If you take off 25 cents, the new price is 75 cents. Right?"
Sam: "Right."
Esab: "Well, 25 cents over 75 cents is one-third of the new price. You've got a one-third, or 33% sale."
Sam: "Fantastic! I've been practically giving it away without knowing it."
Esab: "It's worse than you think, Sam. The cost to you is 50 cents, right? Nobody can expect to stay in business if you ever went below that price, right?"
Sam: "Right."
Esab: "So we should really be advertising in terms of reduction in profits. See?"
Sam: "I'm beginning to see. At $1.00 our profit would be 50 cents per unit. We have reduced our profit by 25/50 or 50 percent. It's really a 50% off sale!"
Esab: "Yes, if you want to be completely honest about it . . ."
Sam: "Of course I want to be honest. But out of idle curiosity . . ."
Esab: "Well, if you don't mind bending the truth ever so slightly . . ."
Sam: "Bending slightly is honest. Bending big is dishonest."
Esab: "Well, your profit is 25 cents. Your reduction is 25 cents. If you take the reduction and divide by the profit . . ."
Sam: "You get a 100% reduction. Why those damned customers, ripping me off like that."

* Runyon, R. P., *Winning with Statistics*. Addison-Wesley, Reading, Mass., 1977.

Esab: "Yes, we'll call it a 'Rip Off Sam Sale.' Oh, yes, by the way, have you ever considered putting a price tag of $1.25 on that item?"
Sam: "No. Should I?"
Esab: "Well, when you have your sale and reduce the price to 75 cents, you can claim . . . let's see now . . . 50/25 × 100 . . . that's a 200% reduction!"
Sam: "And Joe Snopes down the block can only advertise that he sells below cost."
Esab: "You can do better than that. The usual cost to the customer is $1.25, right?"
Sam: "With our new price tag, that's right."
Esab: "Well you're going to reduce it to 75 cents, 50 cents below cost. That's 50/125 × 100 = 40%. Just think of it, 40% below cost!"
Sam: "I like 200% reduction better."
Esab: "Why?"
Sam: "It sounds like I'm giving it away, which I am."

2.8.1 Types of Statistical Ratios

The preceding discussion should make it clear that there are several different types of statistical ratios, and that statements involving ratios, proportions, and percentages should be carefully crutinized to avoid the confusion to which we have alluded.

The following are some of the statistical ratios most commonly employed in elementary statistical analyses.

1. The **distribution ratio** is defined as the ratio of a part to a total which includes that part. Symbolically, in the case of a two-category variable, the distribution ratio may be represented as

$$\frac{x}{x + y}.$$

To illustrate, if there are 300 women (x) in a class which includes 900 males (y), the **frequency** of women is 300. The proportion of women to the total is

$$\frac{300}{300 + 900},$$

expressed decimally, $p = 0.25$. To express this proportion as a percentage, you simply multiply by 100, i.e., the percentage of women is 25% of the total.

The **proportion** of men (y) to the total is $\frac{900}{1200}$ or $p = 0.75$. The percentage of men is, therefore, 75% of the total.

You will note that the sum of all the proportions is 1.00 and the sum of the percentages is 100%. The calculation of the distribution ratios when more than two categories are involved is a simple extension of the above procedures. Thus if we were interested in describing the proportions of the 1000 students studying in each of five major areas of concentration, we might obtain the following: 300 in business administration, 150 in humanities, 250 in life sciences, 100 in physical sciences, and 200 in social sciences. The respective proportions would be $\frac{300}{1000} = 0.30$, $\frac{150}{1000} = 0.15$, $\frac{250}{1000} = 0.25$, $\frac{100}{1000} = 0.10$, and $\frac{200}{1000} = 0.20$. You will note that the sum of these proportions is 1.00.

2. The **interclass ratio** is defined as a *ratio of a part in a total to another part in the same total.* Thus, in the class of 1200 students, the ratio of female students to male students is 1:3. You will note that we might also express the ratio in terms of male to female students, in which case the interclass ratio would be 3:1.

3. The **time ratio**, or *time relative*, is a measure which expresses the change in a series of values arranged in a time sequence and is typically shown as percentages.

There are two main classes of time ratios: (a) those which employ a **fixed-base** period and (b) those employing a **moving base** as, for example, the preceding year.

The data in Table 2.2 show the total elementary and secondary school enrollment in the period 1869–1974. Employing the total enrollment in 1939–1940 as a fixed base, time ratios have been calculated. In addition, the last column presents **moving-base time ratios** which employ the preceding year.

Time ratios are often referred to as index numbers. The following excerpt from Runyon[†] demonstrates ways in which index numbers may be employed to confuse the reader.

Table 2.2

Total elementary and secondary school enrollment for selected school years between 1869 and 1974, fixed-base time ratios (base: 1939–40), and moving-base time ratios (base: preceding year)*

Year	Total elementary and secondary school enrollment	Time ratios: Fixed base (1939–1940)	Time ratios: Moving base (preceding year)
1869–1870	6,872	27.0	—
1899–1900	15,503	61.0	225.6
1909–1910	17,814	70.0	114.9
1919–1920	21,578	84.8	121.1
1929–1930	25,678	101.0	119.0
1939–1940	25,434	100.0	99.0
1949–1950	25,111	98.7	98.7
1959–1960	36,087	141.9	143.7
1969–1970	45,619	179.4	126.4
1971–1972	46,081	181.2	101.0
1973–1974	45,409	178.5	98.5

* *The 1976 World Almanac*, p. 197; *1976 Information Please Almanac*, p. 740.

But who's to select the base year? Aye, there's the rub. The particular year that you choose to represent your base year can have a profound effect on the appearance of all of your subsequent figures. For example, if you want to show a period of very rapid growth in some particular industry, you would pick as your index year one in which the performance of that industry was at its very lowest. If, on the other hand, you wanted to give just the opposite impression that the industry was not doing well, you would purposely select the year in which the industry had

† Runyon, R. P., *Winning with Statistics.* Addison-Wesley, Reading, Mass., 1977.

Table 2.3

Violent crime rates per 100,000 population for selected years between 1960 and 1974, and time ratios using two different years as the fixed base. (*Source: The American Almanac for 1976, The Statistical Abstract of the U.S.* New York: Grosset & Dunlap, 1975.)

		Time ratio (percentage change)	
Year	Rate/100,000 pop.	Fixed base, year 1960	Fixed base, year 1970
1960	160	100	44
1965	199	124	55
1967	251	157	69
1968	297	186	82
1969	327	204	90
1970	362	226	100
1971	394	246	109
1972	299	187	83
1973	415	259	115
1974	459	287	127

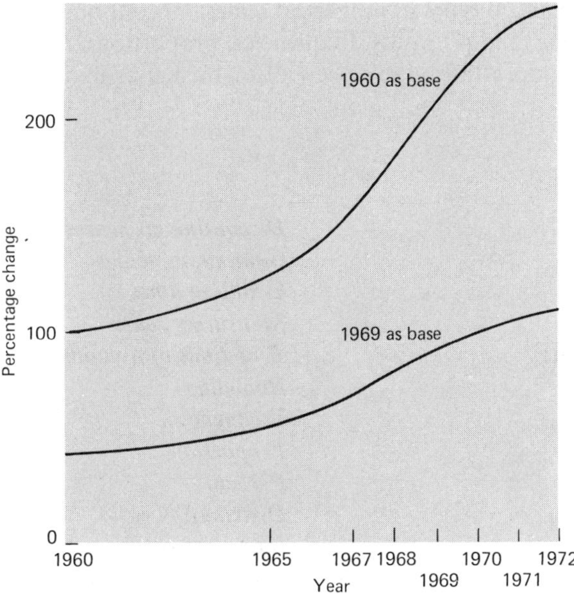

Fig. 2.3 Rate of growth in violent crimes. The impression given by the graph depends upon the year selected as the base year. When 1960 is used as fixed base, the crime rate increase appears explosive. When 1969 is used as base year, the growth rate appears more modest.

its best performance. Abracadabra, you would now convey—with the very same data—a distinct impression of relatively poor performance over recent years.

Now let's see how this is done. Shown in Table 2.3 are the violent crime rates per 100,000 population for selected years from 1960 to 1974. If you simply look at the crime rate, it is quite obvious that the rates are increasing rather alarmingly. Now one can produce quite different impressions if one simply selects different years as the fixed base year and calculates percentages of change from that year. If you look at Fig. 2.3, you will see two plots. One plots the time ratio for the violent crime rates for the selected years of 1960 to 1974, and we use 1960 as the base year. In the second we use precisely the same data but we shift to 1970 as the base year. Inspection of the two line drawings clearly shows a huge rate of violent crime increase when 1960 is used as the base year. When 1970 is used as the base year, the rate of increase looks much more modest.

CHAPTER SUMMARY

In this chapter, we pointed out that advanced knowledge of mathematics is not a prerequisite for success in this course. A sound background in high school mathematics plus steady application to assignments should be sufficient to permit mastery of the fundamental concepts put forth in this text.

To aid the student who may not have had recent contact with mathematics, we have attempted to review some of the basic concepts of mathematics. Included in this review are: (1) the grammar of mathematical notation, (2) summation rules, (3) types of numbers, (4) types of numerical scales, (5) continuous and discontinuous scales, (6) rounding, and (7) ratios, frequencies, proportions, and percentages. Students requiring a more thorough review of mathematics may refer to Appendix I.

Terms to Remember:

$\sum$
Nominal numeral
Ordinal numeral
Cardinal number
Variable or Variate
Value of variable
Nominal scale
Frequency data
Enumerative data
Attributive data
Categorical data
Ordinal scale
Interval scale
Ratio scale
Discontinuous scales
Continuous scales
Counting units
Measuring units
True limit of a number
Rounding
Frequency
Proportion
Percent
Distribution ratio
Interclass ratio
Fixed-base time ratio
Moving-base time ratio
Index number

BOX 2.2 SUMMARY OF SCALES OF MEASUREMENT

QUALITATIVE AND QUANTITATIVE VARIABLES

There are two broad types of variables with which researchers work—qualitative and quantitative. Qualitative variables differ in kind, rather than by "how much." They are classified according to the attributes by which they differ, rather than in terms of the degree to which they share a given attribute. In contrast, a quantitative variable is one which expresses a difference in terms of *how much* individuals or objects possessing an attribute differ from one another.

Qualitative Variables

Qualitative variables are the result of classifying individuals or objects that share an attribute in common as "the same" and those that do not share this attribute as "different."

Examples: An American is anyone holding United States citizenship. A person holding citizenship in another country is not a United States citizen. (Note that this is true for the overwhelming majority of people. There is an occasional person who holds a dual citizenship).
A red automobile is different from a green automobile.

With respect to the attribute in question, qualitative variables are either-or.

A person is either a male or a female.
An animal is either a dog, or a cat, or an elephant, or . . . .
An automobile is either a Ford, a Chevrolet, a Fiat, a Toyota, or . . . .

In items 2.2.1 to 2.2.6, check the qualitative variable in each pair.

2.2.1	Sex	Temperature
2.2.2	Height	Color
2.2.3	Zip code	Weight
2.2.4	Distance	Hair color
2.2.5	Speed of travel	Religious preference
2.2.6	Time	Occupation

Data obtained with qualitative variables involves counting and enumeration.

Examples: The statistics class contains 15 male and 22 female students.
The number of registered Republican voters in Yasa City is 8,245; Democrat, 7,993; Independent, 2,372.

In items 2.2.7 to 2.2.9, which member of each pair (italicized) involves qualitative data?

2.2.7 Number of *blonds* in class.
Height of the tallest person in class.

2.2.8 Number of *Datsuns* sold in Yasa City during 1976.
The *weight* of a Datson 210.

2.2.9 The low *temperature* in Gunnison, Colorado on January 5, 1977.
The population of the *original 13 colonies*, December 31, 1776.

Nominal and Ordinal Scales

Nominal scales are qualitative variables in which no order, direction, or magnitude is implied. Ordinal scales are qualitative variables in which some order, direction, or magnitude is implied.

Examples: Car color is a nominal scale; rank in military (Private through General) is ordinal.
Political affiliation is a nominal scale; order of finish in "most valuable player award" is ordinal.

In items 2.2.10 to 2.2.12, select the member of each pair which involves ordinal scales.

2.2.10	Rank in leadership qualities	Hair color
2.2.11	Worst natural disaster in history	Make of automobile
2.2.12	Post position in horse races	Type of animal

Classification into ordinal scales involves algebra of inequalities.

Examples: $a > b$ may mean: a is greater than b, higher rank than b, more prestigious than b; prettier than b, etc.;
$a < b$ may mean a is less than b, lower rank than b, less prestigious than b, etc.

In items 2.2.13 to 2.2.18, using symbols $>$ or $<$, complete the statements.

2.2.13	General	Private
2.2.14	Corporal	Sergeant
2.2.15	Leader	Follower
2.2.16	Third post position	Eighth post position
2.2.17	Winner of beauty pageant	Also ran
2.2.18	Player sent down to minor leagues	Most valuable player on team

The most common descriptive statistics with nominal and ordinal variables involve proportions and percentages.

Examples: The proportion of registered voters in Yasa City is 0.44.
The percentage of female students in the statistics class is 59.46.

In items 2.2.19 to 2.2.21, select from each pair the statement which is most likely to involve proportions or percentages.

2.2.19 Mathematics scores of subjects following administration of a drug.
Cars sold in Yasa City by manufacturers.

2.2.20 Number of horses who finished first at each post position at an oval track during 1976.

2.2.21 The heights of American males over 18 years of age.
Salaries of employees in mining industries.
Number of different types of crimes during 1977.

Quantitative Variables

Quantitative variables indicate how much of a given attribute an object or person has.

Examples: Frances Green is 63 inches tall.
John McMahon weighs 235 pounds.

Quantitative variables can be added, subtracted, multiplied and divided.

Examples: Frances Green is 63 inches tall; Morris French stands 70 inches high. Their combined height is 133 inches.
Sirloin roasts in the meat counter at the local supermarket weigh 56, 72, 35, and 46 ounces. The first two weigh 47 ounces more than the last two.

In items 2.2.22 to 2.2.25, indicate which of each pair is a quantitative variable.

2.2.22 Number of blonds in class.
Height of tallest person in class.

2.2.23 Number of Datsuns sold in Yasa City during 1976.
The weight of a Datsun 210.

2.2.24 The low temperature in Gunnison, Colorado on January 5, 1977.
The population of the original 13 colonies, December 31, 1776.

2.2.25 The size of the family.
The colors of the rainbow.

Continuous vs. Discrete Measurements

Some quantitative variables increase or decrease in continuous gradations. No matter how accurately measured, it is always possible to conceive of a more accurate measure. Weight can be 125, 125.01, 125.0063, 125.006347, etc. Such variables are referred to as continuous.

Examples: Height, weight, speed, temperature, altitude.

Other quantitative variables increase by discrete, rather than by continuous, amounts. A family unit may contain 0, 1, 2, 3, etc. children but not 1.43 children.

In items 2.2.26 to 2.2.29, check continuous (C) or discrete (D) opposite each variable.

2.2.26 A child's spinner. C D

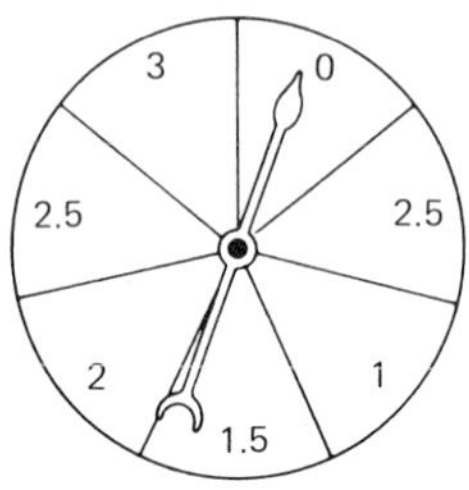

2.2.27 Weight of each horse at the County Fair. C D
2.2.28 Size of family unit. C D
2.2.29 The height of various skyscrapers throughout the world. C D

Descriptive statistics with quantitative variables commonly involve arithmetic procedures such as adding, subtracting, multiplying, and dividing.

Examples: The weights of 4 sixth grade students are 80, 92, 73, 75 pounds. Their mean weight is

$$\bar{X} = \frac{\sum_{i=1}^{N} X_i}{N} = \frac{320}{4} = 80 \text{ pounds.}$$

The range of weights (highest weight minus lowest weight) is $92 - 73 = 19$.

In items 2.2.30 to 2.2.32, select from each pair the one that is most likely to involve arithmetic procedures in calculating descriptive statistics.

2.2.30 Height of various athletes.
Types of cars manufactured in Detroit.
2.2.31 The number of children per family unit.
Eye color of students in statistics class.
2.2.32 Number of winners at each post position on an oval track.
Daily miles driven to work by commuters.

PROPORTIONS AND PERCENTAGES

We previously noted that the descriptive statistics most commonly used with qualitative data involve proportions and percentages.

Interclass Ratio

An interclass ratio is defined as the ratio of a part to a total which includes that part. When a proportion is used as an interclass ratio, it can vary only between 0.00 and 1.00. If a is the number of males in a class, b the number of females, the proportion of males is:

$$\text{Proportion of males} = \frac{a}{a + b}$$

The proportion of females is:

$$\text{Proportion of females} = \frac{b}{a + b}$$

EXERCISES

The following exercises are based on this chapter and Appendix I.

2.1 Solve the following equations for X.

a) $a + X = b - c$ b) $2X + a = c$ c) $aX + bY = c$

d) $\frac{a}{X} = \frac{b}{c}$ b) $\frac{b}{c/(X/a)} = Y$ f) $\frac{a + X}{Y/(b/c)} = c$

g) $\frac{a}{b + X} = c$ h) $(X^3)(X^4)(X^2) = Y$ i) $\frac{X^5}{X^8} = aX^2$

2.2 Substitute in the following equations and solve for X.

a) $aX + bY = c$, in which $a = 3$, $b = 2$, $Y = 8$, $c = 25$

b) $a^5 - a^3 = X$, in which $a = 2$

c) $(a^2)(b^3) = X$, in which $a = \frac{1}{6}$, $b = \frac{5}{6}$

d) $\frac{a^5}{a^2} = X$, in which $a = 3$

e) $a^3 - \frac{b}{c} = X$, in which $a = 2$, $b = \frac{2}{3}$, $c = \frac{3}{4}$

2.3 Round the following numbers to the second decimal place.

a) 99.99500 b) 46.40501
c) 2.96500 d) 0.00501
e) 16.46500 f) 1.05499
g) 86.2139 h) 10.0050

2.4 Determine the answers to the following problems to as many places as is standard procedure.

a) 0.275 times 0.111 b) 0.3811 times 0.2222
c) 0.999 times 0.121 d) 150 divided by 400
e) 0.1 divided by 0.9 f) 0.006 divided by 0.007

2.5 Determine the proportion and percentage of the following.
   a) Male students in your statistics class.
   b) Number of face cards in a conventional 52-card deck.
   c) 25 items correct on a quiz consisting of 33 items.
   d) 4652 voters out of a total registration of 9686.

2.6 Determine the value of the following expressions in which $X_1 = 4$, $X_2 = 5$, $X_3 = 7$, $X_4 = 9$, $X_5 = 10$, $X_6 = 11$, $X_7 = 14$.

a) $\sum_{i=1}^{4} X_i =$ b) $\sum_{i=1}^{7} X_i =$ c) $\sum_{i=3}^{6} X_i =$

d) $\sum_{i=2}^{5} X_i =$ e) $\sum_{i=1}^{N} X_i =$ f) $\sum_{i=1}^{N} X_i =$

2.7 Express the following in summation notation.
   a) $X_1 + X_2 + X_3$ b) $X_1 + X_2 + \cdots + X_N$
   c) $X_3^2 + X_4^2 + X_5^2 + X_6^2$ d) $X_4^2 + X_5^2 + \cdots + X_N^2$

2.8 The answers to the following questionnaire items are based on what scale of measurement?
   a) What is your height?
   b) What is your weight?
   c) What is your occupation?
   d) How does this course compare with others you have taken?
   e) What is your name?

2.9 In the following examples, identify the scale of measurement, and determine whether the italicized variable is continuous or discontinuous.
   a) *Distance* traveled from home to school.
   b) Number of infants born at *varying times of the day.*
   c) Number of votes compiled by each of *three candidates* for a political office.
   d) Number of servicemen at varying *ranks in the U.S. Army.*

2.10 Determine the square roots of the following numbers to two decimal places.
   a) 160   b) 16   c) 1.60   d) 0.16   e) 0.016

2.11 State the true limits of the following numbers.
   a) 0   b) 0.5   c) 1.0   d) 0.49   e) −5   f) −4.5

2.12 Using the values of X_i given in Exercise 2.6 above, show that

$$\sum_{i=1}^{N} X_i^2 \neq \left(\sum_{i=1}^{N} X_i\right)^2.$$

2.13 An appliance salesman sold 40 color television sets in one month. The preceding month he had sold 80. What is the percentage decrease? (a) $33\frac{1}{3}\%$, (b) 50%, (c) $66\frac{2}{3}\%$, (d) 200%, (e) 300%.

2.14 An item costs a retailer \$1.50 and she sells it for \$2.00. The percentage of profit using her cost as a base is (a) 25%, (b) $33\frac{1}{3}\%$, (c) 75%, (d) 300%, (e) 400%.

2.15 Referring to the preceding problem, the percentage of profit using selling price as a base is (a) 25%, (b) $33\frac{1}{3}$%, (c) 75%, (d) 300%, (e) 400%.

2.16 A local merchant has reduced the price of an item from $1.50 to $0.75. She advertises: "Price slashed by 200%!" Comment on her ad. What is the true percentage reduction in price?

2.17 Brand *A* cigarettes, which has been on the market for a year, increased its sales in one month by 40%. Brand B, an old standby of many years, increased its sales by 5% over the same one-month period. Can we conclude that Brand A is more popular than Brand B? Explain.

2.18 A vendor sells a package of cigarettes for $0.45. If he claims a 50% profit on his investment, how much did he pay for the cigarettes?

2.19 A salesperson sold 200 pairs of shoes this week. How many pairs must she sell next week to show a 25% increase in sales?

2.20 The birth of the computer age has ushered in a new type of crime, unimaginable several decades ago. Table 2.4 reports verified instances of computer crime or computer abuse between 1969 and 1973.

Table 2.4

Cases of computer abuse reported and verified

Year	Financial fraud	Theft of information or property	Unauthorized use	Vandalism	Total
1969	3	6	0	3	12
1970	7	5	9	8	29
1971	22	18	6	6	52
1972	12	15	16	12	55
1973	21	15	8	9	53

Source: Parker, D. B., S. Nyeum, and S. S. Oura. 1973. *Computer Abuse.* Menlo Park, Ca., Stanford Research Institute.

a) Sum all of the entries in the Total column and express each year as a percentage of the whole.
b) Construct a fixed-base time ratio using 1969 as the base year; 1971 as the base year.
c) Convert each entry to percentages, using the sum obtained in part (a) as the denominator.

2.21 Using the figures shown in the table below, answer the following questions:
a) Of all students majoring in each academic area, what percentage is female?
b) Considering only the males, what percentages are found in each academic area?
c) Considering only the females, what percentages are found in each academic area?
d) Of all students majoring in the five areas, what percentage is male? What percentage is female?

The number of students, by sex, majoring in each of five academic areas:

	Male	Female
Business administration	400	100
Education	50	150
Humanities	150	200
Science	250	100
Social science	200	200

2.22 A large discount house advertises, "Prices reduced below cost." What does this claim mean to you? Ask your friends to interpret the claim. Are the interpretations all in agreement?

2.23 Indicate which of the following variables represent discrete or continuous series.

a) The time it takes you to complete these problems.
b) The number of newspapers sold in a given city on August 7, 1976.
c) The amount of change in weight of 5 women during a period of 4 weeks.
d) The number of home runs hit by 10 pitchers, selected at random, during the 1977 baseball season.
e) The number of stocks on the New York Stock Exchange that increased in selling price on December 19, 1976.

2.24 Following is a list showing the number of births in the United States (expressed in thousands) between 1969 and 1973 (Source: National Center for Health Statistics, Dept. of Health, Education and Welfare, 1976 *Information Please Almanac*).

Year	Males	Females
1969	1,847	1,754
1970	1,915	1,816
1971	1,823	1,733
1972	1,670	1,588
1973	1,608	1,529

a) Calculate the percentage of males and females for each year.
b) Convert (separately for males and females) into time ratios employing a
  i) fixed base (1969), ii) moving base.

2.25 Referring to Exercise 1.15 calculate the following:

a) The distribution ratio and percentage of the number of X albums sold to the total number of albums. (*Hint*: Total = X + Y + Z in this example.)
b) The distribution ratio and percentage of the number of X plus Y albums to the total.
c) The distribution ratio and percentage of Z albums to the total.
d) The interclass ratio of the number of X albums to the number of Y albums sold.
e) The interclass ratio of the number of X plus Y albums to Z albums.
f) The interclass ratio of X to Z.

2.26 "In 1975, about 33,000,000 persons owned shares in American corporations, compared to 8,630,000 in 1956." (*The 1976 World Almanac and Book of Facts*. New York: Newspaper Enterprise Association, Inc., 1975, p. 84.) Assuming that these estimated numbers are exact, calculate:

a) The percent increase from 1956 to 1975.
b) What percent the number of persons owning shares in 1956 is of the number in 1975.

2.27 Withholding—weekly payroll period

Single person—including head of household:

If the amount of wages is:	The amount of income tax to be withheld shall be:	
Not over $20	0	
Over — But not over		of excess over
$ 20—$ 31	14%	$ 20
$ 31—$ 50	$ 1.54, plus 17%	$ 31
$ 50—$100	$ 4.77, plus 20%	$ 50
$100—$135	$14.77, plus 18%	$100
$135—$212	$21.07, plus 21%	$135
$212	$37.24, plus 24%	$212

(From *The 1972 World Almanac*, p. 61.)

a) Calculate the amount withheld per week for single persons earning the following amounts:
i) $75/week ii) $120/week iii) $200/week iv) $300/week
b) What percent of the total weekly earnings is withheld for each of the above earnings?

2.28

Indexes of retail prices of foods for the years 1969–1975 (1967 = 100).

Year and month		Food price index
1969		108.9
1970		114.9
1971		118.4
1972		123.5
1973		141.4
1974		161.7
1975	January	170.9
	April	171.2
	July	178.6

(From Bureau of Labor Statistics, United States Department of Labor, *The 1976 World Almanac*, p. 46.)

a) What is the percent increase in food:
   i) from 1969 to January 1975? ii) from 1969 to 1970?
   iii) from 1970 to January 1975?
b) If a food item cost $0.50 in 1967, estimate what the price would have been in:
   i) 1969, ii) 1971, iii) January 1975, iv) July 1975.
c) Are these estimates accurate? Explain your answer.
d) Calculate the moving base and price for each year from 1969 to 1974 for an item costing $0.50 in 1967.

2.29 Suppose production cost of a certain vegetable is $0.25 and a farmer sells it for $0.50. Determine his percentage of profit:
a) Using production cost as a base.
b) Using selling price as a base.

2.30 **Table 2.5**

Employment status of persons 16 years or older

Labor force (in thousands) All persons	Total	Employed	March 1970 Unemployed
Men	51,621	48,379	2,081
Married, wife present	39,138	37,103	1,020
Married, wife absent	1,065	983	70
Widowed	673	624	48
Divorced	1,200	1,117	74
Single	9,545	8,552	869
Women	31,233	29,581	1,652
Married, husband present	18,377	17,497	880
Married, husband absent	1,411	1,325	97
Widowed	2,542	2,463	79
Divorced	1,927	1,823	104
Single	6,965	6,473	492

(From Bureau of Labor Statistics, U.S. Department of Labor, *The 1972 World Almanac*, p. 151. The sum of individuals in the employed and unemployed categories will not always equal the total because of rounding errors obtained during the course of collecting these data.)

a) Determine the percent of employed men and the percent of employed women of the total labor force of men and women, respectively.
b) Determine the percent of unemployed men and the percent of unemployed women of the total labor force of men and women whose spouses are present.
c) Determine the percent of employed men and women of the labor force who are divorced.
d) Determine the percent of unemployed men and women of the labor force who are single.

2.31 Referring to the data presented in Box 1.1, calculate the percentage of cars that average 20 miles per gallon or better in city driving
a) for all cars.
b) for cars weighing 4,000 pounds or more.
c) for cars with manual transmissions.
d) for cars with automatic transmissions.

2.32 Repeat Exercise 2.31 for highway driving. Compare the results.

2.33 Shown below is the annual incidence of alcohol-related collisions among London, Ontario motorists in various age groups, 1968–1973 (as of July 1 each year).

Year	Age group		
	16–17	18–20	24
1968–1969	17	47	5
1969–1970	26	39	7
1970–1971	25	48	9
1971–1972	33	133	14
1972–1973	23	153	27

Source: Adapted from Whitehead, P. C., J. Craig, N. Langford, C. MacArthur, B. Stanton, and R. Ferrence. Collision Behavior of Young Drivers. *J. Studies Alcohol.*, **36**: 1208–1223, 1975.

a) Construct fixed-base ratios for each of the three age groups using 1968–1969 as the base year.
b) At the beginning of 1971–1972, the legal age for purchasing and consuming alcoholic beverages was changed from 21 to 18. For each age group, determine the percentage of alcohol-related collisions prior to 1971–1972 and from 1971–1972 through 1972–1973.
c) Time ratios and percentages can sometimes lead to deceptive results, particularly when the base year has a low value and the total number of observations is small. With this in mind, contrast the change in alcohol-related collisions among drivers 18–20 and drivers 24 years old.

2.34 How often have you walked away from a multiple-choice exam exclaiming in distress, "Every time I change an answer, I go from right to wrong."? It is almost a cherished article of faith that when in doubt, you should always stick with your first answer. At last, the validity of this belief has been subjected to research verification. Professor James J. Johnston recorded the number of students who made the unfortunate decision of changing from right to wrong answers and the number who made the correct decision of changing from wrong to right. Surprisingly, 71 students made a greater number of wrong-to-right responses whereas only 31 made a larger number of right-to-wrong changes. Determine the proportion of students who made the "correct" decision and the proportion who took the primrose path from right to wrong. Does the evidence appear to support the popular student belief? (Source: James J. Johnston. Sticking with first responses on multiple-choice exams: for better or for worse? *Teaching of Psychology*, **2**, #4, 1975).

2.35 Red dye number 2 became a source of great controversy in early 1976. A report was issued by the Food and Drug Administration showing the number of rats afflicted with cancer under low and high dosage levels of this common food additive. The results were as follows: among 44 rats receiving low daily dosages of red dye number 2, 4 contracted cancer. Among 44 receiving high dosages, 14 fell prey to cancer. Determine the proportion of rats in each group that contracted cancer; determine the proportion that did not contract cancer.

3 Frequency Distributions

BOX 3.1

Look again at the data presented in Box 1.1. What is the best highway mileage achieved by any of the cars? The poorest? What mileage occurs most frequently? These are difficult questions to answer when confronted with a mass of unorganized data. However, there are simple procedures for transforming chaos into order. In this chapter we will see how to obtain a **grouped frequency distribution** such as the one below. Note that the questions we asked are now readily answered. For example, the most commonly obtained highway mileage is between 15 and 16 mpg.

Class interval (highway mpg)	Real limits	f
41–42	40.5–42.5	1
39–40	38.5–40.5	0
37–38	36.5–38.5	1
35–36	34.5–36.5	9
33–34	32.5–34.5	3
31–32	30.5–32.5	9
29–30	28.5–30.5	14
27–28	26.5–28.5	3
25–26	24.5–26.5	9
23–24	22.5–24.5	6
21–22	20.5–22.5	14
19–20	18.5–20.5	2
17–18	16.5–18.5	20
15–16	14.5–16.5	28
13–14	12.5–14.5	5

3.1 GROUPING OF DATA

Let us imagine that you have just accepted a position as curriculum director in a large senior school. Your responsibility is to develop curricula which are in agreement with the needs and the motivations of the students and, at the same time, provide maximum challenges to their intellectual capacities. Now, it is beyond the scope of the text to provide a solution to this very complex and provoking problem. However, it is clear that no steps toward a solution can be initiated without some assessment of the intellectual capacities of the student body. Accordingly, you go to the Guidance Office and "pull out" **at random** (i.e., in such a way that every member of the population shares an equal chance of being selected) 110 student dockets containing a wealth of personal and scholastic information. Since your present interest is to assess intellectual ability, you focus your attention upon the entry labeled, "I.Q. estimate." You write these down on a piece of paper, with the results listed in Table 3.1.

Table 3.1

I.Q. scores of 110 high school students selected at random

154	131	122	100	113	119	121	128	112	93
133	119	115	117	110	104	125	85	120	135
116	103	103	121	109	147	103	113	107	98
128	93	90	105	118	134	89	143	108	142
85	108	108	136	115	117	110	80	111	127
100	100	114	123	126	119	122	102	100	106
105	111	127	108	106	91	123	132	97	110
150	130	87	89	108	137	124	96	111	101
118	104	127	94	115	101	125	129	131	110
97	135	108	139	133	107	115	83	109	116
110	113	112	82	114	112	113	142	145	123

As you mull over these figures, it becomes obvious to you that you cannot "make heads or tails" out of them unless you organize them in some systematic fashion. It occurs to you to list all the scores from highest to lowest and then place a slash mark alongside of each score every time it occurs (Table 3.2). The number of slash marks, then, represents the frequency of occurrence of each score.

When you have done this, you have constructed an ungrouped frequency distribution of scores. Note that, in the present example, the scores are widely spread out, a number of scores have a frequency of zero, and there is no "visually" clear indication of central tendency. Under these circumstances, it is customary for most researchers to "group" the scores into what is referred to as **class intervals** and then obtain a frequency distribution of "grouped scores."

Table 3.2

Frequency distribution of I.Q. scores of 110 high school students selected at random

<table>
<tr><th>X</th><th>f</th><th>X</th><th>f</th><th>X</th><th>f</th><th>X</th><th>f</th></tr>
<tr><td>154</td><td>|</td><td>135</td><td>||</td><td>116</td><td>||</td><td>97</td><td>||</td></tr>
<tr><td>153</td><td></td><td>134</td><td>|</td><td>115</td><td>||||</td><td>96</td><td>|</td></tr>
<tr><td>152</td><td></td><td>133</td><td>||</td><td>114</td><td>||</td><td>95</td><td></td></tr>
<tr><td>151</td><td></td><td>132</td><td>|</td><td>113</td><td>||||</td><td>94</td><td>|</td></tr>
<tr><td>150</td><td>|</td><td>131</td><td>||</td><td>112</td><td>|||</td><td>93</td><td>||</td></tr>
<tr><td>149</td><td></td><td>130</td><td>|</td><td>111</td><td>|||</td><td>92</td><td></td></tr>
<tr><td>148</td><td></td><td>129</td><td>|</td><td>110</td><td>|||||</td><td>91</td><td>|</td></tr>
<tr><td>147</td><td>|</td><td>128</td><td>||</td><td>109</td><td>||</td><td>90</td><td>|</td></tr>
<tr><td>146</td><td></td><td>127</td><td>|||</td><td>108</td><td>||||||</td><td>89</td><td>||</td></tr>
<tr><td>145</td><td>|</td><td>126</td><td>|</td><td>107</td><td>||</td><td>88</td><td></td></tr>
<tr><td>144</td><td></td><td>125</td><td>||</td><td>106</td><td>||</td><td>87</td><td>|</td></tr>
<tr><td>143</td><td>|</td><td>124</td><td>|</td><td>105</td><td>||</td><td>86</td><td></td></tr>
<tr><td>142</td><td>||</td><td>123</td><td>|||</td><td>104</td><td>||</td><td>85</td><td>||</td></tr>
<tr><td>141</td><td></td><td>122</td><td>||</td><td>103</td><td>|||</td><td>84</td><td></td></tr>
<tr><td>140</td><td></td><td>121</td><td>||</td><td>102</td><td>|</td><td>83</td><td>|</td></tr>
<tr><td>139</td><td>|</td><td>120</td><td>|</td><td>101</td><td>||</td><td>82</td><td>|</td></tr>
<tr><td>138</td><td></td><td>119</td><td>|||</td><td>100</td><td>||||</td><td>81</td><td></td></tr>
<tr><td>137</td><td>|</td><td>118</td><td>||</td><td>99</td><td></td><td>80</td><td>|</td></tr>
<tr><td>136</td><td>|</td><td>117</td><td>||</td><td>98</td><td>|</td><td></td><td></td></tr>
</table>

3.1.1 Grouping into Class Intervals

Grouping into class intervals involves a sort of "collapsing the scale" in which we assign scores to **mutually exclusive*** classes in which the classes are defined in terms of the grouping intervals employed. The reasons for grouping are twofold: (1) It is uneconomical and unwieldy to deal with a large number of cases spread out over many scores unless automatic calculators are available. (2) Some of the scores have such low frequency counts associated with them that it is not warranted to maintain these scores as separate and distinct entities.

On the negative side is, of course, the fact that grouping inevitably results in the loss of information. For example, individual scores lose their identity when we group into class intervals and some small errors in statistics based upon grouped scores are unavoidable.

The question now becomes, "On what basis do we decide upon the grouping intervals which we will employ?" Obviously, the interval selected must not be so gross that we lose the discrimination provided by our original measurement. For example, if we were to divide the previously collected I.Q. scores into two classes;

* We refer to the classes as *mutually exclusive* because it is impossible for a subject's score to belong to more than one class.

those below 100 and those 100 and above, practically all the information inherent in the original scores would be lost. On the other hand, the class intervals should not be so fine that the purposes served by grouping are defeated. In answer to our question, there is, unfortunately, no general prescription which can be applied to all data. Much of the time the choice of the number of class intervals must represent a judgment based upon a consideration of the relative effects of grouping upon discriminability and presentational economy. However, it is generally agreed that most data in the behavioral sciences can be accommodated by 10 to 20 class intervals. For uniformity, we shall aim for approximately 15 class intervals for the data that we shall discuss in this text.

Having decided upon the number of class intervals that is appropriate for a set of data, the procedures for assigning scores to class intervals are quite straightforward. Although one of several different techniques may be used, we shall employ only one for the sake of consistency. The procedures to be employed are as follows:

Step 1. Find the difference between the highest and the lowest score values contained in the original data. Add 1 to obtain the total number of scores or potential scores. In the present example, this result is $(154-80) + 1 = 75$.

Step 2. Divide this figure by 15 to obtain the number of scores or potential scores in each interval. If the resulting value is not a whole number, and it usually is not, the authors prefer to round to the nearest odd number so that a whole number will be at the middle of the class interval. However, this practice is far from universal and you would not be wrong if you rounded to the *nearest number*. In the present example, the number of scores for each class interval is $\frac{75}{15}$, or 5. We shall designate the class interval by the symbol i. In the example, $i = 5$.

Step 3. Take the lowest score in the original data as the minimum value in the lowest class interval. Add to this $i - 1$ to obtain the maximum score of the lowest class interval. Thus the lowest class interval of the data on hand is 80–84.

Step 4. The next higher class interval begins at the integer following the maximum score of the lower class interval. In the present example, the next integer is 85. Follow the same steps as in step 3 to obtain the maximum score of the second class interval. Follow these procedures for each successive higher class interval until all the scores are included in their appropriate class intervals.

Step 5. Assign each obtained score to the class interval within which it is included. The **grouped frequency distribution** appearing in Table 3.3 was obtained by employing the above procedures.

You will note that by grouping we may obtain an immediate "picture" of the distribution of I.Q. scores among our high school students. For example, we note that there is a clustering of frequencies in the class intervals between the scores of 100 and 119. It is also apparent that the number of scores in the extremes tends to dwindle off. Thus we have achieved one of our objectives in grouping, to provide an economical and manageable array of scores.

Table 3.3

Grouped frequency distribution of I.Q. scores based upon data appearing in Table 3.2

Class interval	f	Class interval	f
150–154	2	110–114	17
145–149	2	105–109	14
140–144	3	100–104	12
135–139	5	95–99	4
130–134	7	90–94	5
125–129	9	85–89	5
120–124	9	80–84	3
115–119	13		
			$N = 110$

Excellent examples of employing overlapping classes (i.e., not mutually exclusive) to produce statistical absurdities are found in Huff's book.*

A good deal of bumbling and chicanery have come from adding together things that don't add up but merely seem to. Children for generations have been using a form of this device to prove that they don't go to school.

You probably recall it. Starting with 365 days to the year you can subtract 122 for the one-third of the time you spend in bed and another 45 for the three hours a day used in eating. From the remaining 198 take away 90 for summer vacation and 21 for Christmas and Easter vacations. The days that remain are not even enough to provide for Saturdays and Sundays. [See Fig. 3.1.]

Using the same logic, a dean of a college faculty might prove that her teachers never work . . . which might be true, but not for the reasons given.

3.1.2 The True Limits of a Class Interval

In our prior discussion of the "true limits" of a cardinal number, we pointed out that the "true" value of a number is equal to its apparent value plus and minus one-half of the unit of measurement. Of course, the same is true of these values even after they have been grouped into class intervals. Thus, although we write the limits of the lowest class interval as 80–84, the **true limits of the interval** are 79.5–84.5 (i.e., the lower real limit of 80 and the upper real limit of 84, respectively).

It is important to keep in mind that the true limits of a class interval are not the same as the *apparent limits*. When calculating the median and percentile ranks for grouped data, we shall make use of the *true limits* of the class interval.

* Reprinted from *How to Lie with Statistics*, by Darrell Huff. Pictures by Irving Geis. By permission of W. W. Norton & Company, Inc. Copyright 1954 by Darrell Huff and Irving Geis.

© 1955 United Feature Syndicate, Inc.

Fig. 3.1 However, when you subtract 2928 hours a year spent in bed, 1080 hours spent in eating, 2160 hours for summer vacation, 504 hours for Christmas and Easter vacations. . . .

BOX 3.2 SMOOTHING FREQUENCY DISTRIBUTIONS

It is often desirable to smooth the graphic representations of frequency distributions. One of the most commonly employed techniques is referred to as the method of moving averages. The procedures consist of the following steps:

a) Add two class intervals to each end of the frequency distribution. All four of these intervals will have zero frequency associated with them.
b) Add a third column and label it "smoothed frequency." Enter a zero in the third column of the lowest class interval.
c) To find the smoothed frequency of the second class interval from the bottom, add together the frequency in that interval, the interval above, and the frequency in the interval below, and divide by 3. Enter this value in the third column corresponding to the second from the lowest class interval.
d) Obtain the moving average for each class interval by first summing together the frequency within that interval and the frequencies in the intervals directly above and below that interval and then by dividing by 3.

Table 3.4 gives the smoothed frequency distribution for the data given in Table 3.3.

Table 3.4
Smoothed frequency distribution

Class interval	f	Smoothed frequency
160–164	0	0.00
155–159	0	0.67
150–154	2	1.33
145–149	2	2.33
140–144	3	3.33
135–139	5	5.00
130–134	7	7.00
125–129	9	8.33
120–124	9	10.33
115–119	13	13.00
110–114	17	14.67
105–109	14	14.33
100–104	12	10.00
95–99	4	7.00
90–94	5	4.67
85–89	5	4.33
80–84	3	2.67
75–79	0	1.00
70–74	0	0.00

3.2 CUMULATIVE FREQUENCY AND CUMULATIVE PERCENTAGE DISTRIBUTIONS

It is often desirable to arrange the data from a frequency distribution into a **cumulative frequency distribution**. Besides aiding in the interpretation of the frequency distribution, a cumulative frequency distribution is of great value in obtaining the median and the various percentile ranks of scores, as we shall see in Chapter 5.

The cumulative frequency distribution is obtained in a very simple and straightforward manner. Let us attend to the data in Table 3.5.

The entries in the frequency distribution indicate the number of high school students falling within each of the class intervals. Each entry within the cumulative frequency distribution indicates the number of all cases or frequencies *below the upper real limit* of that interval. Thus, in the third class interval from the bottom in Table 3.5, the entry "13" in the cumulative frequency distribution indicates that a total of 13 students scored lower than the upper real limit of that interval, which is 94.5. The entries in the cumulative frequency distribution are obtained by a simple process of successive addition of the entries in the frequency column. Thus the cumulative frequency of the interval 104.5–109.5 is obtained by successive addition of $3 + 5 + 5 + 4 + 12 + 14 = 43$. Note that the top entry in the cumulative frequency column is always equal to N. If you fail to obtain this result, you know that you have made an error in cumulating frequencies and should check your work.

The **cumulative percentage distribution**, also shown in Table 3.5, is obtained by dividing each entry in the cumulative frequency column by N and multiplying by 100. Note that the top entry must be 100% since all cases fall below the upper real limit of the highest interval.

Table 3.5

Grouped frequency distribution and cumulative frequency distribution based upon data appearing in Table 3.3. $N = 110$

Class interval	f	Cumulative f	Cumulative %
150–154	2	110	100
145–149	2	108	98
140–144	3	106	96
135–139	5	103	94
130–134	7	98	89
125–129	9	91	83
120–124	9	82	75
115–119	13	73	66
110–114	17	60	55
105–109	14	43	39
100–104	12	29	26
95–99	4	17	15
90–94	5	13	12
85–89	5	8	7
80–84	3	3	3

CHAPTER SUMMARY

This chapter was concerned with "making sense" out of a mass of data by the construction of frequency distributions of scores or values of the variable. When the scores are widely spread out, many have a frequency of zero, and when there is no clear indication of central tendency, it is customary to group scores into class intervals. The resulting distribution is referred to as a grouped frequency distribution.

The basis for arriving at a decision concerning the grouping units to employ and the procedures for constructing a grouped frequency distribution were discussed and demonstrated. It was seen that the true limits of a class interval are obtained in the same way as the true limits of a score. Finally, the procedures for converting a frequency distribution into a cumulative frequency distribution and a cumulative percentage distribution were demonstrated.

Terms to Remember:

Random	***True limits of an interval***
Frequency distribution	***Apparent limits of an interval***
Grouped frequency distribution	***Cumulative frequency distribution***
Class intervals	***Cumulative percentage distribution***
Mutually exclusive	

EXERCISES

3.1 Give the true limits, the midpoints, and the width of interval for each of the following class intervals.

a) 8–12 b) 6–7 c) 0–2 d) 5–14
e) (−2)–(−8) f) 2.5–3.5 g) 1.50–1.75 h) (−3)–(+3)

3.2 For each of the following sets of measurements, state (a) the best width of class interval (i), (b) the apparent limits of the lowest interval, (c) the true limits of that interval, (d) the midpoint of that interval.

i) 0 to 106 ii) 29 to 41 iii) 18 to 48
iv) −30 to +30 v) 0.30 to 0.47 vi) 0.206 to 0.293

3.3 Given the following list of scores in a statistics examination, use $i = 5$ for the class intervals and (a) set up a frequency distribution; (b) list the true limits and the midpoint of each interval; (c) prepare a cumulative frequency distribution; and (d) prepare a cumulative percentage distribution.

Scores on a Statistics Examination

63	88	79	92	86	87	83	78	41	67
68	76	46	81	92	77	84	76	70	66
77	75	98	81	82	81	87	78	70	60
94	79	52	82	77	81	77	70	74	61

Table 3.6

High prices per share of 250 stocks

55	$90\frac{7}{8}$	$22\frac{1}{2}$	$40\frac{1}{2}$	81	47	$48\frac{3}{8}$	9	$37\frac{1}{2}$	$26\frac{3}{8}$	$33\frac{1}{2}$	86	$40\frac{5}{8}$	$16\frac{3}{4}$	$15\frac{5}{8}$	$41\frac{5}{8}$	$78\frac{1}{4}$	$10\frac{7}{8}$	$14\frac{7}{8}$	$32\frac{1}{8}$
$16\frac{1}{2}$	$21\frac{5}{8}$	$30\frac{1}{2}$	$68\frac{3}{4}$	68	$29\frac{5}{8}$	$31\frac{3}{4}$	$44\frac{1}{8}$	$20\frac{3}{4}$	$29\frac{1}{2}$	$36\frac{1}{2}$	$42\frac{1}{2}$	$33\frac{1}{8}$	$114\frac{3}{4}$	$55\frac{1}{8}$	32	$31\frac{5}{8}$	$31\frac{1}{2}$	13	$50\frac{3}{8}$
$73\frac{5}{8}$	$42\frac{3}{4}$	$26\frac{3}{8}$	22	$58\frac{1}{8}$	$55\frac{3}{4}$	$18\frac{1}{2}$	$31\frac{1}{2}$	$43\frac{1}{2}$	$30\frac{5}{8}$	$42\frac{1}{4}$	56	$14\frac{1}{4}$	$24\frac{1}{8}$	$29\frac{1}{8}$	$64\frac{5}{8}$	$17\frac{1}{4}$	69	$12\frac{1}{2}$	$29\frac{5}{8}$
$39\frac{3}{4}$	$21\frac{5}{8}$	$37\frac{7}{8}$	$25\frac{7}{8}$	$185\frac{1}{4}$	$47\frac{1}{8}$	$34\frac{3}{4}$	$34\frac{1}{2}$	$32\frac{1}{2}$	42	$33\frac{1}{4}$	$6\frac{1}{2}$	$14\frac{1}{2}$	$34\frac{3}{8}$	$27\frac{1}{4}$	$31\frac{1}{8}$	$80\frac{1}{2}$	$27\frac{7}{8}$	34	41
$10\frac{3}{8}$	$69\frac{5}{8}$	$64\frac{1}{4}$	$29\frac{1}{2}$	230	18	38	$62\frac{1}{8}$	$21\frac{3}{4}$	$106\frac{3}{4}$	$21\frac{7}{8}$	$28\frac{1}{2}$	$107\frac{1}{4}$	24	77	70	$60\frac{5}{8}$	91	$54\frac{1}{2}$	$37\frac{3}{4}$
$76\frac{1}{2}$	35	$44\frac{3}{4}$	$32\frac{7}{8}$	$97\frac{1}{4}$	26	34	$36\frac{1}{2}$	$24\frac{1}{8}$	$17\frac{1}{2}$	$28\frac{3}{4}$	$16\frac{7}{8}$	101	$21\frac{1}{4}$	86	36	$37\frac{1}{4}$	$24\frac{5}{8}$	$54\frac{5}{8}$	$57\frac{5}{8}$
27	$53\frac{7}{8}$	$33\frac{1}{8}$	$34\frac{7}{8}$	76	$14\frac{5}{8}$	26	$72\frac{1}{8}$	$30\frac{3}{4}$	$21\frac{1}{2}$	$11\frac{1}{8}$	$54\frac{1}{4}$	51	$67\frac{3}{4}$	$61\frac{5}{8}$	13	34	48	$15\frac{1}{4}$	$40\frac{1}{8}$
30	$52\frac{3}{4}$	$90\frac{3}{4}$	$31\frac{5}{8}$	$51\frac{1}{2}$	$32\frac{1}{2}$	$33\frac{1}{4}$	$23\frac{3}{4}$	$23\frac{5}{8}$	33	$34\frac{3}{4}$	93	40	$35\frac{3}{8}$	$22\frac{1}{4}$	$50\frac{3}{8}$	8	$32\frac{1}{2}$	$57\frac{1}{8}$	$43\frac{1}{2}$
$8\frac{5}{8}$	$36\frac{7}{8}$	$20\frac{5}{8}$	$28\frac{1}{4}$	$42\frac{5}{8}$	25	$104\frac{1}{2}$	$57\frac{3}{8}$	$53\frac{3}{4}$	63	$41\frac{1}{2}$	$23\frac{5}{8}$	$26\frac{1}{4}$	96	$49\frac{1}{2}$	$32\frac{5}{8}$	$9\frac{3}{4}$	108	85	$29\frac{3}{8}$
$34\frac{1}{2}$	78	$12\frac{3}{4}$	$17\frac{7}{8}$	$43\frac{1}{4}$	$10\frac{3}{4}$	$46\frac{1}{2}$	$35\frac{1}{2}$	67	$39\frac{7}{8}$	$23\frac{3}{4}$	$43\frac{1}{4}$	$64\frac{1}{2}$	$72\frac{5}{8}$	$43\frac{1}{2}$	$68\frac{7}{8}$	43	$163\frac{1}{4}$	$61\frac{7}{8}$	114
$19\frac{7}{8}$	$89\frac{1}{8}$	$27\frac{5}{8}$	$53\frac{3}{4}$	$65\frac{7}{8}$	$10\frac{3}{4}$	$38\frac{1}{2}$	$25\frac{5}{8}$	$78\frac{1}{2}$	16	$16\frac{3}{8}$	$51\frac{1}{2}$	$51\frac{1}{4}$	24	$9\frac{3}{4}$	$92\frac{7}{8}$	$33\frac{5}{8}$	132	53	103
$12\frac{1}{8}$	66	27	$14\frac{7}{8}$	$14\frac{3}{4}$	50	$43\frac{3}{8}$	32	$120\frac{3}{8}$	$26\frac{1}{4}$	$56\frac{1}{4}$	$20\frac{3}{4}$	$45\frac{1}{8}$	84	29	$43\frac{1}{4}$	$29\frac{3}{4}$	107	85	33
$11\frac{7}{8}$	$14\frac{1}{2}$	$55\frac{1}{2}$	$54\frac{1}{2}$	$28\frac{1}{8}$	$26\frac{3}{4}$	$57\frac{1}{2}$	$111\frac{5}{8}$	$131\frac{1}{2}$	22										

3.4 Using the data in Exercise 3.3, set up frequency distributions with the following: (a) $i = 1$ (ungrouped frequency distribution), (b) $i = 3$, (c) $i = 10$, (d) $i = 20$. Discuss the advantages and the disadvantages of employing these widths.

3.5 Given the following list of numbers: (a) Construct a grouped frequency distribution. (b) List the true limits and the midpoint of each interval. Indicate the width employed. (c) Compare the results with those of Exercise 3.3 and 3.4.

6.3	8.8	7.9	9.2	8.6	8.7	8.3	7.8	4.1	6.7
6.8	7.6	4.6	8.1	9.2	7.7	8.4	8.6	7.0	6.6
7.7	7.5	9.8	8.1	8.2	8.1	8.7	7.8	7.0	6.0
9.4	7.9	5.2	8.2	7.7	8.1	7.7	7.0	7.4	6.1

3.6 Several entries in a frequency distribution showing the yield of corn per acre of land are 15–21, 8–14, 1–7. (a) What is the width of the interval? (b) What are the lower and upper real limits of each interval? (c) What are the midpoints of each interval?

3.7 Listed in Table 3.6 are the high prices per share of 250 stocks sold on the New York Stock Exchange on January 27, 1968, as shown in The New York Times, January 27, 1968.

a) Round each price to the nearest dollar and group the results into a frequency distribution.
b) Try several grouping intervals and note changes on the form of the resulting frequency distributions.
c) Identify the apparent and real limits of the lowest class interval for each of the resulting frequency distributions.

3.8 Listed in Table 3.7 are the prices per share of 30 stocks sold on the New York Stock Exchange on January 1, 1967, the prices on January 1, 1968, and the net change. (Data are based on The New York Times, January 8, 1968).

a) Convert the data to percentage of change, employing the 1967 figures as a base.
b) Arrange the stocks in order of the 1967 prices, from the lowest to the highest. Examine the percentage-of-increase figures. Does there appear to be any relationship between the cost of the stock and the percentage of change during the year?

Table 3.7

1967	1968	Net change	1967	1968	Net change	1967	1968	Net change	1967	1968	Net change
13	$16\frac{3}{8}$	$3\frac{3}{8}$	$18\frac{1}{2}$	$42\frac{1}{2}$	24	$29\frac{1}{8}$	$18\frac{7}{8}$	$-10\frac{1}{4}$	33	40	7
46	47	1	$27\frac{3}{4}$	$26\frac{3}{8}$	$-1\frac{3}{8}$	35	64	29	$15\frac{1}{8}$	$27\frac{1}{4}$	$12\frac{1}{8}$
$28\frac{3}{8}$	$28\frac{3}{4}$	$\frac{3}{8}$	$7\frac{3}{4}$	$13\frac{3}{4}$	6	$32\frac{1}{4}$	$41\frac{1}{2}$	$9\frac{1}{4}$	$53\frac{1}{4}$	$48\frac{3}{4}$	$-4\frac{1}{2}$
$38\frac{2}{8}$	$44\frac{7}{8}$	$6\frac{5}{8}$	$25\frac{1}{2}$	46	$20\frac{1}{2}$	$103\frac{1}{4}$	124	$20\frac{3}{4}$	$27\frac{1}{4}$	$57\frac{1}{4}$	30
$39\frac{7}{8}$	$34\frac{1}{2}$	$-5\frac{3}{8}$	$57\frac{3}{4}$	$70\frac{3}{4}$	13	$29\frac{5}{8}$	$34\frac{5}{8}$	5	$22\frac{1}{2}$	$36\frac{7}{8}$	$14\frac{3}{8}$
$26\frac{7}{8}$	$31\frac{3}{8}$	$4\frac{1}{2}$	$27\frac{3}{4}$	$22\frac{1}{4}$	$-5\frac{1}{2}$	$3\frac{1}{2}$	10	$6\frac{1}{2}$	$70\frac{1}{2}$	$64\frac{1}{8}$	$-6\frac{3}{8}$
$15\frac{7}{8}$	$62\frac{7}{8}$	47	102	95	-7	$30\frac{7}{8}$	34	$3\frac{1}{8}$	$10\frac{7}{8}$	$19\frac{1}{8}$	$8\frac{1}{4}$
$44\frac{7}{8}$	$79\frac{3}{8}$	$29\frac{1}{2}$	$21\frac{3}{8}$	$25\frac{1}{2}$	$4\frac{1}{8}$						

3.9 If all the net-change figures in the preceding problem were placed in a hat and drawn at random, what is the likelihood that (round all figures to the nearest integer):

a) a stock would be selected which changed by as much as 10 points (*Note*: this statement implies *both* positive and negative changes)?
b) a given stock would change by as much as +10 points?
c) a given stock would change between +5 and +10 points?
d) a given stock would change between −1 and −10 points?

3.10 Construct a grouped frequency distribution, using 5–9 as the lowest class interval, for the following list of numbers. List the width, midpoint, and real limits of the highest class interval.

67	63	64	57	56	55	53	53	54	54
45	45	46	47	37	23	34	44	27	44
45	34	34	15	23	43	16	44	36	36
35	37	24	24	14	43	37	27	36	26
25	36	26	5	44	13	33	33	17	33

3.11 Do Exercise 3.10 again, using 3–7 as the lowest class interval. Compare the resulting frequency distribution with those of Exercises 3.10, 3.12, and 3.13.

3.12 Repeat Exercise 3.10, using $i = 2$. Compare the results with those of Exercises 3.10, 3.11, and 3.13.

3.13 Repeat Exercise 3.10, using $i = 10$. Compare the results with those of Exercises 3.10, 3.11, and 3.12.

3.14 Refer to Table 3.5, and answer the following. If all the scores were placed in a hat and selected at random, what is the likelihood that:
a) The score would be in the interval 145–149?
b) The score would be between 130 and 138?
c) The score would be 94 or below?
d) The score would be 145 or above?

3.15 Referring to Exercise 1.17, calculate the distribution ratio of the times the investigator recorded each of the following time periods over the total number of observations:

a) 1.0	b) 1.1	c) 1.2	d) 1.3
e) 1.4	f) 1.5	g) 0.9	h) 0.8
i) 0.7	j) 0.6	k) 0.5	

3.16 Assume that the scores in Exercise 1.17 are grouped in six class intervals.
a) Give the true limits, midpoints, and width of each interval.
b) Set up the frequency distribution.
c) Obtain the cumulative frequency distribution and the cumulative percentage distribution.

3.17 Suppose a person investigates the price of a certain item at forty different stores and finds that the prices are listed as follows:

$60	$75	$89	$77	$65	$80	$63	$72
$87	$64	$73	$75	$67	$74	$75	$74
$68	$73	$75	$75	$74	$76	$71	$76
$86	$82	$70	$71	$68	$78	$83	$77
$74	$67	$88	$80	$72	$78	$85	$84

a) Construct the frequency distribution, using 15 class intervals.
b) List the true limits and midpoints of each interval.

3.18 Referring to Exercise 3.17, set up the frequency distribution for:
a) $i = 3$ b) $i = 10$

3.19 Referring again to Exercise 3.17, suppose a person randomly selects one of the stores and buys the item there. What is the likelihood that it would cost:
a) \$60 b) \$89 c) \$75 d) from \$72–\$73
e) from \$78–\$79 f) from \$74–\$75 g) more than \$75 h) less than \$67

3.20 A manager of a company found that the number of sick days taken by his 50 employees last year was:

10	35	12	8	44	6	15	20	5	7
5	11	17	8	4	7	25	9	2	10
12	12	3	10	9	13	5	16	31	9
0	4	7	11	3	18	2	10	6	22
2	9	8	29	6	4	7	10	0	1

a) Using $i = 3$, set up the frequency distribution.
b) Obtain the cumulative frequency distribution and cumulative percentage distribution.

3.21 If the lowest class interval on a frequency distribution is 64–72, and there are 15 class intervals, what are the midpoint, the width, the apparent limits, and the true limits of the highest interval?

3.22 A manager of a company recorded the units completed in a certain clerical job by each employee. Fifty employees perform the same tasks, but are seated in two different rooms. In room A, the number of units completed in one day is:

21	22	20	15	25
0	28	9	28	30
24	29	27	34	38
24	35	36	31	41
32	43	44	53	50

In room B, the number of units completed is:

6	21	13	36	18
24	32	16	38	20
28	25	33	26	30
26	29	35	45	59
32	31	30	40	30

a) Combine all the scores and obtain the frequency distribution for $i = 4$.
b) Obtain the cumulative frequency distribution and the cumulative percentage distribution.

3.23 Referring to Exercise 3.22, suppose the manager introduces piped music, carpeting, and higher illumination into room B and then records new scores for room B:

36	41	40	44	47
45	44	42	48	46
48	49	50	49	51
51	53	54	54	56
52	55	52	59	58

a) Set up a frequency distribution, $i = 4$, for these scores and compare with the distribution found in Exercise 3.22.
b) Combine these scores with those of room A and set up the frequency distribution with $i = 4$.
c) Compute the cumulative frequency distribution and the cumulative percentage distribution.

3.24 Referring to Exercises 3.22 and 3.23, obtain and compare the frequency distributions for room A and both sets of data for room B.

3.25 Set up a frequency distribution for city miles per gallon using the data presented in Box 1.1.

3.26 Table 3.8 shows by state, the number of families receiving welfare in February 1975 and the average monthly payment per family.
a) Set up a frequency distribution for the number of families receiving welfare. Round to the nearest thousand and use $i = 25$.
b) Set up a frequency distribution for the average payment per family. Round to the nearest dollar and use $i = 15$.

Table 3.8

Welfare for families with dependent children (February 1975)

State	Number of recipient families	Average payment per family
Alabama	49,642	$97.06
Alaska	4,257	260.35
Arizona	20,593	127.15
Arkansas	31,635	127.11
California	440,493	238.43
Colorado	30,842	201.24
Connecticut	38,785	266.06
Delaware	10,182	165.59
Dist. of Columbia	31,737	242.83
Florida	80,967	118.47
Georgia	114,174	101.74
Hawaii	14,632	305.93
Idaho	6,361	210.28
Illinois	219,080	320.88
Indiana	51,810	149.77
Iowa	26,747	281.10
Kansas	22,310	205.84
Kentucky	50,980	179.40
Louisiana	65,965	120.42
Maine	24,522	172.38
Maryland	67,890	168.95
Massachusetts	108,062	283.44
Michigan	198,623	269.58

(Continued)

Table 3.8 *(Continued)*

State	Number of recipient families	Average payment per family
Minnesota	43,550	248.46
Mississippi	53,811	49.74
Missouri	83,140	121.29
Montana	7,447	161.92
Nebraska	11,891	180.36
Nevada	4,723	144.89
New Hampshire	8,515	235.94
New Jersey	130,129	273.34
New Mexico	18,793	137.55
New York	349,689	325.04
North Carolina	58,841	161.29
North Dakota	4,509	199.78
Ohio	171,870	174.76
Oklahoma	30,916	180.93
Oregon	35,704	202.87
Pennsylvania	182,475	263.83
Puerto Rico	47,561	45.85
Rhode Island	16,009	244.43
South Carolina	42,254	88.67
South Dakota	7,889	202.47
Tennessee	63,878	106.12
Texas	113,536	107.77
Utah (January)	11,865	219.64
Vermont	6,530	255.86
Virginia	56,588	192.66
Washington	48,937	234.14
West Virginia	21,210	174.26
Wisconsin	53,702	265.60
Wyoming	2,320	153.08

Source. Social and Rehabilitation Service, Department of Health Education and Welfare.

3.27 Table 3.9 shows the number of traffic deaths, by state, per 100 million vehicle miles during 1973 and 1974. Plot separate frequency distributions of rates of traffic deaths for 1973 and 1974, using a class interval of 0.3.

3.28 Table 3.10 shows the worldwide suicide rates per 100,000 population. Prepare a grouped frequency distribution of worldwide suicide rates using $i = 2.4$.

3.29 Table 3.11 shows typical tuition fees at selected colleges and universities (for state or other public institutions the figure shown is the tuition fee for residents). (Source: *The 1976 World Almanac and Book of Facts.*) Using $i = 300$, set up a frequency distribution.

Table 3.9

Motor-vehicle traffic deaths by states, per 100 million vehicle miles, 1973–1974

State	1974 rate	1973 rate	State	1974 rate	1973 rate
Alabama	4.1	6.2	Montana	5.1	5.8
Alaska	4.1	4.0	Nebraska	3.5	3.9
Arizona	4.7	6.0	Nevada	5.2	6.4
Arkansas	3.9	5.0	New Hampshire	3.4	2.8
California	3.1	3.8	New Jersey	2.4	2.8
Colorado	3.8	4.2	New Mexico	5.7	6.7
Connecticut	2.2	2.8	New York	4.0	4.6
Delaware	3.3	3.7	North Carolina	4.5	5.3
District of Columbia	2.6	2.5	North Dakota	3.7	4.8
Florida	3.7	4.5	Ohio	3.0	3.7
Georgia	4.4	5.3	Oklahoma	3.5	3.7
Hawaii	3.3	3.4	Oregon	4.4	4.0
Idaho	5.9	6.5	Pennsylvania	3.2	3.7
Illinois	3.4	3.9	Rhode Island	1.8	2.4
Indiana	3.3	4.2	South Carolina	4.4	4.7
Iowa	3.6	4.1	South Dakota	4.5	5.6
Kansas	3.4	4.1	Tennessee	4.1	4.9
Kentucky	3.3	4.6	Texas	3.9	4.6
Louisiana	4.4	6.0	Utah	3.1	5.0
Maine	3.3	3.6	Vermont	4.2	4.7
Maryland	3.1	3.2	Virginia	3.1	3.5
Massachusetts	3.4	3.4	Washington	3.4	3.3
Michigan	3.4	3.8	West Virginia	4.3	4.7
Minnesota	3.5	4.1	Wisconsin	3.3	4.0
Mississippi	4.7	6.4	Wyoming	5.6	5.6
Missouri	3.5	4.7			

Source. National Safety Council.

Table 3.10

Suicide rates around the world (Deaths per 100,000 population, 1972)

Country	Rate	Country	Rate	Country	Rate
Angola[1]	1.1	Chile[2]	5.0	France[5]	16.0
Australia	12.5	Colombia[4]	2.9	Germany, East[2]	30.5
Austria	23.3	Costa Rica[1]	3.0	Germany, West[2]	21.2
Barbados	1.7	Czechoslovakia[1]	24.2	Greece[1]	3.3
Belgium[2]	16.5	Denmark[1]	24.7	Guatemala[2]	3.2
Bolivia[3]	1.4	Ecuador[1]	2.3	Hong Kong	10.6
Bulgaria	11.4	El Salvador[1]	8.7	Hungary	36.9
Canada	12.2	Finland[1]	21.7	Iceland	8.6

(*Continued*)

Table 3.10 *(Continued)*

Country	Rate	Country	Rate	Country	Rate
Ireland[1]	2.7	Paraguay[1]	1.5	Switzerland[1]	18.6
Israel	6.7	Philippines[2]	0.6	Trinidad & Tobago[4]	5.6
Italy[2]	5.8	Poland	12.0	Turkey[1]	1.9
Jamaica[1]	1.0	Portugal[1]	8.2	United Kingdom:	
Japan	17.0	Puerto Rico	10.1	England and Wales	7.7
Luxembourg	15.0	Singapore	10.9	Northern Ireland	3.0
Mauritius	2.3	South Africa:[2]		Scotland	8.1
Mexico	0.7	Colored	3.4	United States[5]	11.7
Netherlands[5]	7.9	White	12.5	Uruguay[1]	10.4
New Zealand[1]	8.3	Spain[1]	4.3	Venezuela[1]	6.4
Norway[1]	8.1	Sri Lanka[7]	17.2		
Panama	2.1	Sweden[1]	20.4		

[1]1971. [2]1970. [4]1969. [5]Supplied by World Health Organization. [6]1967. [7]1968.

Sources. *United Nations Demographic Yearbook, 1973*; World Health Organization.

Table 3.11

Typical Tuition fees

School	Tuition	School	Tuition
Adrian	$2,118	Carnegie-Mellon	2,900
Akron, Univ. of	705	Case Western Reserve Univ.	3,120
Alabama, Univ. of	595	Clemson Univ.	640
Amherst	3,795	Colorado, Univ. of	353
Anderson	1,860	Connecticut, Univ. of	765
Arizona, Univ. of	450	Dakota State	559
Auburn	549	Dana	2,120
Avila	1,700	Dayton, Univ. of	2,000
Baldwin-Wallace	2,679	Delaware, Univ. of	795
Ball State	720	Denver, Univ. of	3,150
Baylor	1,362	Doane	2,000
Black Hills State	574	Drake Univ.	2,750
Bob Jones	1,020	Duke Univ.	2,980
Boston	2,950	Duquesne Univ.	2,550
Bowdoin	3,385	Evansville, Univ. of	1,995
Bowling Green State	780	Fisk Univ.	1,950
Brigharn Young	340	Fordham Univ.	2,600
Brown	3,930	Furman Univ.	2,048
Bryn Mawr	3,700	Georgia, Univ. of	612
Canisius	2,300	Gonzaga Univ.	2,100
Carleton	4,600	Hampton Inst.	1,895

School	Tuition	School	Tuition
Harvard Univ.	3,400	Pan American Univ.	270
Haverford	2,975	Pennsylvania, Univ. of	3,450
Holy Cross	3,200	Pennsylvania State Univ.	960
Houston, Univ. of	340	Peru State	588
Idaho, Univ. of	400	Pittsburgh, Univ. of	1,024
Idaho State Univ.	400	Portland, Univ. of	3,204
Indiana Univ.	722	Princeton Univ.	4,000
Indiana State Univ.	720	Providence	2,250
Iowa, Univ. of	780	Purdue Univ	2,020
Iowa State Univ. of Sci. & Tech.	720	Redlands, Univ. of	2,975
Jacksonville Univ.	1,970	Rhode Island, Univ. of	796
John Carroll Univ.	3,200	Rice Univ.	2,180
Kansas, Univ. of	573	Richmond, Univ. of	2,425
Kansas State Univ.	532	Roger Willams	1,015
Kent State Univ.	810	Rutgers Univ.	725
Kentucky, Univ. of	480	St. Bonaventure Univ.	2,300
Leland Stanford Univ.	3,810	St. Francis	2,490
Marquette Univ.	2,450	St. Olaf	3,600
Maryland, Univ. of	698	Santa Clara, Univ. of	2,508
Memphis State Univ.	406	Sarah Lawrence	4,150
Michigan, Univ. of	800	Seattle Univ.	2,160
Minnesota, Univ. of	714	Seton Hall Univ.	1,900
Mississippi State Univ.	$511	Southern Calif., Univ. of	2,910
Mississippi, Univ. of	553	Southern Methodist Univ.	2,450
Montana, Univ. of	529	Swarthmore	3,450
Montana State Univ.	519	Syracuse Univ.	3,438
Morgan State	701	Temple Univ.	525
Muskingum	3,730	Texas A & M Univ.	300
Nevada, Univ. of	608	Tufts Univ.	3,600
New Hampshire, Univ. of	2,472	Tulane Univ.	3,000
New Mexico, Univ. of	456	Utah, Univ. of	480
New York Univ.	3,300	Vanderbilt Univ.	3,220
Niagara Univ	2,190	Vassar	3,275
North Carolina, Univ. of	453	Vermont, Univ. of	1,088
North Dakota, Univ. of	527	Wake Forest Univ.	2,200
North Dakota State Univ.	435	Washburn Univ. of Topeka	700
Notre Dame, Univ. of	2,782	Way College of Emporia	2,000
Oberlin	3,300	Wayne State Univ.	332
Ohio State Univ.	780	Wellesley	5,150
Oklahoma, Univ. of	445	West Virginia Univ.	318
Oklahoma State Univ.	464	Wyoming, Univ. of	410
Old Dominion Univ.	570	Yale Univ.	4,050
Oral Roberts Univ.	1,300	Youngstown State Univ.	630

4 Graphing Techniques

BOX 4.1

It has often been said, "A picture is worth a thousand words." In statistics, one might say, a graph is worth a thousand data points. As a case in point, consider our discussion in the previous chapter concerning the size of the class interval. If the interval is too broad, much of the information inherent in the data is lost.

Figures 4.1 to 4.3 present the mileage data using three different interval widths.

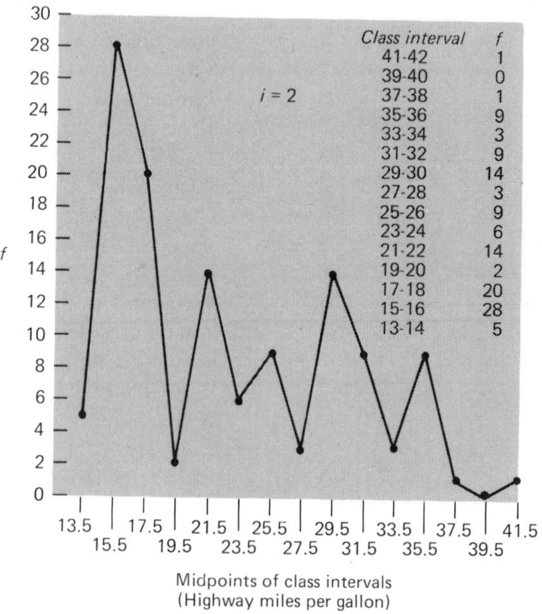

Figure 4.1

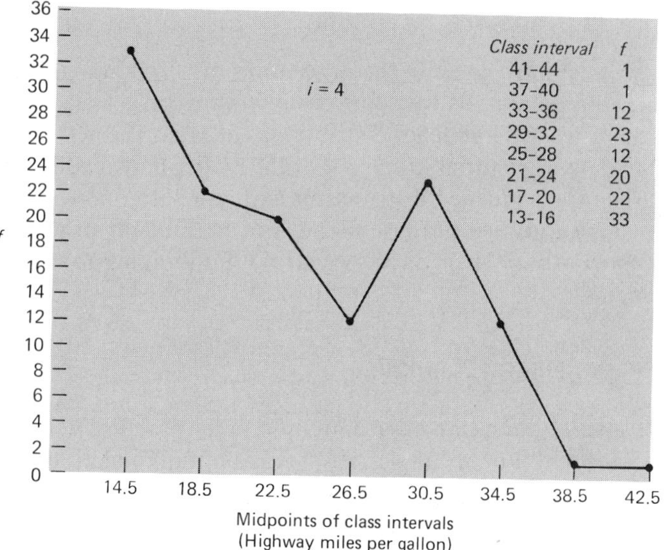

36
34
32
30
28
26
24
22
20
f 18
16
14
12
10
8
6
4
2
0
i = 4
Class interval f
41-44 1
37-40 1
33-36 12
29-32 23
25-28 12
21-24 20
17-20 22
13-16 33
14.5 18.5 22.5 26.5 30.5 34.5 38.5 42.5
Midpoints of class intervals
(Highway miles per gallon)

Figure 4.2

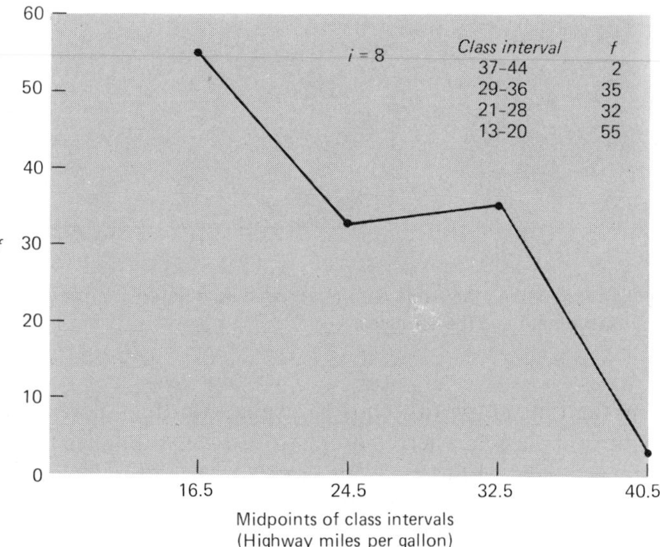

60
50
40
f 30
20
10
0
i = 8
Class interval f
37-44 2
29-36 35
21-28 32
13-20 55
16.5 24.5 32.5 40.5
Midpoints of class intervals
(Highway miles per gallon)

Figure 4.3

4.1 INTRODUCTION

In Chapter 3, we examined some of the procedures involved in making sense out of a mass of unorganized data. As we pointed out, your work is usually just beginning when you have constructed frequency distributions of data. The next step, commonly, is to present the data in pictorial form so that the reader may readily apprehend the essential features of a frequency distribution and compare one with another if he desires. Such pictures, called graphs, should not be thought of as substitutes for statistical treatment of data but rather as visual aids for thinking about and discussing statistical problems.

4.2 MISUSE OF GRAPHING TECHNIQUES

As you are well aware, graphs are often employed in the practical world of commerce to mislead the reader. For example, by the astute manipulation of the vertical (**ordinate** or Y-axis) and horizontal (**abscissa** or X-axis) axes of a graph, it is possible to convey almost any impression that is desired. Figure 4.4 illustrates this misapplication of graphing techniques. In it are shown two bar graphs of the same data in which the ordinate and the abscissa are successively elongated to produce two distinctly different impressions.

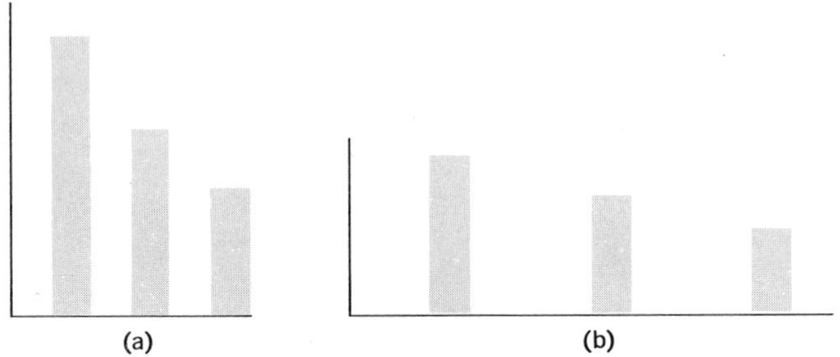

Fig. 4.4 Bar graphs representing the same data but producing different impressions by varying the relative lengths of the ordinate and the abscissa.

It will be noted that graph (a) tends to exaggerate the difference in frequency counts among the four classes, whereas graph (b) tends to minimize these differences.

The differences might be further exaggerated by use of a device called the "oh boy" chart (from *Winning with Statistics*, Richard P. Runyon). This procedure consists of eliminating the zero frequency from the vertical axis and beginning with a frequency count greater than zero. Figure 4.5 is borrowed from Dr. Runyon's book and illustrates quite dramatically the way in which graphs may be employed for purposes of deception.

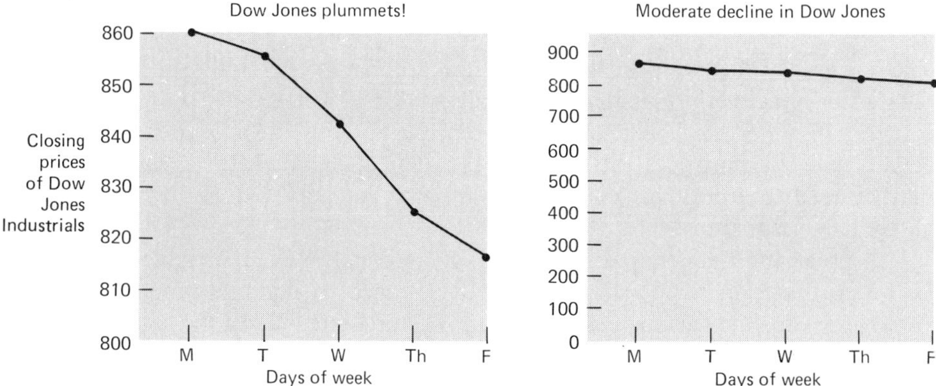

Fig. 4.5 The use of the "Oh Boy" chart to exaggerate differences along the ordinate. (Reprinted from *Winning with Statistics* by Richard P. Runyon. Reading, Mass.: Addison-Wesley, 1977.)

It is obvious that the use of such devices is inimical to the aim of the statistician, which is to present data with such clarity that misinterpretations are reduced to a minimum. We may overcome the second source of error illustrated above by making the initial entry on the Y-axis a zero frequency. The first problem, however, which is the selection of scale units to represent the horizontal and vertical axes, remains. Clearly, the choice of these units is an arbitrary affair, and anyone who decides to make the ordinate twice the length of the abscissa is just as correct as one who decides upon the opposite representation. It is clear that, in order to avoid graphic anarchy, it is necessary to adopt a convention.

4.2.1 Three-Quarter High Rule

For graphic representations of frequency distributions, most statisticians have agreed upon a convention known as the "**three-quarter high rule**" which is expressed as follows.

In plotting the frequencies, the vertical axis should be laid out so that the height of the maximum point (representing the score with the highest associated frequency) is approximately equal to three-quarters the length of the horizontal axis.

The advantage of this convention is that it prevents subjective factors and, possibly, personal biases from influencing decisions concerning the relative proportion of the abscissa and the ordinate in graphic representations. The use of the three-quarter rule is illustrated in the forthcoming section dealing with bar graphs. However, this rule has also been applied to all the graphs appearing in the remainder of the chapter.

4.3 NOMINALLY SCALED VARIABLES

The **bar graph**, illustrated in Fig. 4.6, is the graphic device employed to represent data which are either nominally or ordinally scaled. A vertical bar is drawn for each category in which the height of the bar represents the number of members of that class. If we arbitrarily set the width of each bar at one unit, the *area* of each bar may be used to represent the frequency for that category. Thus the total area of all the bars is equal to N.

In preparing frequency distributions of nominally scaled variables, you must keep two things in mind: (1) No order is assumed to underlie nominally scaled variables. Thus the various categories can be represented along the abscissa in any order you choose. The authors prefer to arrange the categories alphabetically in keeping with their desire to eliminate any possibility of personal factors entering into the decision. (2) The bars should be separated rather than contiguous so that any implication of continuity among the categories is avoided.

The use of the three-quarter rule is perfectly straightforward with graphs of nominally scaled variables. Parenthetically, in implementing the three-quarter rule, we recommend that you acquaint yourself with the use of the metric ruler. You will soon find that the metric system of measurement is far superior to the English system when it comes to dealing with fractions and decimals.

The first decision which must be made concerning the length of the abscissa will be based on such factors as the amount of space available for the graphic representation. Once this decision is made, the remainder follows automatically.

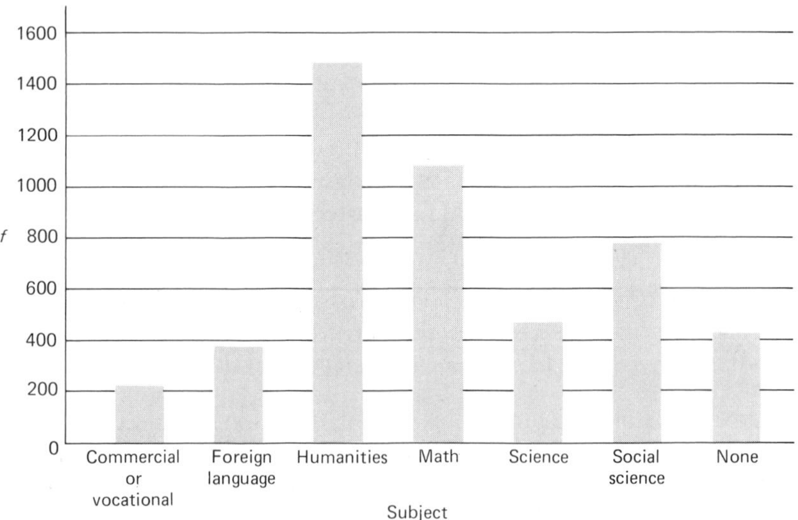

Fig. 4.6 High-school subject disliked most by males 14 to 24 years of age enrolled in high school. (*Source*. Adapted from Manpower Research Monograph, *Career Thresholds*, 1970, U.S. Department of Labor/Manpower Administration.)

Let us say that available space permits us to represent the horizontal axis with a line approximately 100 millimeters (mm) in length. The height of the vertical axis would then be $\frac{3}{4} \times 100$ or 75 mm. In Fig. 4.6, the maximum frequency is approximately 1500. Since 75 mm are to be shared by 1500 frequencies, the number of millimeters representing each frequency is $\frac{75}{1500}$ or 0.05 mm. In other words, each 5 mm along the Y-axis represents 100 frequencies.

Finally, the seven categories represented along the abscissa will occupy 100 mm or approximately 14 mm per class. It will be recalled that the vertical bars should be separated to avoid the implication of continuity. Consequently, the vertical bars should be somewhat less than 14 mm in width to permit this separation.

4.4 ORDINALLY SCALED VARIABLES

It will be recalled that the scale values of ordinal scales carry the implication of an ordering which is expressible in terms of the algebra of inequalities (greater than, less than). In terms of our preceding discussion, ordinally scaled variables should be treated in the same way as nominally scaled variables except that the categories should be placed in their naturally occurring order along the abscissa. Figure 4.7 illustrates the use of the bar graph with an ordinally scaled variable, presenting graphically the data previously shown in Section 2.5.2.

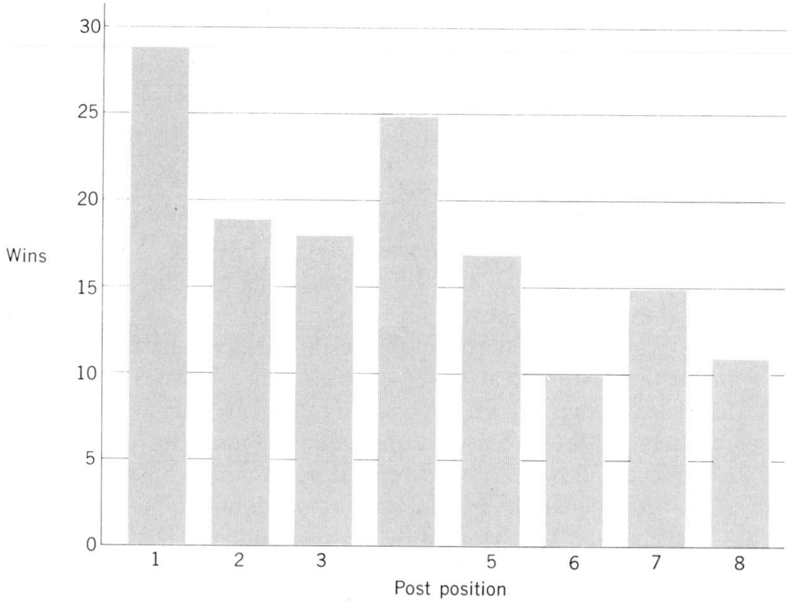

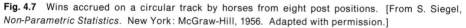

Fig. 4.7 Wins accrued on a circular track by horses from eight post positions. [From S. Siegel, *Non-Parametric Statistics*. New York: McGraw-Hill, 1956. Adapted with permission.]

4.5 INTERVAL AND RATIO SCALED VARIABLES

4.5.1 Histogram

It will be recalled that interval and ratio scaled variables differ from ordinally scaled variables in one important way, i.e., equal differences in scale values are equal. This means that we may permit the vertical bars to touch one another in graphic representations of interval or ratio scaled frequency distributions. Such a graph is referred to as a **histogram** and replaces the bar graph employed with nominal and ordinal variables. Figure 4.8 illustrates the use of the histogram with a discretely distributed ratio scaled variable.

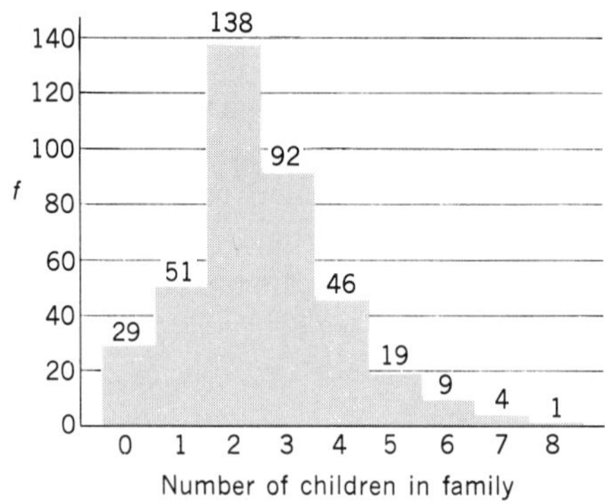

Fig. 4.8 Frequency distribution of the number of children per family among 389 families surveyed in a small suburban community (hypothetical data).

We previously noted (Section 4.3) that frequency may be represented either by the area of a bar or by its height. However, there are many graphic applications in which the height of the bar, or the ordinate, may give misleading information concerning frequency. Consider Fig. 4.9, which shows the data grouped into *unequal* class intervals and the resulting histogram.

If you think of frequency in terms of the height of the ordinate, you might erroneously conclude that the interval 15–25 includes only two cases. However, if we represent each score by one unit on the scale of frequency and an equal unit on the scale of scores, the total area for each score is equal to one. In the interval 15–25 there are 22 frequency units distributed over 11 score units; thus, for this interval, the height of the ordinate will be $\frac{22}{11}$ or 2 score units. Similarly, in the interval 6–8 there are 24 frequency units distributed over 3 score units. Thus the height of the ordinate must equal 8 units.

In general, it is advisable that we consider frequency in terms of area whenever we are dealing with variables in which an underlying continuity may be assumed.

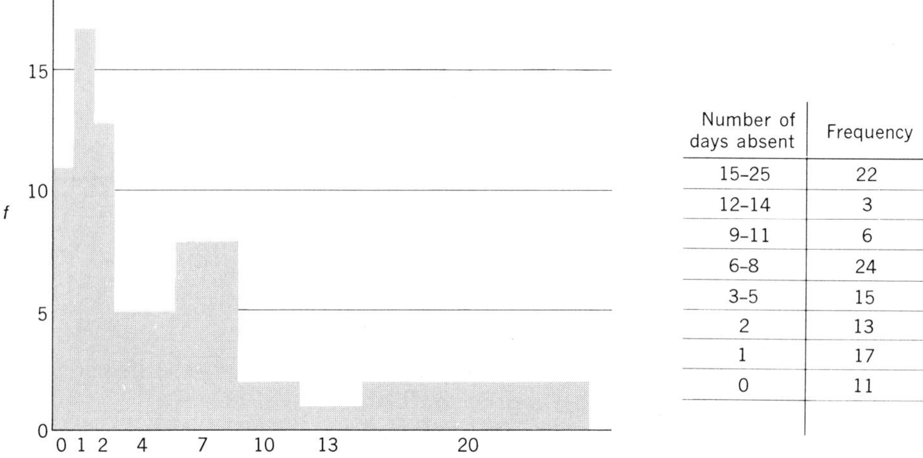

Number of days absent	Frequency
15–25	22
12–14	3
9–11	6
6–8	24
3–5	15
2	13
1	17
0	11

Fig. 4.9 Histogram employing unequal class intervals (hypothetical data).

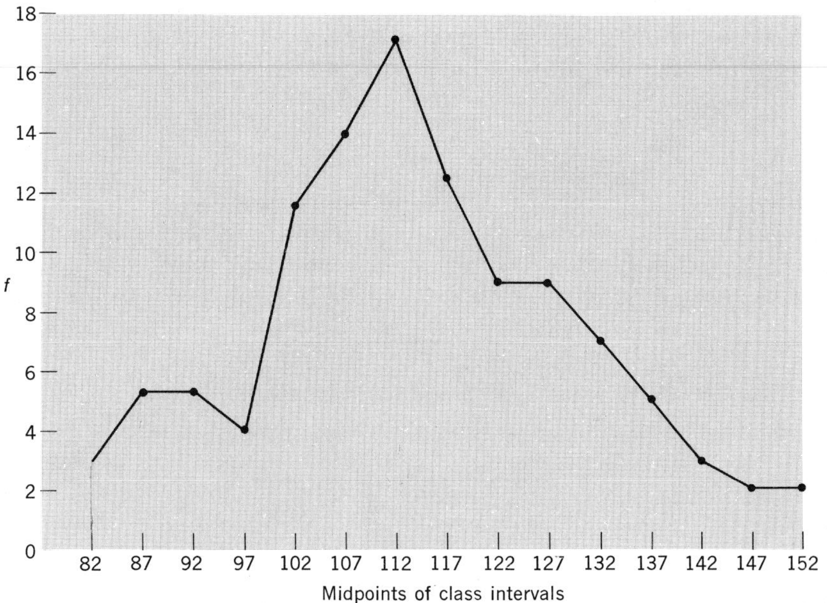

Fig. 4.10 Frequency polygon based on the data appearing in Table 3.4.

4.5.2 Frequency Polygon

We can readily convert the histogram into another commonly employed form of graphic representation, the **frequency polygon**, by joining the midpoints of the bars with straight lines. However, it is not necessary to construct a histogram prior to the construction of a frequency polygon. All you need to do is place a dot where the tops of the bars would have been and join these dots. In practice, the authors prefer to reserve the use of the histogram for discrete distributions and the frequency polygon for distributions in which underlying continuity is explicit or may be assumed. When two or more frequency distributions are compared, the frequency polygon provides a clearer picture. Figure 4.10 shows a frequency polygon based upon the grouped frequency distribution appearing in Table 3.3.

In frequency polygons, frequency *cannot* be represented by the height of the ordinate, but only by the area between the ordinates.

4.6 FORMS OF FREQUENCY CURVES

Frequency polygons may take on an unlimited number of different forms. However, many of the statistical procedures discussed in the text assume a particular form of distribution, namely, the "bell-shaped" normal curve.

In Fig. 4.11, several forms of bell-shaped distributions are shown. Curve (a), which is characterized by a piling up of scores in the center of the distribution, is referred to as a **leptokurtic** distribution. In curve (c), in which the opposite condition prevails, the distribution is referred to as **platykurtic**. And finally, curve (b) takes on the ideal form of the normal curve and is referred to as a **mesokurtic** distribution.

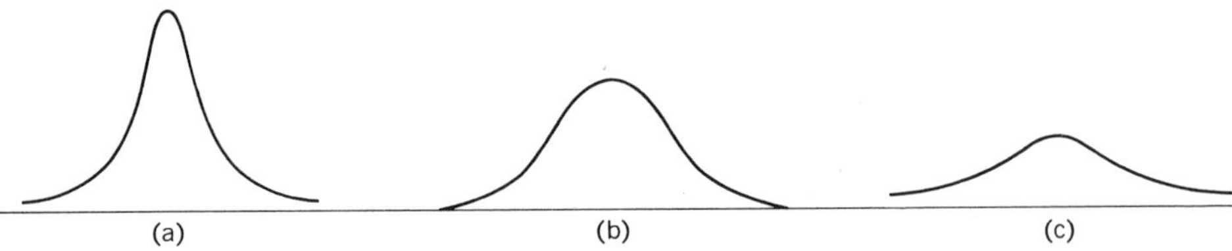

Fig. 4.11 Three forms of bell-shaped distributions: (a) leptokurtic, (b) mesokurtic, and (c) platykurtic.

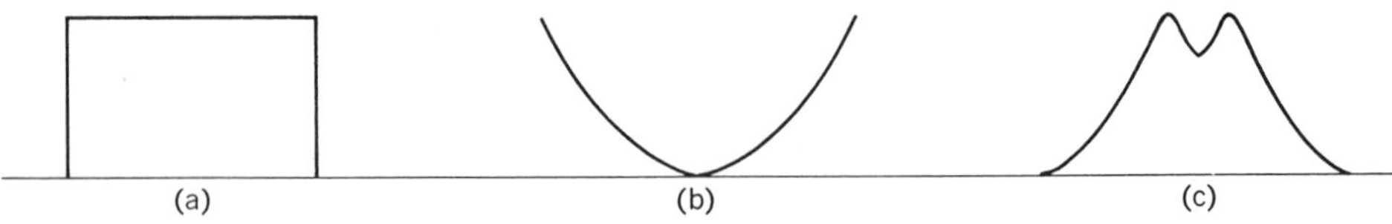

Fig. 4.12 Several symmetrical frequency curves that are not bell-shaped.

The **normal curve** is referred to as a symmetrical distribution, since, if it is folded in half, the two sides will coincide. Not all symmetrical curves are bell-shaped, however. A number of different symmetrical curves are shown in Fig. 4.12.

Certain distributions have been given names, e.g., that in Fig. 4.12(a) is called a **rectangular** distribution and that in Fig. 4.12(b) a **U-distribution**. Incidentally, a bimodal distribution such as appears in Fig. 4.12(c) is found when the frequency distributions of two different populations are represented in a single graph. For example, a frequency distribution of male and female adults of the same age would probably yield a curve similar to Fig. 4.12(c) on a strength of grip task, body weight, or height.

When a distribution is not symmetrical, it is said to be **skewed**. If we say that a distribution is positively skewed, we mean that the distribution tails off at the high end of the horizontal axis and there are relatively fewer frequencies at this end. If, on the other hand, we say that the distribution is negatively skewed, we mean that there are relatively fewer scores associated with the left-hand, or low, side of the horizontal axis. Figure 4.13 presents several forms of skewed distributions.

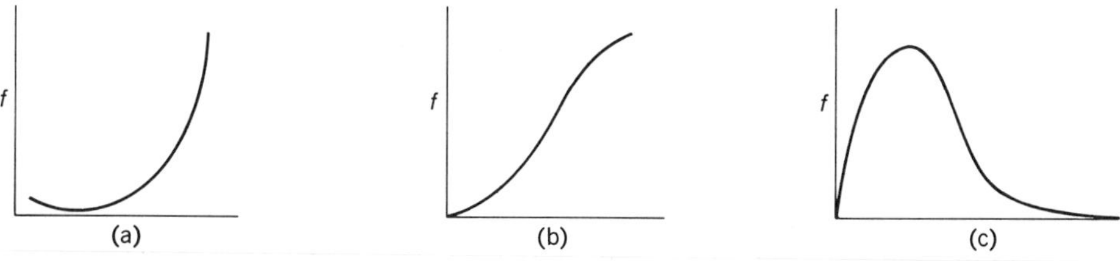

Fig. 4.13 Skewed frequency curves.

Figure 4.13(a) is referred to as a **J-curve** and Fig. 4.13(b) is referred to as an **ogive**. A cumulative frequency distribution of normally distributed data will yield an ogive or S-shaped distribution. Figure 4.13(c) is **positively skewed**. Incidentally, Fig. 4.13(a) illustrates an **extreme negative skew**.

It is not always possible to determine by inspection whether or not a distribution is skewed. There is a precise mathematical method for determining both direction and magnitude of skew. However, it is beyond the scope of this book to go into a detailed discussion of this topic. In Chapter 5, however, we shall outline the procedure for determining the direction, if not the magnitude, of skew.

4.7 PIE CHARTS

Pie charts, also referred to as **circle charts**, are commonly employed to represent distribution ratios. The total is cut into component parts by drawing radii in the circles, thereby dividing it into wedge-shaped parts. The name *pie chart* is derived from the resemblance of these wedges to pieces of pie.

Figure 4.14 shows the data, previously presented in Fig. 4.6, arranged into a pie chart. The circle represents the sum of the set of distribution ratios (100%). Each wedge represents one ratio in the set.

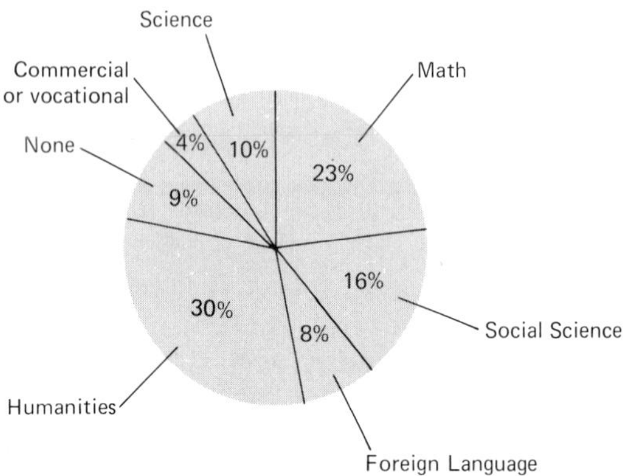

Fig. 4.14 Percentage of male high-school students who indicate dislike for a particular high school subject. (Adapted from Manpower Research Monograph, *Career Thresholds*, 1970, U.S. Department of Labor/Manpower Administration.)

To construct a circle chart, you multiply each percentage by 360 (the number of degrees in a circle) to obtain the number of degrees to assign to each component. Then, starting at the top, the degrees for each component are successively counted in a clockwise direction. For this work, a protractor and compass are, of course, necessary.

Figure 4.15 shows the use of the **slit chart** (a variant of the pie chart) to present the President's budget estimate for the 1969 fiscal year. Pie charts and their variants are particularly effective means of showing distributions of money. Note, for example, that the figures are readily converted to percentages by merely changing cents to percent.

As an interesting sidelight, many statistical textbooks include in their discussion of pie charts the statement that circle charts cannot be read as rapidly or as accurately as corresponding bar graphs. Thus the recommended use of pie charts is usually reserved for special situations in which one is as much interested in attracting attention as conveying information. The interesting fact is that research simply does not support the view that pie charts are more difficult to read. In one study involving the comparison of eight different graphic methods for representing distribution ratios, it was found that the circle graph is read most accurately (Peterson and Schramm, 1954). Numerous other studies support this conclusion (Mathews, 1926; Eels, 1926).

Budget Receipts and Outlays
(in millions of dollars)

Description	1967 actual	1968 estimate	1969 estimate
Receipts by source:			
Individual income taxes	61,526	67,700	80,900
Corporation income taxes	33,971	31,300	34,300
Employment taxes	27,823	29,730	34,154
Unemployment insurance	3,652	3,660	3,594
Premiums for other insurance and retirement	1,853	2,049	2,275
Excise taxes	13,719	13,848	14,671
Estate and gift taxes	2,978	3,100	3,400
Customs	1,901	2,000	2,070
Other receipts	2,168	2,443	2,744
Total, receipts	149,591	155,830	178,108
Outlays by function:			
National defense	70,092	76,489	79,789
International affairs and finance	4,650	5,046	5,153
Space research and technology	5,423	4,803	4,573
Agriculture and agricultural resources	4,377	5,311	5,609
Natural resources	2,132	2,432	2,490
Commerce and transportation	7,446	7,853	8,121
Housing and community development	2,285	3,954	2,784
Health, labor and welfare	40,084	46,417	51,407
Education	4,047	4,541	4,699
Veterans benefits and services	6,898	7,168	7,342
Interest	12,548	13,535	14,400
General government	2,454	2,578	2,790
Allowances for:			
Civilian and military pay increase			1,600
Contingencies		100	350
Undistributed intragovernmental payments:			
Government contributions for employee retirement	−1,735	−1,913	−2,007
Interest received by trust funds	−2,287	−2,678	−3,042
Total, outlays	158,414	175,635	186,062
Budget deficit (−)	−8,823	−19,805	−7,954

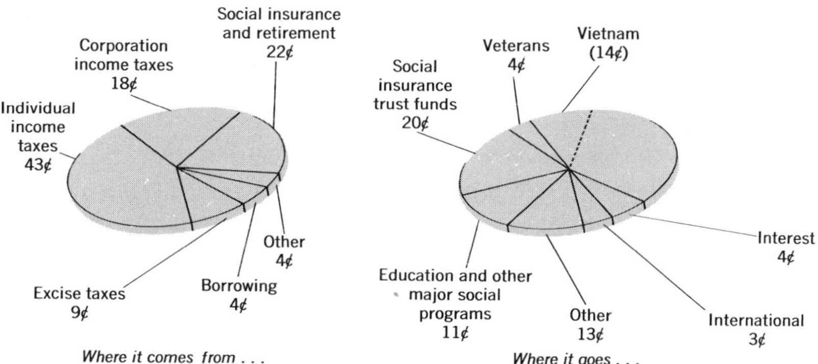

Fig. 4.15 Use of the slit chart to present monetary information.

4.8 OTHER GRAPHIC REPRESENTATIONS

Throughout this chapter we have been discussing graphic representations of frequency distributions. However, other types of data are frequently collected by behavioral scientists. We shall briefly discuss a few graphic representations of such data.

We are frequently interested in comparing various groups or conditions with respect to a given characteristic. These groups or conditions constitute the independent, or experimental, variable, whereas the characteristic we are measuring is referred to as the criterion or dependent variable.

Figure 4.16 contrasts city and highway mean miles per gallon among three weight classes of 1976 automobiles. It should be noted that the experimental variable is represented along the X-axis and the criterion or dependent variable on the Y-axis.

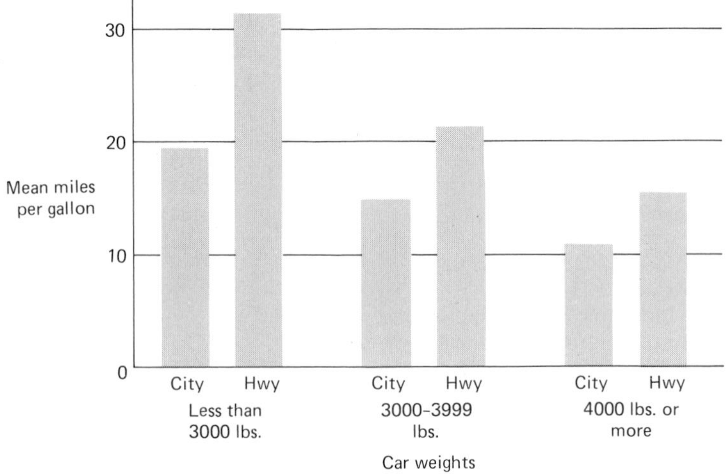

Fig. 4.16 Mean miles per gallon for city and highway driving for three different weight classes. (*Source*. Environmental Protection Agency; 1976 Automobiles.)

In recent years, much has been made of the growing energy crisis our nation faces. In Fig. 4.17 we see a line graph showing the amount of energy consumed during selected years from 1947 through 1972. It can be seen from the graph that our energy consumption more than doubled over this period of 25 years.

Figure 4.18 presents another line graph, this time showing the incidence of alcohol-related collisions in three age groups both before and after reducing the legal age from 21 to 18 for alcohol purchase and consumption. It is interesting to note that the rise in alcohol-related collisions is greatest in the age group (18-20) directly affected by this law.

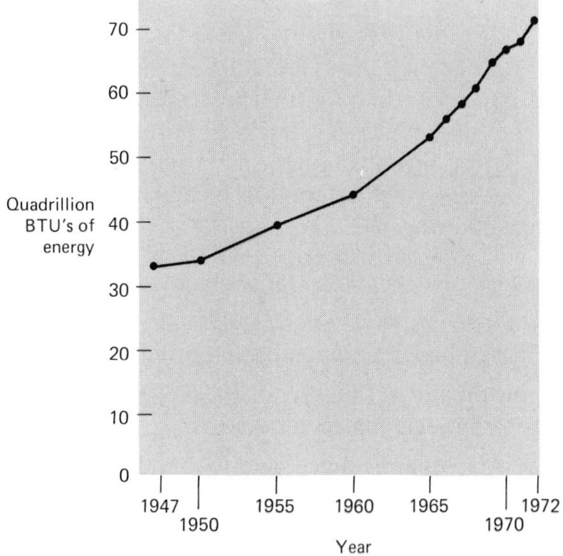

Fig. 4.17 The gross national expenditure of energy, in quadrillions of btu's, during selected years between 1947 and 1972. (*Source*. U.S. Department of the Interior, *U.S. Energy through the year 2000*.)

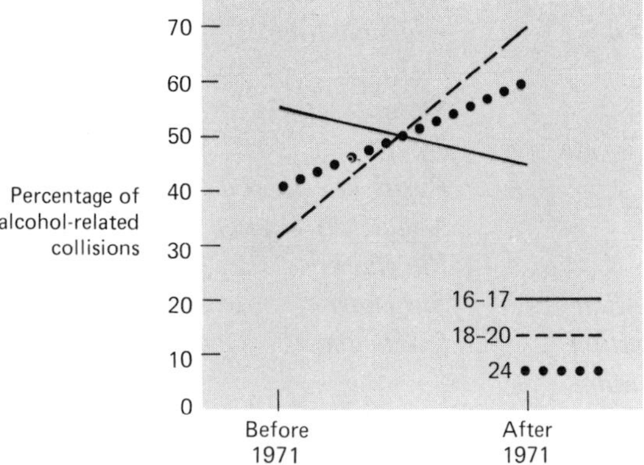

Fig. 4.18 Percentage of alcohol-related collisions in three age groups before and after the legal age for purchasing and consuming alcoholic beverages was changed from 21 to 18. (*Source*. Adapted from Whitehead, P. C., J. Craig, N. Langford, C. MacArthur, B. Stanton, and R. Terrence, "Collision behavior of young drivers," *J. Studies Alcohol*, **36**, 1208–1223, 1975.)

CHAPTER SUMMARY

In this chapter, we reviewed the various graphing techniques employed in the behavioral sciences. The basic purpose of graphical representation is to provide visual aids for thinking about and discussing statistical problems. The primary objective is to present data in a clear, unambiguous fashion so that the reader may apprehend at a glance the relationships which we want to portray.

We discussed the following:

1. Devices employed by unscrupulous individuals to mislead the unsophisticated reader.
2. The use of the bar graph with nominally and ordinally scaled variables.
3. The use of the histogram and the frequency polygon with continuous and discontinuous ratio or interval-scaled variables.
4. The use of the "three-quarter high" rule to standardize the representation of the abscissa and ordinate in graphic techniques.
5. Various forms of bell-shaped and other symmetrical distributions, and asymmetrical or skewed distributions.
6. The use of the pie (circle) chart to represent distribution ratios.
7. Finally, we discussed and demonstrated several graphic representations of data, other than frequency distributions, commonly employed in the behavioral sciences.

Terms to Remember:

Abscissa (X-axis)
Ordinate (Y-axis)
Bar graph
"Oh boy" chart
"Three-quarter high rule"
Histogram
Frequency polygon
Normal curve
Platykurtic distribution
Leptokurtic distribution
Mesokurtic distribution
Rectangular distribution
U-distribution
Ogive
Skew
J-curve
Positively skewed distribution
Negatively skewed distribution
Pie chart
Slit chart
Circle chart

EXERCISES

4.1 Given the following frequency distribution of the weights of 96 female students, draw a histogram.

Class interval	f	Class interval	f
160–164	1	130–134	17
155–159	3	125–129	11
150–154	10	120–124	8
145–149	6	115–119	3
140–144	14	110–114	1
135–139	22		

4.2 Draw the appropriate graphic representations for each of the four examples in Exercise 2.9, using your own hypothetical data.

4.3 Take a pair of dice, toss 100 times and record the sum (on the face of the two dice) for each toss. Prepare a bar graph, showing the number of times each sum occurs.

4.4 Given the following monthly sales by five salespeople in a large appliance store:

Salespeople	Sales, $	Salespeople	Sales, $
Mr. Arthur	22,500	Ms. Emily	22,100
Mr. Stanley	17,900	Mr. William	20,700
Ms. Laura	21,400		

Draw graphs to perpetrate the lies stated below.

a) The sales manager wants to convince the owner of the store that all members of her sales force are functioning at a uniformly high level.

b) The sales manager wants to spur Mr. Stanley to greater efforts.

c) The store owner wants to spur the sales manager to greater efforts.

4.5 Draw a graph of the data in Exercise 4.4 which represents the true state of affairs.

4.6 Below are the scores of two groups of fourth-grade students on a test of reading ability.

Class interval	Group *A* f	Group *B* f	Class interval	Group *A* f	Group *B* f
50–52	5	2	29–31	9	22
47–49	12	3	26–28	6	11
44–46	18	5	23–25	4	9
41–43	19	8	20–22	3	6
38–40	26	12	17–19	1	4
35–37	19	24	14–16	2	2
32–34	13	35			
				137	143

a) Construct a frequency polygon for each of these groups on the same axis.
b) Describe and compare each distribution.

4.7 Describe the types of distributions you would expect if you were to graph each of the following.
a) Annual incomes of American families.
b) The heights of adult American males.
c) The heights of adult American females.
d) The heights of American males and females combined in one graph.

4.8 Given the following frequency distribution of the results of 73 students on a midterm exam, draw a frequency polygon.

Class interval	f	Class interval	f
95–99	2	65–69	11
90–94	2	60–64	6
85–89	5	55–59	3
80–84	9	50–54	0
75–79	16	45–49	1
70–74	18		

4.9 For the data in Exercise 4.8, draw a cumulative frequency polygon.

4.10 Obtain the annual budget estimate of the President of the United States for this year. Show the allocation of funds by category, using both a bar graph and a pie chart.

4.11 Carefully note the forms of graphical representation employed in the financial section of your Sunday newspaper. Collect examples of "good" graphics, i.e., clear to interpret and free of misleading elements. Collect examples of "poor" graphics. What makes them poor? Begin to list characteristics which determine whether graphs are effective or ineffective. Note also graphs employed in advertising. How do they compare in quality to the graphs illustrating financial statistics? Are they as complete? Are the ordinate and abscissa always labeled? Are any misleading? In what way?

4.12 Draw a pie chart to represent the data in Exercise 4.4, expressing the sales figures of each salesperson in terms of percentage of the total.

4.13 Present the data in Fig. 4.15 employing a bar graph.

4.14 Employing the line graph shown in Section 4.8, plot the percentage of change figures calculated for Exercise 3.8 against the 1967 prices of the stocks. (Note that, since the percentage of change is sometimes negative, it will be necessary to represent negative as well as positive changes along the ordinate).

4.15 Group the stock prices (from Exercise 3.8) into approximately eight categories and plot the percentage of change for each stock within that category. Find, for each category, the figure which represents the percentage of change. This figure is approximately in the center of the column representing the category. Join these midpoints. Is there any suggestion of a relationship between the initial price of stock and the percentage of change during the year?

4.16 Following are the number of quarts of milk sold at a supermarket on 52 consecutive Saturdays.

67	56	64	78	88	57
65	70	66	67	67	71
61	60	72	69	73	
69	75	68	65	66	
65	62	71	72	61	
78	63	58	63	64	
75	62	71	92	74	
76	65	73	77	75	
63	65	58	64	62	
64	88	61	69	64	

a) The manager of the dairy department decides to limit the number of quarts of milk on sale each Saturday to 70. Assuming that the sales figures will be the same the following year, what is the likelihood (i.e., the percentage of time) that his department will be caught short?

b) Group the sales figures into a frequency distribution with the lower class limits of 56–58.

c) Prepare a cumulative frequency distribution and cumulative percentage distribution based on the above frequency distribution.

d) Group the sales figures into a frequency distribution with the lower class limit of 55–59. Prepare a cumulative frequency distribution and cumulative percentage distribution based on this frequency distribution. Compare the resulting distributions with those obtained above.

4.17 Draw a frequency polygon for the frequency distribution obtained in Exercise 3.10.

4.18 Draw a frequency polygon for the frequency distribution obtained in Exercise 3.11. Compare this distribution with that of Exercise 4.17.

4.19 Draw a cumulative frequency polygon for the frequency distribution obtained in Exercise 3.10.

4.20 Draw a cumulative frequency polygon for the frequency distribution obtained in Exercise 3.11. Compare this distribution with that of Exercise 4.19.

4.21 Referring to Exercise 1.15, construct a histogram which shows the total number of each album sold.

a) Reconstruct the histogram, making the ordinate twice the height and the abscissa half as long.

b) Again reconstruct the histogram, making the ordinate half as high as the first drawing, and the abscissa twice as long.

c) Which graph would suggest that there was a large difference in the number of albums sold? Which would suggest a small difference?

4.22 Draw a polygon for Exercise 3.20. Is the form symmetrical or skewed? If skewed, is it positively or negatively skewed?

4.23 Graph the frequency distribution and the cumulative frequency distribution determined in Exercise 3.22.

4.24 Graph the frequency distribution of Exercise 3.23. Compare the form of this graph with that of the above problem.

4.25 Graph the frequency distribution and the cumulative frequency distribution in Exercise 3.15. What type of polygon does this represent?

4.26 Using Table 4.1, round the numbers to the nearest billion and construct the appropriate graph.

Table 4.1

Gross National Product (in millions of dollars)

Year	GNP	Year	GNP
1950	284,769	1972	1,155,155
1960	503,734	1973	1,294,919
1970	977,080	1974	1,397,400

Source. Department of Commerce, *The 1976 World Almanac*. New York: Newspaper Enterprise Association, Inc.

4.27 Referring to Table 4.2, construct a bar graph showing the side-by-side comparisons of exports and imports for each commodity.

Table 4.2

U.S. exports and imports of leading commodities, 1974

Commodity	Value (in millions of dollars) Exports	Imports
Food and live animals	13,983	9,379
Beverages and tobaccos	1,247	1,321
Crude materials, inedible other than fuels	10,934	5,915
Animal and vegetable oils and fats	1,423	544
Chemicals	8,822	3,991
Machinery	38,189	24,713
Transport equipment	13,871	12,630
Other manufactured goods	16,516	27,507
Other transactions	2,587	2,252

Source. Bureau of International Commerce, Dept. of Commerce. *The 1976 World Almanac*, p. 111.

4.28 Draw the three forms of normal distributions. Assign values along the ordinate and abscissa to each graph, and determine the frequency distribution from these values.

4.29 Repeat the steps in Exercise 4.28 with a negatively skewed polygon and with one that is positively skewed.

4.30 The Federal Reserve Board reported percentages of income going to various categories of families:

Group	1970	1969	1965	1950
Lowest fifth	5.5	5.6	5.3	4.5
Second fifth	12.0	12.3	12.1	12.0
Middle fifth	17.4	17.6	17.7	17.4
Fourth fifth	23.5	23.5	23.7	23.5
Highest fifth	41.6	41.0	41.3	42.6
Top 5%	14.4	14.0	15.8	17.0

Source. *The Wall Street Journal*, Feb. 28, 1972, p. 1:5.

a) Represent the six groups and four years on one graph (see Fig. 4.18).
b) Reconstruct the graph to emphasize the differences among the groups.
c) Reconstruct the graph to minimize the differences among groups.

4.31 For every dollar paid in car insurance, \$0.43 is spent on overhead and \$0.57 is paid back to the consumer. Of the \$0.43, \$0.36 is spent on administration and selling and \$0.07 for claimants' legal costs. Of the amount paid back, \$0.30 is spent for car repairs, \$0.21 for bodily injuries, and \$0.06 for all other payments. Draw a histogram and a pie chart representing these amounts. Which appears easier to read? (Source: *Best's Aggregates and Averages*, 1970 and *U.S. Department of Transportation Studies*. Ford advertisement, *The Wall Street Journal*, March 3, 1972, p. 7:4–5.)

4.32 Using Table 4.3, construct a pie chart.

Table 4.3
How consumers spend their dollar

Group	1973 % of total
Food and tobacco	22.2
Clothing, accessories and jewelry	10.1
Personal care -	1.8
Housing	14.5
Household operation	14.6
Medical care	7.8
Personal business	5.6
Transportation	1.4
Recreation	6.5
Private education and research	1.6
Religious and welfare activities	1.3
Foreign travel and other	0.7

Source. Department of Commerce.

Table 4.4

	Age	Number of words per sentence
Females	2	2.1
	3	3.8
	4	4.6
	5	5.1
	6	5.6
Males	2	1.4
	3	3.4
	4	4.5
	5	4.9
	6	5.1

4.33 It has frequently been reported that females are more advanced than males in early language development. Data reported in one study are presented in Table 4.4. Represent this information on a graph.

4.34 The deployment of energy in the United States is as follows: industrial, 37.2%; transportation, 24.6%; residential and commercial, 22.4%; heat loss, 15.8%.
a) Construct a pie chart to represent this deployment of energy.
b) Construct a bar graph for these same data.

4.35 In Exercise 2.20, we presented the data of Parker *et al.* showing the breakdown of various types of computer abuses from 1969 to 1973.
a) Construct a graph of a fixed-base time ratio of the Total column using 1969 as the base year; 1971 as the base year.
b) In which of the two graphs does the increase in computer abuse appear to be greater?
c) Using the three-quarter high rule, prepare a graph of the fixed-base time ratio with 1971 as the base year. Contrast the appearance of this graph with the one obtained in part (a).

4.36 The following data show our nation's gross energy expenditure, in quadrillion btu's, during selected years between 1947 and 1972.

Year	Energy expenditure	Year	Energy expenditure	Year	Energy expenditure
1947	33.0	1965	53.3	1969	65.0
1950	34.0	1966	56.4	1970	67.1
1955	39.7	1967	58.3	1971	68.7
1960	44.6	1968	61.7	1972	72.1

Source. U.S. Dept. of the Interior, *United States Energy Through the Year 2000*

a) Prepare a histogram showing the growth in energy expenditures.
b) Prepare an "oh boy" chart showing the growth in energy expenditure.

4.37 The following table shows the approximate frequencies of numbers of hours per week spent on homework by white and black males, 14–24 years of age.

Hours per week doing homework	Whites	Blacks
None	366	41
1–4	2654	330
5–9	3295	495
10–14	2014	357
15–19	641	96
20 or More	183	69
Total	9153	1388

Source. Adapted from Manpower Research Monograph, *Career Thresholds*, *1*, 1970. U.S. Dept. of Labor/Manpower Administration.

a) Convert the entries in each of the columns to percentages.
b) Draw a graph comparing the percentage of whites and blacks doing homework various numbers of hours per week.

4.38 Refer back to Exercise 3.27 and prepare frequency polygons of traffic deaths for 1973 and 1974, superimposing one upon the other.

*4.39 Imagine that you have a population consisting of seven scores: 0, 1, 2, 3, 4, 5, 6. You write each of these numbers on a paper tab and place them in a hat. Then you draw a number, record it, place it back in the hat, and draw a second number. You add the second number to the first to obtain a sum. You continue drawing pairs of scores and obtaining their sums until you have obtained all possible combinations of these seven scores, taken two at a time.

The following table shows all possible results from drawing samples of $n = 2$ from this population of seven scores. The values within each cell show the sum of the two scores.

First draw

Second draw	0	1	2	3	4	5	6
0	$0+0=0$	$1+0=1$	$2+0=2$	$3+0=3$	$4+0=4$	$5+0=5$	$6+0=6$
1	$0+1=1$	$1+1=2$	$2+1=3$	$3+1=4$	$4+1=5$	$5+1=6$	$6+1=7$
2	$0+2=2$	$1+2=3$	$2+2=4$	$3+2=5$	$4+2=6$	$5+2=7$	$6+2=8$
3	$0+3=3$	$1+3=4$	$2+3=5$	$3+3=6$	$4+3=7$	$5+3=8$	$6+3=9$
4	$0+4=4$	$1+4=5$	$2+4=6$	$3+4=7$	$4+4=8$	$5+4=9$	$6+4=10$
5	$0+5=5$	$1+5=6$	$2+5=7$	$3+5=8$	$4+5=9$	$5+5=10$	$6+5=11$
6	$0+6=6$	$1+6=7$	$2+6=8$	$3+6=9$	$4+6=10$	$5+6=11$	$6+6=12$

a) Obtain an ungrouped frequency distribution of the 49 sums.
b) Construct a histogram from the frequency distribution.
c) Note that the original distribution of scores was rectangular since each score occurred with the same frequency. Compare the form of the original distribution with the form of the frequency distribution of sums.

5 Percentiles

BOX 5.1

Suppose you were thinking of buying a Chevrolet Chevette with manual transmission. You observe that the average highway mileage performance is 33 miles per gallon. How does this performance figure compare with all of the others listed in Box 1.1? A glance at the figures indicates that this is a pretty good mileage rating. As we shall see in this chapter, there are more precise ways of determining the relative performance of a specific automobile.

Shown in Table 5.1 are the grouped cumulative frequency and percentage distributions of highway mileage for the 124 cars listed in Box 1.1. Accompanying

Table 5.1

Class interval	Real limits	f	Cumulative f	Cumulative %
41–42	40.5–42.5	1	124	100
39–40	38.5–40.5	0	123	99
37–38	36.5–38.5	1	123	99
35–36	34.5–36.5	9	122	98
33–34	32.5–34.5	3	113	91
31–32	30.5–32.5	9	110	89
29–30	28.5–30.5	14	101	81
27–28	26.5–28.5	3	87	70
25–26	24.5–26.5	9	84	68
23–24	22.5–24.5	6	75	60
21–22	20.5–22.5	14	69	56
19–20	18.5–20.5	2	55	44
17–18	16.5–18.5	20	53	43
15–16	14.5–16.5	28	33	27
13–14	12.5–14.5	5	5	4

these distributions (Fig. 5.1) is a cumulative percentage graph from which we can determine with reasonable accuracy the relative standing of the Chevette. Note that we have constructed a vertical line at 33 mpg until it intercepts the polygon. The horizontal line from this point tells us that the percentile rank of 33 mpg is approximately 90. Stated another way, the Chevette's mileage performance exceeded that of 90% of the 124 cars.

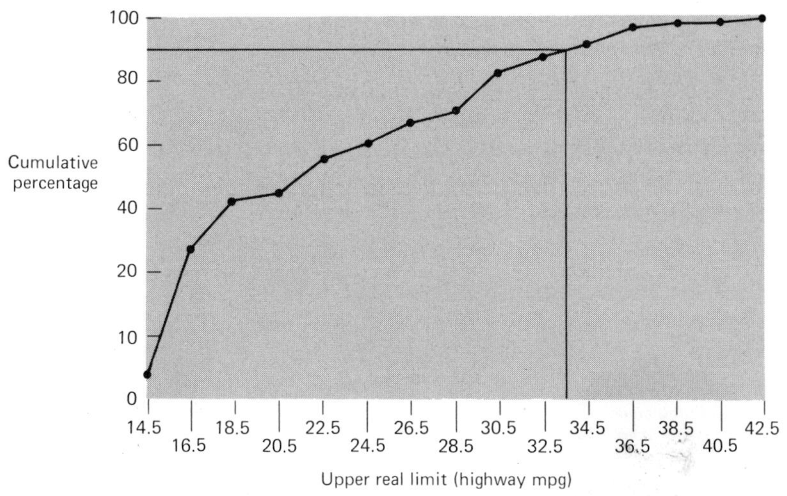

Fig. 5.1 Cumulative percentage distribution.

5.1 INTRODUCTION

Let us suppose that a younger brother came home from school and announced, "I received a score of 127 on my scholastic aptitude test." What would be your reaction? Commend the child for obtaining such a fine score? Criticize for not getting a higher one? Or reserve judgment until you learned more about the distribution of scores within your sibling's class or group? If you have passed the course up to this point, you have undoubtedly selected the last of these three alternatives.

It should be clear that a score by itself is meaningless. It takes on meaning only when it can be compared to some standard scale or base. Thus if your younger sibling were to volunteer the information, "79% of the students scored lower than I," he would be providing some frame of reference for interpreting the score. Indeed, he would have been citing the percentile rank of his score. The **percentile rank** of a score, then, *represents the percent of cases in a comparison group which achieved scores lower than the one cited.* Thus to say that a score of 127 has a percentile rank of 79 is to indicate that 79% of the comparison group scored below 127. Incidentally, each score is considered to be a hypothetical point without dimension, so that it would be equally meaningful to say that 21% of the comparison group scored higher than 127.

5.2 CUMULATIVE PERCENTILES AND PERCENTILE RANK

5.2.1 Obtaining the Percentile Rank of Scores from a Cumulative Percentage Graph

In Chapter 3, we learned how to construct cumulative frequency and cumulative percentage distributions. If we were to graph a cumulative percentage distribution, we could read the percentile ranks directly from the graph.* Figure 5.2 presents a graphic form of the cumulative percentage distribution presented in Table 3.5.

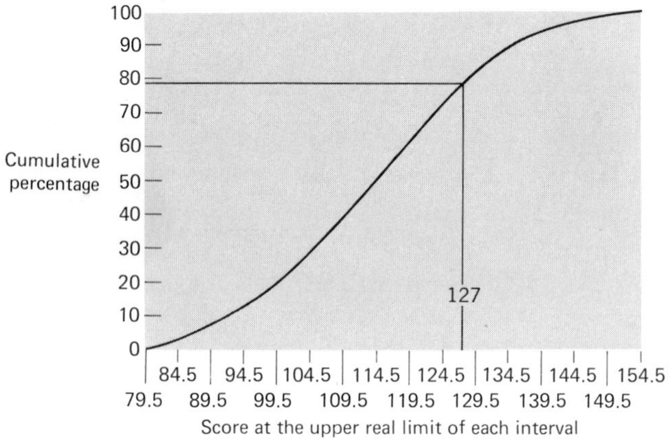

Fig. 5.2 Graphic representation of a cumulative percentage distribution.

To illustrate, let us imagine that we wanted to determine the percentile rank of a score of 127. We locate 127 along the abscissa and construct a perpendicular at that point so that it intercepts the curve. Reading directly across on the scale to the left, we see that the percentile rank is approximately 79. On the other hand, if we wanted to know the score at a given percentile, we could reverse the procedure. For example, what is the score at the 90th percentile? We locate the 90th percentile on the ordinate, we read directly to the right until it intercepts the curve, we construct a line perpendicular to the abscissa, and we read the value on the scale of scores. In the present example, it can be seen that the score at the 90th percentile is approximately 135.

* Note that the reverse is also true, i.e., given a percentile rank, we could read the corresponding scores.

5.2.2 Obtaining the Percentile Rank of Scores Directly

We are often called upon to determine the percentile rank of scores without the assistance of a cumulative percentage polygon or with greater precision than is possible with a graphical representation. To do this, it is usually necessary to interpolate within the cumulative frequency column to determine the precise cumulative frequency corresponding to a given score.

Using the grouped frequency distribution found in Table 5.2, let us determine directly the percentile rank of a score of 127 which we previously approximated by the use of the cumulative percentage polygon. The first thing we should note is that a score of 127 falls within the interval 125–129. The total cumulative frequency below that interval is 82. Since a percentile rank of a score is defined symbolically as

$$\text{Percentile rank} = \frac{\text{cum } f}{N} \times 100, \tag{5.1}$$

it is necessary to find the precise cumulative frequency corresponding to a score of 127. It is clear that the cumulative frequency corresponding to a score of 127 lies somewhere between the 82nd and the 91st cases, the cumulative frequencies at both extremes of the interval. What we must do is to interpolate within the interval 124.5–129.5 to find the exact cumulative frequency of a score of 127. In doing this, we are actually trying to determine the proportion of distance that we must move into the interval in order to find the number of cases included up to a score of 127.

Table 5.2

Grouped frequency distribution and cumulative frequency distribution of scores in an educational test (hypothetical data)

Class interval	Real limits	f	Cumulative f
150–154	149.5–154.5	2	110
145–149	144.5–149.5	2	108
140–144	139.5–144.5	3	106
135–139	134.5–139.5	5	103
130–134	129.5–134.5	7	98
125–129	124.5–129.5	9	91
120–124	119.5–124.5	9	82
115–119	114.5–119.5	13	73
110–114	109.5–114.5	17	60
105–109	104.5–109.5	14	43
100–104	99.5–104.5	12	29
95–99	94.5–99.5	4	17
90–94	89.5–94.5	5	13
85–89	84.5–89.5	5	8
80–84	79.5–84.5	3	3
		$N = 110$	

A score of 127 is 2.5 score units above the lower real limit of the interval (that is, $127 - 124.5 = 2.5$). Since there are 5 score units within the interval, a score of 127 is 2.5/5 of the distance through the interval. We now make a very important assumption, i.e., *that the cases or frequencies within a particular interval are evenly distributed throughout that interval*. Since there are 9 cases within the interval, we may now calculate that a score of 127 is $2.5/5 \times 9$, or the 4.5th case within the interval. In other words, the frequency 4.5 in the interval corresponds exactly to a score of 127. We have already seen, however, that 82 cases fall below the lower real limit of the interval. Adding the two together, we find that the score of 127 has a cumulative frequency of exactly 86.5. Substituting 86.5 into formula (5.1), we obtain the following:

$$\text{Percentile rank of } 127 = \frac{86.5}{110} \times 100 = 78.64.$$

You will note that this answer, when rounded to the nearest percentile, agrees with the approximation obtained by the use of the graphical representation of a cumulative percentage distribution (Fig. 5.2).

Figure 5.3 summarizes graphically the procedures involved in finding the cumulative frequency of a given score. You will note that the interval 124.5–129.5 is divided into 5 equal units corresponding to the scores within that interval, whereas the frequency scale is divided into 9 equal units corresponding to the 9 frequencies within that interval. What we are accomplishing, in effect, in finding the frequency corresponding to a score, is a *linear transformation* from a scale of scores to a scale of frequencies, which is analogous to converting Fahrenheit readings to values on a centigrade scale and conversely; or we are converting percentage of profit figures based on selling price to percentage of profit based on production cost (see Section 2.7).

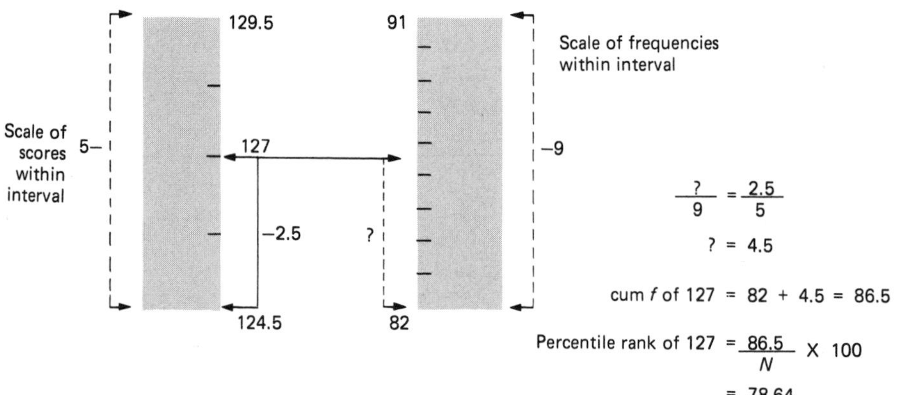

Fig. 5.3 Graphic representation of the procedures involved in finding the cumulative frequency corresponding to a given score.

5.2.3 Finding the Score Corresponding to a Given Percentile Value

Let us imagine that your younger brother, instead of apprising you of his score, reported, instead, his percentile rank. He tells you his score was at the 96th percentile. What was his score?

To obtain the answer, we must interpolate in the reverse direction, from the cumulative frequency scale to the scale of scores. The first thing we must learn is the cumulative frequency corresponding to the 96th percentile. It follows algebraically from formula (5.1) that

$$\text{cum } f = \frac{\text{percentile rank} \times N}{100}. \tag{5.2}$$

Since we are interested in a score at the 96th percentile and our N is 110, the cumulative frequency of a score at the 96th percentile is

$$\text{cum } f = \frac{96 \times 110}{100} = 105.6.$$

Referring to Table 5.2, we see that the frequency 105.6 is in the interval with the real limits of 139.5–144.5. Indeed, it is 2.6 frequencies into the interval since the cum f at the lower real limit of the interval is 103, which is 2.6 less than 105.6. There are three cases in all within the interval. Thus the frequency 105.6 is 2.6/3 of the way through an interval with a lower real limit of 139.5 and an upper real limit of 144.5. In other words, it is 2.6/3 of the way through 5 score units. Expressed in terms of score units, then, it is $(2.6/3) \times 5$ or 4.33 score units above the lower real limit of the interval. By adding 4.33 to 139.5, we obtain the score at the 96th percentile, which is 143.83.

Figure 5.4 represents, graphically, the procedures involved in the linear transformation from units of the frequency scale to units of the scale of scores.

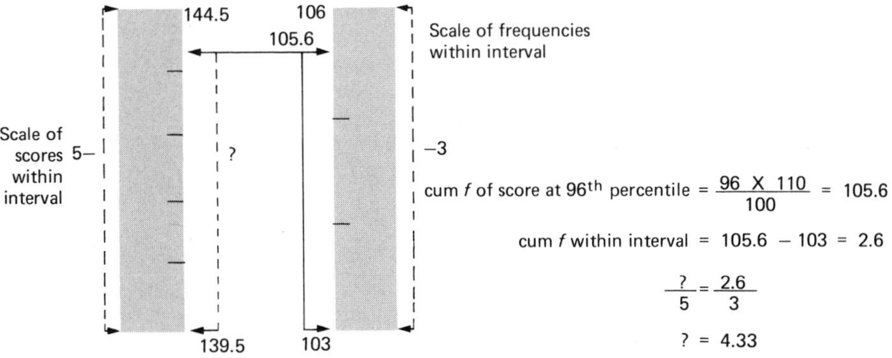

Fig. 5.4 Graphic representation of the procedures involved in finding the score corresponding to a given frequency within the interval.

It is generally a good idea to check the accuracy of our calculations. In other words, whenever you find the percentile rank of a score, you may take that answer and determine the score corresponding to that percentile value. You should obtain the original score. Similarly, whenever you obtain a score corresponding to a given percentile rank you may take that answer and determine the percentile rank of that score. You should always come back to the original percentile rank. Failure to do so indicates that you have made an error. It is preferable to repeat the solution without reference to your prior answer rather than to attempt to find the mistake in your prior solution. Such errors are frequently of the "proofreader" type which defy detection, and are time consuming and highly frustrating to locate.

In Section 5.2.2 we found that a score of 127 had a corresponding percentile rank of 78.64. Using the procedures described above, find the score at the 78.64th percentile. You should obtain a score of 127.

5.3 PERCENTILE RANK AND REFERENCE GROUP

Just as a score is meaningless in the abstract, so also is a percentile rank. A percentile rank must always be expressed in relation to some reference group. Thus if a friend claims that he obtained a percentile rank of 93 in a test of mathematical aptitude, you might not be terribly impressed if the reference group were made up of individuals who completed only the eighth grade. On the other hand, if the reference group consisted of individuals holding a doctorate in mathematics, you attitude would unquestionably be quite different.

Many standardized tests employed in education and industry publish separate norms for various reference groups. Table 5.3 shows the raw-score equivalents for selected percentile points on a test widely employed for graduate-school admissions. You will note that a person obtaining a raw score of 50 on this test, would obtain a percentile rank of 25, 35, 40, and 50 when compared successively with reference groups in the biological sciences, medical science, agriculture, and social work.

CHAPTER SUMMARY

In this chapter we saw that a score, by itself, is meaningless unless it is compared to a standard base or scale. Scores are often converted into units of the percentile rank scale in order to provide a readily understandable basis for their interpretation and comparison.

We saw that:

1. Percentile ranks of scores and scores corresponding to a given percentile may be approximated from a cumulative percentage graph.
2. Direct computational methods permit a more precise location of the percentile rank of a score and the score corresponding to a given percentile.

These methods were discussed and demonstrated in the text. They are summarized in Tables 5.4 and 5.5.

Table 5.3

Raw score equivalents of selected percentile points on the Miller Analogies Test for eight graduate and professional school groups. (Reproduced by permission. Copyright © 1970 by The Psychological Corporation, New York, N.Y. All rights reserved.)

Percentile	Physical sciences	Agriculture	Medical science	Biological sciences	Social sciences	Social work	Languages and literature	Law school freshmen
99	93	89	92	88	90	81	87	84
95	91	86	83	87	85	76	84	79
90	88	77	78	86	82	67	80	73
85	85	72	76	80	79	64	76	66
80	82	67	74	76	76	61	74	63
75	80	64	71	70	74	60	73	60
70	78	61	67	68	69	58	68	58
65	76	59	64	67	67	57	66	55
60	74	57	60	65	64	54	65	53
55	70	56	58	63	63	52	61	51
50	68	54	57	61	61	50	59	49
45	65	51	55	58	58	47	56	47
40	63	50	53	55	56	46	53	45
35	60	48	50	53	53	45	51	42
30	58	43	47	52	51	41	46	40
25	55	40	45	50	49	39	43	37
20	51	37	43	48	46	37	41	35
15	47	34	41	47	44	32	38	32
10	43	31	34	41	39	27	35	30
5	39	26	30	37	32	22	29	25
1	28	5	24	28	18	9	7	18
N	251	125	103	84	229	116	145	558
Mean	66.7	53.6	57.6	61.5	60.2	49.4	57.7	49.6
SD	16.6	17.3	16.2	15.6	16.0	15.2	17.4	16.1

Table 5.4

Summary of procedures: obtaining the score corresponding to a given percentile rank

Class interval	Real limits	f	Cum f
41–43	40.5–43.5	1	98
38–40	37.5–40.5	2	97
35–37	34.5–37.5	4	95
32–34	31.5–34.5	9	91
29–31	28.5–31.5	10	82
26–28	25.5–28.5	13	72
23–25	22.5–25.5	18	59
20–22	19.5–22.5	14	41
17–19	16.5–19.5	11	27
14–16	13.5–16.5	8	16
11–13	10.5–13.5	5	8
8–10	7.5–10.5	2	3
5–7	4.5–7.5	1	1
	$i = 3$	98	

Problem: Find the score corresponding to a percentile rank of 75.17.

Steps:

1. Sum f column to obtain N. $N = 98$.
2. Cumulate frequencies by successive addition of frequencies from lowest to highest class intervals. Last cumulative frequency must equal N.
3. Multiply 75.17 by N and divide by 100. Thus $(75.17/100) \times 98 = 73.67$. This represents the cumulative frequency corresponding to the unknown score.
4. Find the class interval containing the 73.67th cumulative frequency. This is the class interval 28.5 to 31.5.
5. Subtract from the value obtained in step 4 the cumulative frequency at the upper real limit of the *adjacent lower* interval, i.e., $73.67 - 72 = 1.67$. This value shows that the desired score is 1.67 frequencies within the interval 28.5–31.5.
6. Divide the value found in step 5 by the frequency within the interval and multiply by the width of the interval ($i = 3$). Thus $(1.67/10) \times 3 = 0.50$.
7. Add the value found in step 6 to the score at the lower real limit of the interval containing the 73.67th cumulative frequency. Thus $28.5 + 0.5 = 29.0$ (answer).

Terms to Remember:

Percentile rank
Percentile

EXERCISES

5.1 Estimate the percentile rank of the following scores, employing Fig. 5.2.
a) 104.5 b) 112 c) 134

5.2 Calculate the percentile rank of the scores in Exercise 5.1, employing Table 5.2.

5.3 Estimate the scores corresponding to the following percentiles, employing Fig. 5.2.
a) 25 b) 50 c) 75

5.4 Calculate the scores corresponding to the percentiles in Exercise 5.3, employing Table 5.2.

Table 5.5

Summary of procedures: obtaining percentile rank of a score

Class interval	Real limits	f	Cum f
41–43	40.5–43.5	1	98
38–40	37.5–40.5	2	97
35–37	34.5–37.5	4	95
32–34	31.5–34.5	9	91
29–31	28.5–31.5	10	82
26–28	25.5–28.5	13	72
23–25	22.5–25.5	18	59
20–22	19.5–22.5	14	41
17–19	16.5–19.5	11	27
14–16	13.5–16.5	8	16
11–13	10.5–13.5	5	8
8–10	7.5–10.5	2	3
5–7	4.5–7.5	1	1
	$i = 3$	98	

Problem: Find the percentile rank of a score of 29.

Steps:

1. Sum f column to obtain N. $N = 98$.
2. Cumulate frequencies by successive addition of frequencies from lowest to highest class intervals. Last cumulative frequency must equal N.
3. Identify interval containing the score: 29.
4. Subtract the score at the *lower real limit* of that interval from 29. $29 - 28.5 = 0.5$.
5. Note the frequency within that interval: 10.
6. Divide the value found in step 4 by the width of the interval and multiply by the frequency found in step 5. Thus $(0.5/3) \times 10 = 1.67$. This value represents the frequency within the interval 28.5–31.5 corresponding to a score of 29.
7. Note the cumulative frequency appearing in the *adjacent lower interval*: Cum $f = 72$.
8. Add the value found in step 6 to the value found in step 7. Thus $1.67 + 72 = 73.67$.
9. Divide the value found in step 8 by N and multiply by 100. Thus $(73.67/98) \times 100 = 75.17$ (answer).

5.5 A younger sibling informs you he obtained a score of 130 on a standard vocabulary test. What additional information might you seek in order to interpret his score?

5.6 Refer back to Exercise 4.6.

a) A student in group A obtained a score of 45 on the test of reading ability. What is his percentile rank in the group?

b) What is the percentile rank of a student in group B who also obtained a score of 45?

c) Combine both groups into an overall frequency distribution and obtain the percentile rank of a score of 45. What happens to the percentile rank of the student in group A? group B? Why?

5.7 If we were to place all the scores shown in Table 5.2 into a hat, what is the likelihood that, selecting at random, we would obtain:

a) a score equal to or higher than 131?

b) a score equal to or lower than 131?

c) a score equal to or below 84?

d) a score equal to or above 154?
e) a score between 117 and 133?
f) a score equal to or greater than 148, or equal to or less than 82?

5.8 The following questions are based on Table 5.3.
a) John G. proudly proclaims that he obtained a "higher" percentile ranking than his friend, Howard. Investigation of the fact reveals that his score on the test was actually lower. Must it be concluded that John G. was lying, or is some other explanation possible?
b) Fran F. obtained a percentile rank of 65 on the social work scale. What was her score? What score would she have had to obtain to achieve the same percentile rank on the physical sciences scale?
c) World Law School employs the Miller Analogies Test as an element of the admissions procedure. No applicant obtaining a percentile rank below 75 is considered for admission regardless of other qualifications. Thus the 75th percentile might be called a "cutoff" point. Which reference group is most likely involved in the decision? What score constitutes the cutoff point for this distribution?
d) Selip Medical School also employs the 75th percentile as a cutoff point. John G. obtained a raw score of 62. What are his chances of being considered for admission?

5.9 Employing the frequency distribution obtained in Exercise 3.10, calculate the percentile ranks of the following scores:
a) 9.5 b) 17.5 c) 24 d) 32 e) 34.5 f) 52.5

5.10 Repeat Exercise 5.9, employing the frequency distribution obtained in Exercise 3.11. Compare the results with those of Exercises 5.9, 5.11, and 5.12.

5.11 Repeat Exercise 5.9, employing the frequency distribution obtained in Exercise 3.12. Compare the results with those of Exercises 5.9, 5.10, and 5.12.

5.12 Repeat Exercise 5.9, employing the frequency distribution obtained in Exercise 3.13. Compare the results with those of Exercises 5.9, 5.10, and 5.11.

5.13 Employing the frequency distribution obtained in Exercise 3.10, find the scores corresponding to the following percentiles:
a) 10 b) 20 c) 50
d) 60 e) 75

5.14 Repeat Exercise 5.13, employing the frequency distribution obtained in Exercise 3.11. Compare the results with those of Exercises 5.13, 5.15, and 5.16.

5.15 Repeat Exercise 5.13, employing the frequency distribution obtained in Exercise 3.12. Compare the results with those of Exercises 5.13, 5.14, and 5.16.

5.16 Repeat Exercise 5.13, employing the frequency distribution obtained in Exercise 3.13. Compare the results with those of Exercises 5.13, 5.14, and 5.15.

5.17 Employing the frequency distribution obtained in Exercise 3.22, calculate the percentile ranks of the following scores:
a) 59 b) 43 c) 25
d) 18 e) 13 f) 0

5.18 Employing the frequency distribution obtained in Exercise 3.22, find the scores corresponding to the following percentiles:
a) 10 b) 25 c) 35
d) 50 e) 75 f) 90

5.19 Referring to the ungrouped data in Exercise 3.22, count the number of scores above and below the score at the 25th percentile.

a) What is the ratio of the number of scores below the 25th percentile to the total number of scores?

b) What is the ratio of the number of scores above the 25th percentile to the total number of scores?

5.20 Answer the above questions for the 50th and 75th percentiles.

5.21 Employing the two frequency distributions obtained for room B in Exercise 3.24, calculate the percentile ranks of the following scores for both sets of data for room B:

a) 59 b) 45 c) 41 d) 39

5.22 Employing the two frequency distributions obtained for room B in Exercise 3.24, calculate the scores corresponding to the following percentiles:

a) 10 b) 50 c) 75 d) 80

5.23 Referring to Exercise 3.20, determine how many sick days a person in the 50th percentile would have taken in a year.

5.24 Referring to Exercise 3.17, calculate the cost of the item that would correspond to the 50th percentile.

a) What is the difference between the costs at the 50th and 99th percentiles?

b) What is the difference between the costs at the 50th and 25th percentiles?

c) What is the difference between the costs at the 75th and 25th percentiles?

5.25 Suppose a manager were interested in comparing the scores of 25 employees on a mathematical test. Assigning one person to each letter, the following scores are recorded:

A. 55	F. 40	K. 50	P. 40	U. 40
B. 50	G. 60	L. 45	Q. 35	V. 30
C. 35	H. 45	M. 25	R. 25	W. 45
D. 45	I. 50	N. 35	S. 20	X. 35
E. 40	J. 30	O. 55	T. 30	Y. 40

Calculate the percentile rank for each person, employing $i = 5$ and assuming that the lowest class interval is 20–24.

5.26 In addition, the manager studies the number of mathematical problems the employees solve per day. He finds the following number of units completed per day for the same 25 employees (the employees were designated by the same letters as above):

A. 45	F. 30	K. 40	P. 30	U. 30
B. 40	G. 50	L. 35	Q. 25	V. 20
C. 25	H. 35	M. 15	R. 15	W. 35
D. 35	I. 40	N. 25	S. 10	X. 25
E. 30	J. 20	O. 45	T. 20	Y. 30

Calculate the percentile rank for each person and compare it to the ranks in Exercise 5.25. Employ $i = 5$ and a lowest class interval of 10–14.

5.27 Referring back to Boxes 1.1 and 5.1, calculate the percentile ranks for highway miles per gallon of the following automobiles.

a) Buick Century Regal, automatic

b) Cadillac Seville, automatic

c) Chevrolet Monza, manual
d) Lincoln Continental, automatic
e) Plymouth Valiant/Duster, manual
f) Volkswagen Beetle, manual
g) Volvo 260, manual

5.28 Referring back to Box 5.1, calculate the miles per gallon (highway) for cars achieving the following percentile ranks.
a) 88 b) 30 c) 98 d) 5 e) 12 f) 61

5.29 Refer back to Exercise 3.27.
a) Construct a cumulative frequency and cumulative percentage distribution for 1973 and 1974.
b) Find the percentile ranks for each of the following states for 1973 and 1974:
i) Alabama ii) Arizona iii) California
iv) Illinois v) New York vi) your own state

5.30 Refer to Exercise 4.38 and construct both a cumulative frequency and cumulative percentage distribution of the 49 sums.

5.31 Review the sampling experiment demonstrated in Exercise 4.38. Then return to the cumulative percentage distribution in Exercise 5.29. Answer the following questions:
a) What is the likelihood (what is the percentage of times?) that a sum equal to or less than 11 would have been drawn in a sample of $n = 2$ from the original population of seven numbers?
b) What is the likelihood (what percentage of times?) that a sum greater than 5 or less than 10 would have been obtained? [*Hint*: The cumulative percentage to the upper real limit of 9 is 88; below the upper limit of 5 is 43.]
c) What percentage of times would a sum equal to or less than 2 be obtained?
d) What percentage of times would a sum equal to or greater than 10 be obtained?
e) What percentage of times would a sum equal to or less than 2 or equal to or greater than 10 be obtained?

6 Measures of Central Tendency

BOX 6.1

The term "average" is commonly used to describe scores or numerical values in the central part of a distribution. Unfortunately, there are three different measures of central tendency which, for a given distribution, may deviate substantially from one another. Therefore it is possible for a person who wants to prove a particular point to cite that measure of central tendency which best supports his or her point. For example, it is conceivable that an automobile manufacturer whose car attains 17 miles per gallon in the highway test to claim that the car's performance is "above average." At the same time, the manufacturer could also discredit a competitor's car, which obtained 22 miles per gallon on the same highway test, as being "below average." How is this possible?

Class interval		Midpoint	f
41–42		41.5	1
39–40		39.5	0
37–38		37.5	1
35–36		35.5	9
33–34		33.5	3
31–32		31.5	9
29–30		29.5	14
27–28		27.5	3
25–26		25.5	9
23–24	mean →	23.5	6
21–22	median	21.5	14
19–20		19.5	2
17–18		17.5	20
15–16	mode	15.5	28
13–14		13.5	5

Shown here is the highway mileage distribution of 124 cars. The mean, the median, and the mode are indicated by arrows. Thus our tricky manufacturer has simply used the mode (15.5) as the standard for judging the company's car, and the mean (22.9) as the benchmark for the competitor's car.

6.1 INTRODUCTION

One of the greatest sources of confusion among lay people and, perhaps, a cause for their suspicions that statistics is more of an art than a science revolves about the ambiguity in the use of the term "average." Unions and management speak of average salaries, and frequently cite numerical values which are in sharp disagreement with each other; television programs and commercials are said to be prepared with the average viewer in mind; politicians are deeply concerned about the views of the average American voter; the average family size is frequently given as a fractional value, a statistical abstraction which is ludicrous to some and an absurdity to others; the term "average" is commonly used as a synonym for the term "normal"; the TV weatherbriefer tells us we had an average day or that rainfall for the month is above or below average. Indeed, the term "average" has so many popular connotations that many statisticians prefer to drop it from the technical vocabulary and refer, instead, to **measures of central tendency**. We shall define a measure of central tendency as an *index of central location employed in the description of frequency distributions.* Since the center of a distribution may be defined in different ways, there will be a number of different measures of central tendency. In this chapter, we shall concern ourselves with three of the most frequently employed measures of central tendency: the mean, the median, and the mode.

6.1.1 Why Describe Central Tendency?

Through the first five chapters of the book, we concerned ourselves primarily with organizing data into a meaningful and useful form. Beyond this, however, we want to describe our data in such ways that quantitative statements can be made. A frequency distribution represents an organization of data but it does not, in itself, permit us to make quantitative statements either describing the distribution or comparing two or more distributions.

There are two features of many frequency distributions which statisticians have noted and have developed quantitative methods for describing: (1) Frequently data cluster around a central value which is between the two extreme values of the variable under study. (2) The data may tend to be dispersed and distributed about the central value in a way which can be specified quantitatively. The first of these features is the topic of the present chapter, whereas the second (dispersion) is discussed in the forthcoming chapter.

Being able to locate a point of central tendency, particularly when coupled with a description of the dispersion of scores about that point, can be very useful to the statisticians. For example, they may be able to reduce a mass of data to a simple quantitative value which may be understood and communicated to other scientists.

We have already stated that the statistician is frequently called upon to compare the measurements obtained from two or more groups of subjects for the purpose of drawing inferences about the effects of an independent variable. Measures of central tendency greatly simplify the task of drawing conclusions.

© 1955 United Feature Syndicate, Inc. SCHULZ

Fig. 6.1 Is Charley Brown above average, whatever that means?

In 1949, the median per-family income was estimated by the Bureau of the Census as \$3100. It was concluded that the "income of the average family was \$3100." Chimpanzees at a given age can learn to solve a difficult problem in approximately twenty trials (the arithmetic mean). It was concluded, "The average chimpanzee of this age can solve the problem in twenty trials." What is wrong with these conclusions? Basically, when you have established that some measure of central tendency—or average—may be used to characterize a population, you do not prove that an object or person who achieves this value is average. In other words, the fact that you may have an "average" score on some variable does not make you an average person. You may be 8′2″, weigh 400 lb, have flat feet, eat 3 lb of spaghetti for dinner, and have an I.Q. of 100 . . . the mean of the population. It would hardly seem accurate for you to proclaim, "I have an average I.Q. Therefore I am an average person."

This error is so subtle and pervasive that we have heard many highly educated and knowledgeable persons make this error without any awareness of a lack of logical connection between having an average score on some trait and "being average," whatever that means. (See Fig. 6.1.)

6.2 THE ARITHMETIC MEAN

6.2.1 Methods of Calculation

You are probably intimately familiar with the arithmetic mean, for whenever you obtain an "average" of grades by summing the grades and dividing by the number of grades, you are calculating the arithmetic mean. In short, the **mean** is *the sum of the scores or values of a variable divided by their number.* Stated in algebraic form:

$$\bar{X} = \frac{X_1 + X_2 + \cdots + X_N}{N} = \frac{\sum X}{N}, \tag{6.1}$$

where

$\bar{X}$ = the mean and is referred to as X bar,*
N = the number of scores, and
$\sum$ = the mathematical verb directing us to sum all the measurements.

Thus the arithmetic mean of the scores 8, 12, 15, 19, 24 is $\bar{X} = \frac{78}{5} = 15.60$.

Obtaining the mean from an ungrouped frequency distribution. You will recall that we constructed a frequency distribution as a means of eliminating the constant repetition of scores that occur with varying frequency in order to permit a single entry in the frequency column to represent the number of times a given score occurs. Thus, in Table 6.1, we know, from the column headed f, that the score of 8 occurred 6 times. In calculating the mean, then, it is not necessary to add 8 six times since we may multiply the score by its frequency and obtain the same value of 48. Since each score

* In Section 1.2 we indicated that italic letters would be employed to represent sample statistics and Greek letters to represent population parameters. The Greek letter μ will be used to represent the population mean.

is multiplied by its corresponding frequency prior to summing, we may represent the mean for frequency distributions as follows:

$$\bar{X} = \frac{\sum fX}{N}. \tag{6.2}$$

Table 6.1

Computational procedures for calculating the mean with ungrouped frequency distributions

X	f	fX	
12	1	12	
11	2	22	$\bar{X} = \frac{\sum fX}{N}$
10	5	50	
9	4	36	
8	6	48	$\bar{X} = \frac{232}{29}$
7	4	28	
6	3	18	$\bar{X} = 8.00$
5	2	10	
4	2	8	
	$N = 29$	$\sum fX = 232$	

Obtaining the mean from a grouped frequency distribution, raw score method. The calculation of the mean from a grouped frequency distribution involves essentially the same procedures that are employed with ungrouped frequency distributions. To start with, the midpoint of each interval is used to represent all scores within that interval. The midpoint of each interval is multiplied by its corresponding frequency, and the product is summed and divided by N. The procedures employed in calculating the mean from a grouped frequency distribution are demonstrated in Table 6.2.

6.2.2 Properties of the Arithmetic Mean

One of the most important properties of the mean is that *it is the point in a distribution of measurements or scores about which the summed deviations are equal to zero.* In other words,

$$\sum(X - \bar{X}) = 0. \tag{6.3}$$

The algebraic proof of this statement is

$$\begin{aligned}\sum(X - \bar{X}) &= \sum X - \sum \bar{X} \\ &= N\bar{X} - N\bar{X} \\ &= 0.\end{aligned}$$

Table 6.2

Computational procedures for calculating the mean from a grouped frequency distribution

1 Class interval	2 Frequency f	3 Midpoint X	4 Frequency × midpoint fX	
125–129	2	127	254	$\bar{X} = \frac{\sum fX}{N}$
120–124	5	122	610	
115–119	8	117	936	$= \frac{10,195}{100}$
110–114	10	112	1120	
105–109	15	107	1605	
100–104	20	102	2040	$= 101.95$
95–99	15	97	1455	
90–94	10	92	920	
85–89	8	87	696	
80–84	4	82	328	
75–79	3	77	231	
	$N = 100$		$\sum fX = 10,195$	

In following this algebraic proof, it is important to note that (1) since

$$\bar{X} = \frac{\sum X}{N},$$

it follows that $\sum X = N\bar{X}$ and (2) summing the mean over all the scores ($\sum \bar{X}$) is the same as multiplying $\bar{X}$ by N, that is, $N\bar{X}$.

Therefore the mean is a score or a potential score which balances all the scores on either side of it. In this sense, it is analogous to the fulcrum on a teeter board. In playing on teeter boards, you may have noticed that it is possible for a small individual to balance a heavy individual by moving the latter closer to the fulcrum. Thus, if you wanted to balance a younger brother or sister (assuming he or she is lighter than you) on a teeter board, you would move yourself toward the center of the board. This analogy leads to a second important characteristic of the mean; that is, *the mean is very sensitive to extreme measurements when these measurements are not balanced on both sides of it.*

Observe the two *arrays* of scores in Table 6.3. An **array** *is an arrangement of data according to their magnitude from the smallest to the largest value.* Note that all the scores in both distributions are the same except for the very large score of 33 in column X_2. This one extreme score is sufficient to double the size of the mean. The sensitivity of the mean to extreme scores is a characteristic which has important implications governing our use of it. These implications will be discussed in Section 6.5, in which we compare the three measures of central tendency.

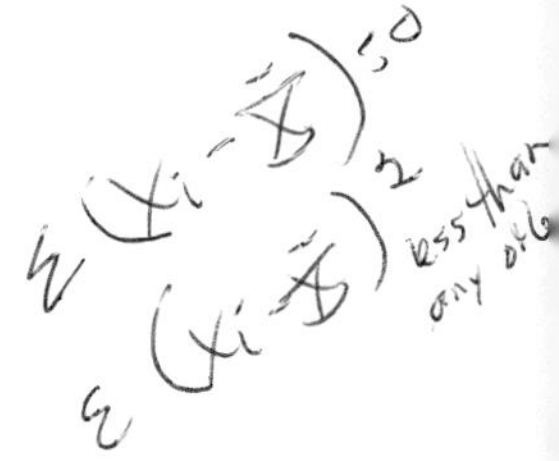

Table 6.3

Comparison of the means of two arrays of scores, one of which contains an extreme value

X_1	X_2
2	2
3	3
5	5
7	7
8	33
$\sum X_1 = 25$	$\sum X_2 = 50$
$\bar{X}_1 = 5.00$	$\bar{X}_2 = 10.00$

A third important characteristic of the mean is that *the sum of squares of deviations from the arithmetic mean is less than the sum of squares of deviations about any other score or potential score.*

To illustrate this characteristic of the mean, Table 6.4 shows the squares and the sum of squares when deviations are taken from the mean and various other scores in a distribution. It can be seen that the sum of squares is smallest in column 4, when deviations are taken from the mean.

Table 6.4

The squares and sum of squares of deviations taken from various scores in a distribution

1 X	2 $(X - 2)^2$	3 $(X - 3)^2$	4 $(X - \bar{X})^2$	5 $(X - 5)^2$	6 $(X - 6)^2$
2	0	1	4	9	16
3	1	0	1	4	9
4	4	1	0	1	4
5	9	4	1	0	1
6	16	9	4	1	0
$\sum$	30	15	10	15	30
$N = 5$					
$\bar{X} = 4$					

This property of the mean provides us with another definition, i.e., *the* **mean** *is that measure of central tendency which makes the sum of squared deviations around it minimal.* The method of locating the mean by finding the minimum sum of squares is referred to as the **least squares** method. The least squares method is of considerable value in statistics, particularly when it is applied to curve fitting.

6.2.3 The Weighted Mean

Let us imagine that, while on a vacation trip, you kept a record of your gasoline purchases. At various stops, you obtained gasoline at 55, 60, 53 and 50 cents a gallon. Could you sum these four prices together and divide by four to obtain a mean cost-per-gallon of 54.5 cents for the trip? This could be done only if you purchased the identical amount of gasoline at each stop. What if, as a matter of fact, you had obtained 15 gallons at the first stop, 20 at the second, 15 at the third, and 10 at the fourth?

The total amount of money you spent to obtain 60 gallons of gas would be

$$15(55) + 20(60) + 15(53) + 10(50) = 3320 \text{ cents.}$$

Since you purchased 60 gallons, the mean cost per gallon is 3320/60 or 55.33 cents.

The **weighted mean** of a set of numbers $X_1, X_2, \ldots, X_n$ can be expressed as a sum of each of these numbers multiplied by their corresponding weights, w. Thus

$$\bar{X}_w = \frac{\sum(w \cdot X)}{\sum w}.$$

For the above problem, the weighted mean cost per gallon becomes

$$\bar{X}_w = \frac{15(55) + 20(60) + 15(53) + 10(50)}{15 + 20 + 15 + 10}$$

$$= \frac{3320}{60} = 55.33 \text{ cents per gallon.}$$

Note that if the weights are equal (i.e., the same amount of gasoline is purchased at each station), the formula reduces to that of the arithmetic mean.

As another example, imagine that during the year you purchase shares of common stock at the following prices: 60 at \$15 per share, 100 at \$25 per share, 150 at \$10 per share, and 200 at \$12 per share. What is the break-even point, i.e., the weighted mean?

$$\bar{X}_w = \frac{60(15) + 100(25) + 150(10) + 200(12)}{60 + 100 + 150 + 200} = \frac{7300}{510} = \$14.31.$$

A special case arises when we have a number of individual means, each based on a different number of cases, and we want to obtain an overall mean. Imagine that you are an instructor in economics teaching three sections of a course. In one section, containing 45 students, the mean grade on the final exam is 74.5. In the remaining two sections, containing 50 and 65 students respectively, the means are 77.4 and 78.3. What is the weighted mean of the means? The weighted mean in this case is given by

$$\bar{X}_w = \frac{\sum(n \cdot \bar{X})}{\sum n}. \tag{6.4}$$

In the present problem, the weighted mean of all three sections is

$$\bar{X}_w = \frac{45(74.5) + 50(77.4) + 65(78.3)}{45 + 50 + 65} = \frac{12{,}312}{160} = 76.95.$$

6.3 THE MEDIAN

With grouped frequency distributions, the **median** is defined as *that score or potential score in a distribution of scores, above and below which one-half of the frequencies fall.* If this definition sounds vaguely familiar to you, it is not by accident. The median is merely a special case of a percentile rank. Indeed, the median is the score at the 50th percentile. It should be clear that the generalized procedures discussed in chapters for determining the score at various percentile ranks may be applied to the calculation of the median.

Let us find the median of the data in Table 5.1 (Section 5.2.2). The cumulative frequency of a score at the 50th percentile is

$$\text{cum } f = \frac{50 \times 110}{100} = 55.$$

The cum f within the interval (109.5 to 114.5) $= 55 - 43 = 12$.

$$\frac{?}{5} = \frac{12}{17},$$

$$? = 3.53,$$

$$\text{median} = 109.5 + 3.53$$
$$= 113.03$$

6.3.1 The Median of an Array of Scores

Occasionally, it will be necessary to obtain the median when the N is not sufficient to justify casting the data into the form of a frequency distribution or a grouped frequency distribution. Consider the following array of scores: 5, 19, 37, 39, 45. Note that the scores are arranged in order of magnitude and that N is an odd number. A score of 37 is the median since two scores fall above it and two scores fall below it.* If N is an even number, the median is the sum of the two middle values divided

* When working with an array of numbers where N is odd, the definition of the median does not quite hold, i.e., in the example above, in which the median is 37, two scores lie below it and two above it, as opposed to one-half of N. If the score of 37 is regarded as falling one-half on either side of the median, this disparity is reconciled.

by two. The two middle values in the array of scores 8, 26, 35, 43, 47, 73 are 35 and 43. Therefore, the median is (35 + 43)/2, or 39.

6.3.2 Characteristics of the Median

An outstanding characteristic of the median is its insensitivity to extreme scores. Consider the following set of scores: 2, 5, 8, 11, 48. The median is 8. This is true, in spite of the fact that the set contains one extreme score of 48. Had the 48 been a score of 97, the median would remain the same. This characteristic of the median makes it valuable for describing central tendency in certain types of distributions. This point will be further elaborated in Section 6.5 when the uses of the three measures of central tendency are discussed.

6.4 THE MODE

Of all measures of central tendency, the mode is the most easily determined since it is obtained by inspection rather than by computation. The **mode** is simply the score which occurs with greatest frequency. For grouped data, the mode is designated as the midpoint of the interval containing the highest frequency count. In Table 6.2 the mode is a score of 102 since it is the midpoint of the interval (100–104) containing the greatest frequency.

In some distributions, which we shall not consider here, there will be two high points which produce the appearance of two humps, as on a camel's back. Such distributions are referred to as being *bimodal.* A distribution containing more than two humps is referred to as being *multimodal.*

6.5 COMPARISON OF MEAN, MEDIAN, AND MODE

We have seen that the mean is a measure of central tendency in which the *sum* of the deviations on one side of it equals the *sum* of the deviations on the other side. The median, on the other hand, divides the area under the curve into two equal halves so that the *number* of scores below the median equals the *number* of scores above the median.

In general, the arithmetic mean is the preferred statistic for representing central tendency because it possesses several desirable properties. To begin with, the mean is a member of a mathematical system which permits its use in more advanced statistical analyses. We have used deviations from the mean to demonstrate two of its most important characteristics, i.e., the sum of deviations is zero and the sum of squares is minimal. Deviations of scores from the mean provide valuable information about any distribution. We shall be making frequent use of deviation scores throughout the remainder of the text. In contrast, deviation scores from the median and the corresponding squared deviations have only limited applications to more advanced statistical considerations. Another important feature of the mean is that it is the more stable or reliable measure of central tendency. If we were to take repeated samples from a given population, the mean would usually show less fluctuation than either the median or the mode. In other words, the mean generally provides a better estimate of the corresponding population parameter.

On the other hand, there are certain situations in which the median is preferred as the measure of central tendency. When the distribution is symmetrical, the mean and the median are identical. Under these circumstances, the mean should be employed. However, as we have seen, when the distribution is markedly skewed, the mean will provide a misleading estimate of central tendency. In column X_2 of Table 6.3, the mean is 10 even though four of the five scores are below this value. Annual family income is a commonly studied variable in which the median is preferred over the mean since the distribution of this variable is distinctly skewed in the direction of high salaries, with the result that the mean overestimates the income obtained by most families.

The median is also the measure of choice in distributions in which there are *indeterminate values.* To illustrate, when running rats in a maze, there will be occasions when one or more rats will simply not run. Their time scores are therefore indeterminate. Their "scores" cannot simply be thrown out since the fact of their not running may be of considerable significance in evaluating the effects of the independent variable. Under these circumstances, the median should be employed as the measure of central tendency.

The mode is the appropriate statistic whenever a quick, rough estimate of central tendency is desired or when we are interested only in the typical case. It is rarely used in the behavioral sciences.

6.6 THE MEAN, THE MEDIAN, AND SKEWNESS

In Chapter 4, we demonstrated several forms of skewed distributions. We pointed out, however, that skew cannot always be determined by inspection. If you have understood the differences between the mean and the median, you should be able to suggest a method for determining whether or not a distribution is skewed and, if so, determine the direction of the skew. The basic fact to keep in mind is that the mean is drawn in the direction of the skew whereas the median, unaffected by extreme scores, is not. Thus, when the mean is higher than the median, the distribution may

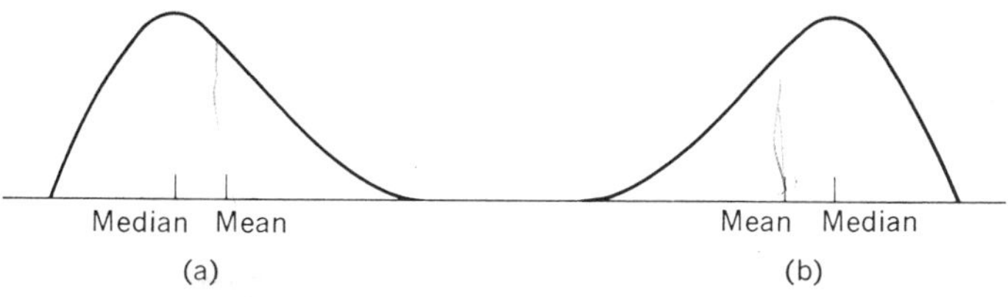

Fig. 6.2 The relation between the mean and median in (a) positively and (b) negatively skewed distributions.

be said to be positively skewed; when the mean is lower than the median, the distribution is negatively skewed. Figure 6.2 demonstrates the relation between the mean and the median in positively and negatively skewed distributions.

CHAPTER SUMMARY

In this chapter we discussed, demonstrated the calculation of, and compared three indices of central tendency that are frequently employed in the description of frequency distributions: the mean, the median, and the mode.

We saw that the mean may be defined variously as the sum of scores divided by their number, the point in a distribution which makes the summed deviations equal to zero, or the point in the distribution which makes the sum of the squared deviations minimal. The median divides the area of the curve into two equal halves so that the number of scores below the median equals the number of scores above it. Finally, the mode is defined as the most frequently occurring score. We demonstrated the method for obtaining the weighted mean of a set of means when each of the individual means is based on a different n.

Because of special properties it possesses, the mean is the most frequently employed measure of central tendency. However, because of its sensitivity to extreme scores which are not balanced on both sides of the distribution, the median is usually the measure of choice when distributions are markedly skewed. The mode is rarely employed in the behavioral sciences.

Finally, we demonstrated that the relationship between the mean and the median is negatively and positively skewed distributions.

Terms to Remember:

Measures of central tendency
Mean
Weighted mean
Median
Mode
Sum of squares
Least squares method
Array

EXERCISES

6.1 Find the mean, the median, and the mode for each of the following sets of measurements. Show that $\sum(X - \bar{X}) = 0$.
a) 10, 8, 6, 0, 8, 3, 2, 2, 8, 0
b) 1, 3, 3, 5, 5, 5, 7, 7, 9
c) 120, 5, 4, 4, 4, 2, 1, 0

6.2 In which of the sets of measurements in Exercise 6.1 is the mean a poor measure of central tendency? Why?

6.3 For each of the sets of measurements in Exercise 6.1, show that the *sum of squares of deviations from the arithmetic mean is less than the sum of squares of deviations about any other score or potential score.*

6.4 You have calculated the maximum speed of various automobiles. You later discover that all the speedometers were set five miles per hour too fast. How will the measures of central tendency based on the corrected data compare with those calculated from the original data?

6.5 You have calculated measures of central tendency on the weights of barbells, expressing your data in terms of ounces. You decide to recompute after you have divided all the weights by 16 to convert them to pounds. How will this affect the measures of central tendency?

6.6 Calculate the measures of central tendency for the data in Exercises 4.1 and 4.8.

6.7 In Exercise 6.1(c), if the score of 120 were changed to a score of 20, how would the various measures of central tendency be affected?

6.8 On the basis of the following measures of central tendency, indicate whether or not there is evidence of skew and, if so, what is its direction?

a) $\bar{X} = 56$ Median $= 62$ Mode $= 68$
b) $\bar{X} = 68$ Median $= 62$ Mode $= 56$
c) $\bar{X} = 62$ Median $= 62$ Mode $= 62$
d) $\bar{X} = 62$ Median $= 62$ Mode $= 30$, Mode $= 94$

6.9 What is the nature of the distributions in Exercise 6.8(c) and (d)?

6.10 Calculate the mean of the following array of scores: 3, 4, 5, 5, 6, 7.

a) Add a constant, say 2, to each score. Recalculate the mean. Generalize: What is the effect on the mean of adding a constant to all scores?
b) Subtract the same constant from each score. Recalculate the mean. Generalize: What is the effect on the mean of subtracting a constant from all scores?
c) Alternately add and subtract the same constant, say 2, from the array of scores (that is, $3 + 2$, $4 - 2$, $5 + 2$, etc). Recalculate the mean. Generalize: What is the effect on the mean of adding and subtracting the same constant an equal number of times from an array of scores?
d) Multiply each score by a constant, say 2. Recalculate the mean. Generalize: What is the effect on the mean of multiplying each score by a constant?
e) Divide each score by the same constant. Recalculate the mean. Generalize: What is effect on the mean of dividing each score by a constant?

6.11 Refer to Exercise 4.6. Which measure of central tendency might best be used to describe group A? Group B? Why?

6.12 In Section 6.5 we stated that the mean is usually *more reliable* than the median, i.e., less subject to fluctuations. We conducted an experiment consisting of 30 tosses of three dice. The results were as follows: 6, 6, 2; 4, 1, 1; 6, 5, 5; 6, 4, 3; 4, 2, 1; 5, 4, 3; 4, 4, 3; 6, 6, 4; 5, 3, 2; 6, 3, 3; 4, 3, 2; 6, 4, 1; 6, 4, 2; 5, 1, 1; 6, 5, 4; 2, 1, 1; 5, 4, 3; 5, 4, 4; 4, 3, 1; 4, 2, 2; 6, 5, 3; 5, 1, 1; 6, 5, 2; 6, 3, 3; 6, 6, 5; 6, 5, 4; 6, 2, 1; 5, 4, 3; 5, 4, 1; 6, 3, 1.

a) Calculate the 30 means and 30 medians.
b) Employing grouping intervals starting with the real limits of the lower interval 0.5–1.5, group the means and medians into separate frequency distributions.
c) Draw histograms for the two distributions. Do they support the contention that the mean is a more stable estimator of central tendency? Explain.
d) Given that we were to place the medians and the means into two separate hats and were to draw one at random from each. What is the likelihood that:
   i) a statistic greater than 5.5 would be obtained?
   ii) a statistic less than 1.5 would be obtained?
   iii) a statistic greater than 5.5 or less than 1.5 would be obtained?

6.13 Shown in Table 6.5 are the unemployment compensation maximums for each state on July 15, 1975. Calculate the mean, the median and the mode. (*Note.* Since both Washington. D.C., and Puerto Rico are included, $N = 52$.)

Table 6.5

State unemployment maximum

State	Maximum weekly benefit amount	State	Maximum weekly benefit amount
Alabama	$90	Montana	$81
Alaska	90	Nebraska	80
Arizona	85	Nevada	88
Arkansas	92	New Hampshire	95
California	90	New Jersey	90
Colorado	108	New Mexico	71
Connecticut	104	New York	95
Delaware	105	North Carolina	90
D.C.	127	North Dakota	86
Florida	82	Ohio	82
Georgia	90	Oklahoma	86
Hawaii	104	Oregon	95
Idaho	90	Pennsylvania	111
Illinois	97	Puerto Rico	55
Indiana	60	Rhode Island	94
Iowa	107	South Carolina	96
Kansas	85	South Dakota	77
Kentucky	80	Tennessee	85
Louisiana	80	Texas	63
Maine	74	Utah	101
Maryland	89	Vermont	91
Massachusetts	95	Virginia	87
Michigan	97	Washington	93
Minnesota	105	West Viginia	115
Mississippi	60	Wisconsin	113
Missouri	85	Wyoming	84

Source. Department of Labor, Manpower Administration, Unemployment Insurance Service.

6.14 During the course of the year you purchased a common stock in the following amounts and prices: 150 at $3 per share, 400 at $2.50 per share, 100 at $4.25 per share, 200 at $3.50 per share.

a) What is the break-even point?

b) If the commissions on the purchases were \$13.50, \$30.00, \$12.75, and \$21.00, respectively, calculate the break-even point, taking the commission into account. (*Hint*: Consider adding the sum of the commissions to $\sum w \cdot X$ prior to dividing).

6.15 In a departmental final exam, the following mean grades were obtained based on n's of 25, 40, 30, 45, 50, and 20: 72.5, 68.4, 75.0, 71.3, 70.6, and 78.1.
a) What is the total mean over all sections of the course?
b) Draw a line graph showing the size of the class along the abscissa and the corresponding means along the ordinate. Does there appear to be a relationship between class size and mean grades on the final exam?

6.16 Give examples of data in which the preferred measure of central tendency would be the
a) mean, b) median, c) mode.

6.17 Find the mean, the median, and the mode for the following array of scores: -9.0, -6.0, -5.0, -5.0, -0.5, 0, 0.1, 2.0, 4.0, 5.0.

6.18 Find the mean, the median, and the mode of the numbers in Exercise 6.17 by first adding 9.0 to each number and then subtracting 9.0 from each of the measures of central tendency.

6.19 List at least three specific instances in which a measure of central tendency was important in describing a group of people.

6.20 List at least three specific instances in which a measure of central tendency was utilized to compare groups of people.

6.21 List at least three specific instances in which a measure of central tendency was estimated for a large group of people from data obtained from a sample of that large group.

6.22 On the basis of examination performance, an instructor identifies the following groups of students:
a) Those with a percentile rank of 90 or higher.
b) Those with a percentile rank of 10 or less.
c) Those with percentile ranks between 40 and 49.
d) Those with percentile ranks between 51 and 60.
Which group would the instructor work with if he wished to raise the *median* performance of the total group? Which group would he work with if he wished to raise the *mean* performance of the total group?

6.23 Referring to Exercise 3.20, calculate the mean, the median, and the mode. Which measure of central tendency would a manager report if he were interested in emphasizing that employees took few sick days? Which measure would suggest a large number of sick days?

6.24 Referring to Exercise 1.17, calculate the mean, the median, and the mode. Which number is closest to the 50th percentile? (See Exercise 5.25.)

6.25 Referring to Exercise 3.17, calculate the mean, the median, and the mode. Show that $\sum(X - \bar{X}) = 0$.

6.26 Suppose that six salesmen are required to sell 50 vacuum cleaners in a month.
Mr. A sells a mean of 1.75 per day in 4 days.
Mr. B sells a mean of 2.0 per day in 5 days.
Mr. C sells a mean of 2.4 per day in 5 days.
Mr. D sells a mean of 2.5 per day in 4 days.
Mr. E sells a mean of 2.0 per day in 3 days.
Mr. F sells a mean of 1.67 per day in 3 days.
What is the mean number of cleaners sold per day?

6.27 Suppose that a given merchant sells the following number of apples from Monday through Saturday.

a) 30, 30, 30, 30, 30, 30
What are the mean, the median, and the mode?

b) 25, 30, 35, 30, 35, 25
What are the mean, the median, and the mode?

c) 10, 25, 30, 35, 25, 30
Calculate the mean, the median, and the mode.

6.28 a) In Exercise 6.27(c), show that $\sum(X - \bar{X}) = 0$.

b) Show that the sum of the deviations from the median and from the mode do not equal zero.

c) Why do the deviations from the mean equal zero, whereas the deviations from the median and mode do not?

6.29 Calculate the mean, the median, and the mode for each of the following sets.

a) 3, 4, 5, 5, 6, 6, 6, 7, 7, 7, 7, 8, 8, 8, 9, 9, 10, 11

b) 5, 5, 6, 6, 6, 7, 7, 7, 7, 7, 7, 7, 7, 8, 8, 8, 9, 9

c) 5, 6, 6, 6, 7, 7, 7, 7, 7, 7, 7, 7, 7, 7, 8, 8, 8, 9

Draw polygons representing the distribution of these scores. Compare their forms.

6.30 Determine the mean, the median, and the mode for the following array:

4, 4, 5, 5, 5, 6, 7, 7, 7, 8, 8

6.31 During a given month, a person buys three dozen eggs for \$0.70, one dozen eggs for \$0.65, two dozen eggs for \$0.75, and one dozen for \$0.60. What is the mean price per dozen?

6.32 A freelance writer earns a mean of \$5.00 per hour for 4 hours of work, \$7.50 per hour for 2 hours work, \$3.33 per hour for 3 hours, and \$4.00 per hour for 1 hour. What are her mean earnings per hour?

6.33 Two manufacturers state that the "average" life of their refrigerators is seven years. However, upon drawing a random sample of durations, a person finds that the life (in years) of 20 machines from manufacturer A is as follows:

5, 5, 5, 6, 6, 6, 6, 7, 7, 7, 7, 7, 7, 8, 8, 8, 8, 9, 9, 9

A sample of machines from manufacturer B shows durations of:

2, 3, 4, 5, 5, 5, 5, 6, 6, 6, 7, 7, 7, 7, 7, 8, 8, 20, 20, 20

What measurement of "average" was each manufacturer reporting? Which machine would probably be the best investment? With which machine would you feel more confident in stating that the "average" life is 7 years?

6.34 At the beginning of 1972, there were 27 nuclear power plants, with a mean output of 436.5 megawatts. An additional 54 plants, with a mean output of 847.8 megawatts, were being built. Finally, 52 additional plants were planned, with a mean output of 991.7 megawatts.

a) What is the anticipated mean output, in megawatts, of all 133 nuclear power plants?

b) What will be the total output of all 133 power plants?

6.35 Shown below are the population and the annual rate of growth for six geographical regions. Calculate the weighted percent of growth over all six regions.

Region	Present population (in millions)	Annual rate of growth (percent)
North America	225	1.1
South America	276	2.9
Europe	456	0.8
USSR	241	1.0
Africa	344	2.4
Asia	1990	2.0

Source. L. Rocks and R. P. Runyon, *The Energy Crisis.* New York: Crown Publishers, 1972.

6.36 Refer to the data presented in Box 1.1. Calculate the mean, the median, and the mode for city miles per gallon.

*6.37 In Chapter 4, Exercise 4.38, we described a sampling experiment in which samples of $n = 2$ were drawn from a population of seven scores. We constructed a table of sums of all possible combinations of seven scores, taken two at a time. Since $n = 2$ in each sample, that sum can be converted to a mean by dividing by 2. Construct a table showing all 49 means.

*6.38 Construct a frequency distribution of the means calculated in Exercise 6.37.

*6.39 Selecting samples of $n = 2$ as we did in Chapter 4, what percentage of the time would we expect to obtain (round to nearest percent):

a) a mean equal to 6?
b) a mean of 5.5 or greater?
c) a mean of 2.5 or less?
d) a mean equal to or greater than 5.5, or equal to or less than 0.5?

*6.40 Calculate the mean of the sample means in Exercise 6.39, above. [*Hint*: Treat each mean as a score, and follow the computational procedures for calculating the mean with ungrouped frequency distributions.] How does the mean of the sample means compare with the original population mean?

Measures of Dispersion

7.1 INTRODUCTION

In the introduction to Chapter 5, we saw that a score by itself is meaningless, and takes on meaning only when it is compared with other scores or other statistics. Thus if we know the mean of the distribution of a given variable, we can determine whether a particular score is higher or lower than the mean. But how much higher or lower? It is clear at this point that a measure of central tendency such as the mean only provides a limited amount of information. To more fully describe a distribution, or to more fully interpret a score, it is clear that additional information is required concerning the **dispersion** of scores about our measure of central tendency.

Consider Fig. 7.1, parts (a) and (b). In both examples of frequency polygons, the mean of the distribution is exactly the same. However, note the difference in the interpretations of a score of 128. In (a), because the scores are widely dispersed about the mean, a score of 128 may be considered only moderately high. Quite a few individuals in the distribution scored above 128, as indicated by the area to the right of 128. In (b), on the other hand, the scores are compactly distributed about the same mean. This is a more homogeneous distribution. Consequently, the score of 128 is

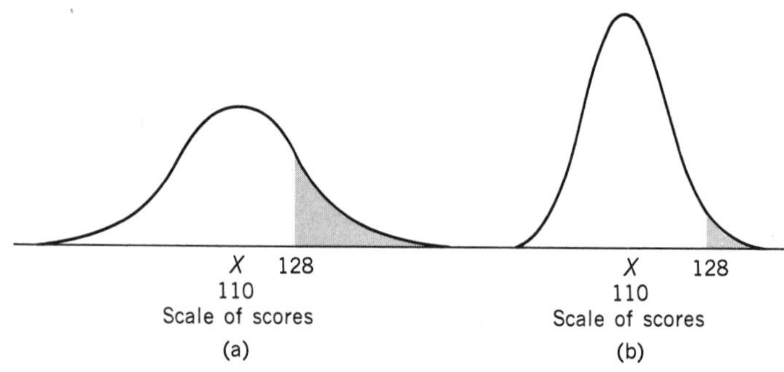

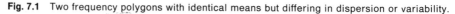

Fig. 7.1 Two frequency polygons with identical means but differing in dispersion or variability.

Fig. 7.2 It is not uncommon for Chambers of Commerce to report the mean annual temperature without reporting variability. A city with an "ideal" annual temperature of, say, 74° may go below zero in winter and above 100° in summer. The term "ideal" becomes a misleading statistical abstraction.

now virtually at the top of the distribution and it may therefore be considered a very high score.

It can be seen, then (Fig. 7.2), that in interpreting individual scores, we must find a companion to the mean or the median. This companion must in some way express the degree of dispersion of scores about the measure of central tendency. We shall discuss five such measures of dispersion or variability: the **range**, the **interquartile**

range, the **mean deviation**, the **variance**, and the **standard deviation**. Of the five, we shall find the standard deviation to be our most useful measure of dispersion in both descriptive and inferential statistics. In advanced inferential statistics, as in analysis of variance (Chapter 16), the variance will become a most useful measure of variability.

The following excerpt from *Winning with Statistics** illustrates the problems which arise when measures of central tendency are reported without any reference to variability.

The truth of the matter is that this world of ours is wonderfully varied. Thank the Creator that we are not all average everything or anything. Of course, if we were it would simplify things. Imagine telephoning the local clothing store, and asking, "Send me an average pair of pants or an average dress," or telling your friendly automobile dealer, "Send me an average lemon." Convenient, yes, but what a crushing bore.

But we are different from one another. Things are different from one another. We are even told that no two snowflakes are ever alike (I'd like to see how anybody would go about proving *that*). Because of the wide variability of all things measured, it is clear that measures of central tendency have only limited value in describing the totality of events that interest us. Clearly some companion to central tendency—a descriptive measure of the spread of scores about central tendency—is desired. "Why?" you ask.

Imagine that you are a home builder and you know that the median family size is 2.19. (That's what it was in 1972; note that, again, we have the statistical abstraction. I am willing to give considerable odds that there are few families of median size.) You are planning to put in a housing development. How many bedrooms should you provide? Our measure of central tendency does not help very much. What you need is some information about how family size is distributed, spread out, or dispersed throughout the population. What proportions of families are two-person, three-person, four-person, five-person, and so on? Data such as the following (which I took from the *American Almanac of 1976: The Statistical Abstract of the U.S.*) are far more useful.

Size of family	Number in thousands	Percent
2 persons	20,592	37.4
3 persons	11,673	21.2
4 persons	10,789	19.6
5 persons	6,386	11.6
6 persons	3,021	5.5
7 or more persons	2,593	4.7

This information tells us that, were we to build all housing units to accommodate two- and three-person families, we would be neglecting about 41 percent of the population. That's a lot of people and a lot of houses.

* Runyon, R. P., *Winning With Statistics.* Addison-Wesley, 1977.

7.2 THE RANGE

When we calculated the various measures of central tendency, we located a *single point* along the scale of scores and identified it as the mean, the median, or the mode. When our interest shifts to measures of dispersion, however, we must look for an index of variability which indicates the *distance* along the scale of scores.

One of the first measures of distance which comes to mind is the so-called **crude range**. The range is by far the simplest and the most straightforward measure of dispersion. It consists simply of the scale distance between the largest and the smallest score.

Although the range is meaningful, it is of little use because of its marked instability. Note that if there is one extreme score in a distribution, the dispersion of scores will appear to be large when, in fact, the removal of that score may reveal an otherwise "compact" distribution. Several years ago, an inmate of an institution for retarded persons was found to have an I.Q. score in the 140's. Imagine the erroneous impression that would result if the range of scores for the inmates was reported as, say, 20–140 or 120! Stated another way, the range reflects only the two most extreme scores in a distribution.

7.3 THE INTERQUARTILE RANGE

In order to overcome the instability of the crude range as a measure of dispersion, the **interquartile range** is sometimes employed. The interquartile range is calculated by simply subtracting the score at the 25th percentile (referred to as the first quartile or Q_1) from the score at the 75th percentile (the third quartile or Q_3). Although this measure of variability of scores is far more meaningful than the crude range, it has two important shortcomings: (1) like the crude range, it does not by itself permit the precise interpretation of a score within a distribution, and (2) like the median, it does not enter into any of the "higher" mathematical relationships that are basic to inferential statistics. Consequently, we shall not devote any more discussion to the interquartile range.

7.4 THE MEAN DEVIATION

In Chapter 6, we pointed out that, when we are dealing with data from normally distributed populations, the mean is our most useful measure of central tendency. We obtained the mean by adding together all the scores and dividing them by N. If we carried these procedures one step further, we could subtract the mean from each score, sum the deviations from the mean, and thereby obtain an estimate of the typical amount of deviation from the mean. By dividing by N, we would have a measure that would be analogous to the arithmetic mean except that it would represent the dispersion of scores from the arithmetic mean.

If you think for a moment about the characteristics of the mean which we discussed in the preceding chapter, you will encounter one serious difficulty. The sum of the deviations of all scores from the mean must add up to zero. Thus if we defined the **mean deviation** (M.D.) as this sum divided by N, the mean deviation would have to be zero. You will recall that we employed the fact that $\sum(X - \bar{X}) = 0$ to arrive at one of several definitions of the mean.

Now, if we were to add all the deviations *without regard to sign* and divide by N, we would still have a measure reflecting the mean deviation from the arithmetic mean. The resulting statistic would, of course, be based upon the **absolute value** of the deviations. The absolute value of a positive number or of zero is the number itself. The absolute value of a negative number can be found by changing the sign to a positive one. Thus the absolute value of $+3$ or -3 is 3. The symbol for an absolute value is $||$. Thus $|-3| = 3$.

The calculation of the mean deviation is shown in Table 7.1.

Table 7.1

Computational procedures for calculating the M.D. from an array of scores

X	$(\lvert X - \bar{X}\rvert)$	
9	$\lvert +4\rvert$	$\text{M.D.} = \dfrac{\sum(\lvert X - \bar{X}\rvert)^*}{N}$ (7.1)
8	$\lvert +3\rvert$	
7	$\lvert +2\rvert$	$\text{M.D.} = \frac{26}{15} = 1.73$
7	$\lvert +2\rvert$	
7	$\lvert +2\rvert$	
5	$\lvert 0\rvert$	
5	$\lvert 0\rvert$	
5	$\lvert 0\rvert$	
5	$\lvert 0\rvert$	
4	$\lvert -1\rvert$	
4	$\lvert -1\rvert$	
3	$\lvert -2\rvert$	
3	$\lvert -2\rvert$	
2	$\lvert -3\rvert$	
1	$\lvert -4\rvert$	
$\sum X = 75$ $N = 15$ $\bar{X} = 5$	$\sum(\lvert X - \bar{X}\rvert) = 26$	

* For scores arranged in the form of a frequency distribution, the following formula for the mean deviation should be used: M.D. = $\sum f(|X - \bar{X}|)/N$.

As a basis for comparison of the dispersion of several distributions, the mean deviation has some value. For example, the greater the mean deviation, the greater the dispersion of scores. However, for interpreting scores within a distribution, the mean deviation is less useful since there is no precise mathematical relationship between the mean deviation, as such, and the location of scores within a distribution.

You may wonder why we have bothered to demonstrate the mean deviation when it is of so little use in statistical analysis. There are two reasons: (1) The standard deviation and the variance, which have great value in statistical analysis, are very close relatives of the mean deviation. In order to calculate the standard deviation and the variance, we shall need only to add one column to Table 7.1. (2) We want you to understand that measures of dispersion represent, in a sense, bases for estimating errors in prediction.

Before taking up the standard deviation, let us examine the second point a bit further by posing a question. In the absence of any specific information, what is our best single basis for predicting a score that is obtained by any given individual? If the data are drawn from a normally distributed population, it turns out that the mean (or *any* measure of central tendency) is our best single predictor. The largest error that we can make in prediction when we employ the mean is the most extreme score minus the mean. On the other hand, if we use *any* other score for prediction, the maximum error possible is the entire range of the distribution. This large an error will occur when we predict the highest score in the distribution for an individual actually obtaining the lowest score, or vice versa.

In our subsequent discussion of the standard deviation and other statistics based on it, we shall stress the fact that these measures provide estimates of error in the prediction of scores.

7.5 THE VARIANCE (s^2) AND STANDARD DEVIATION (s)*

Following a perusal of Table 7.1, you might pose this question, "We had to treat the values in the column headed $(X - \bar{X})$ as absolute numbers because their sum was equal to zero. Why could we not square each $(X - \bar{X})$, and then add the squared deviations? In this way, we would rid ourselves of the minus signs in a perfectly legitimate way while still preserving the information that is inherent in these deviation scores."

The answer is: We could, if, by so doing, we arrived at a statistic which is of greater value in judging dispersion than those we have already discussed. It is most fortunate that the standard deviation, based on the squaring of these deviation scores, is of immense value in three different respects. (1) The standard deviation reflects dispersion of scores so that the variability of different distributions may be compared in terms of the standard deviation (s). (2) The standard deviation permits the *precise* interpretation of scores within a distribution. (3) The standard deviation,

* We remind you that italic letters are used to represent sample statistics, and Greek letters to represent population parameters; e.g., σ^2 represents the population variance and σ represents the population standard deviation. The problem of estimating population parameters from sample values will be discussed in Chapter 13.

like the mean, is a member of a *mathematical system* which permits its use in more advanced statistical considerations. Thus we shall employ measures based upon s when we advance into inferential statistics. We shall have more to say about the interpretive aspects of s after we have shown how it is calculated.

7.5.1 Calculation of Variance and Standard Deviation, Mean Deviation Method, with Ungrouped Scores

The **variance** is defined verbally as *the sum of the squared deviations from the mean divided by N − 1.* Symbolically, it is represented as

$$s^2 = \frac{\sum(X - \bar{X})^2}{N - 1}. \tag{7.2}$$

At times it is more convenient to use the symbol x to represent $X - \bar{X}$. Thus

$$x = X - \bar{X}. \tag{7.3}$$

The variance then becomes

$$s^2 = \frac{\sum x^2}{N - 1}. \tag{7.4}$$

The standard deviation is the square root of the variance and is defined as

$$s = \sqrt{\frac{\sum(X - \bar{X})^2}{N - 1}}, \tag{7.5}$$

or

$$s = \sqrt{\frac{\sum x^2}{N - 1}}. \tag{7.6}$$

The computational procedures for calculating the standard deviation, utilizing the mean deviation method, are shown in Table 7.2.

You will recall that the sum of the $(X - \bar{X})^2$ column [that is, $\sum(X - \bar{X})^2$] is known as the **sum of squares** or the *sum squares* and that this sum is minimal when deviations are taken about the mean. From this point on in the course we shall encounter the sum of squares with regularity. It will take on a number of different forms, depending on the procedures that we elect for calculating it. However, it is important to remember that, whatever the form, the sum squares represents the *sum of the squared deviations from the mean.*

The mean deviation method was shown only to impress you with the fact that the standard deviation is based on the deviation of scores from the mean. This method is extremely unwieldy for use in calculation, particularly when the mean is a fractional value, which is usually the case. Consequently, in the succeeding paragraphs we shall examine a number of alternative ways of calculating the sum squares.

Table 7.2

Computational procedures for calculating s, mean deviation method, from an array of scores

X	$X - \bar{X}$	$(X - \bar{X})^2$	
9	+4	16	$s = \sqrt{\frac{\sum(X - \bar{X})^2}{N - 1}}$*
8	+3	9	
7	+2	4	$= \sqrt{72/14}$
7	+2	4	$= \sqrt{5.14}$
7	+2	4	$= 2.27$
5	0	0	
5	0	0	
5	0	0	
5	0	0	
4	−1	1	
4	−1	1	
3	−2	4	
3	−2	4	
2	−3	9	
1	−4	16	
$\sum X = 75$ $N = 15$ $\bar{X} = 5$	$\sum(X - \bar{X}) = 0$	$\sum(X - \bar{X})^2 = 72$	

* For data cast in the form of a frequency distribution, the formula for the standard deviation is

$$s = \sqrt{\frac{\sum f(X - \bar{X})^2}{N - 1}}.$$

Note that the f appears in the formula to remind you that each $(X - \bar{X})^2$ should be multiplied by its corresponding frequency prior to summing. Even when we are dealing with an array of scores, this is the general formula when the frequency of each score is one, i.e., $f = 1$. For this reason, you should regard the f as implied even when it is not given.

7.5.2 Calculation of Standard Deviation, Raw Score Method, with Ungrouped Scores

It can be shown mathematically that

$$\sum x^2 = \sum X^2 - \frac{(\sum X)^2}{N}, \qquad (7.7)$$

where

$$\sum x^2 = \sum(X - \bar{X})^2 = \sum X^2 - 2\sum X\bar{X} + \sum \bar{X}^2.$$

However, $\sum X = N\bar{X}$ and summing the mean square over all values of $\bar{X}$ is the same as multiplying by N. Thus

$$\begin{aligned}\sum x^2 &= \sum X^2 - 2N\bar{X}^2 + N\bar{X}^2\\ &= \sum X^2 - N\bar{X}^2\\ &= \sum X^2 - N\frac{(\sum X)^2}{N^2}\\ &= \sum X^2 - \frac{(\sum X)^2}{N}.\end{aligned}$$

This leads naturally to the development of a useful raw score formula for s:

$$s = \sqrt{\frac{N\sum X^2 - (\sum X)^2}{N(N-1)}}; \tag{7.8}$$

thus

$$\begin{aligned}s &= \sqrt{\frac{\sum x^2}{N-1}} = \sqrt{\frac{\sum X^2 - (\sum X)^2/N}{N-1}}\\ &= \sqrt{\frac{\sum X^2}{N-1} - \frac{(\sum X)^2}{N(N-1)}} = \sqrt{\frac{N\sum X^2 - (\sum X)^2}{N(N-1)}}.\end{aligned}$$

You will note that the result agrees with the answer which we obtained by the mean deviation method. Table 7.3 summarizes the computational procedure.

7.5.3 Calculation of Standard Deviation, Raw Score Method, with Grouped Data

Before reading the present section, it is recommended that you briefly review the calculation of the mean from grouped data, employing the *raw score method.* It is important to remember that the score taken as representative of each interval is the *midpoint* of the interval.

The raw score formula for calculating s for grouped data is

$$s = \sqrt{\frac{N\sum fX^2 - (\sum fX)^2}{N(N-1)}}. \tag{7.9}$$

Table 7.4 summarizes the computational procedures for calculating s, using the raw score method, for grouped data. It will be noted that the values in the last column, fX^2, may be obtained either by squaring X and multiplying by f or by simply

Table 7.3

Computational procedures for calculating s, raw score method, from an array of scores

X	X^2	
9	81	$s = \sqrt{\dfrac{N\sum X^2 - (\sum X)^2}{N(N-1)}}$
8	64	
7	49	$= \sqrt{\dfrac{15(447) - 75^2}{15(14)}}$
7	49	
7	49	
5	25	$= \sqrt{\dfrac{6705 - 5625}{210}}$
5	25	
5	25	
5	25	$= \sqrt{5.14} = 2.27$
4	16	
4	16	
3	9	
3	9	
2	4	
1	1	
$\sum X = 75$ $N = 15$ $\bar{X} = 5$	$\sum X^2 = 447$	

Table 7.4

Procedures for calculating standard deviation from grouped frequency distributions, raw score method

1 Class interval	2 f	3 Midpoint of interval X	4 fX	5 fX^2	$\bar{X} = \dfrac{990}{66}$ $= 15.00$
26–28	1	27	27	729	$s = \sqrt{\dfrac{N\sum fX^2 - (\sum fX)^2}{N(N-1)}}$
23–25	4	24	96	2304	
20–22	7	21	147	3087	
17–19	12	18	216	3888	$= \sqrt{\dfrac{66(16{,}488) - 990^2}{66(65)}}$
14–16	18	15	270	4050	
11–13	11	12	132	1584	$= \sqrt{\dfrac{108{,}108}{4225}}$
8–10	9	9	81	729	
5–7	3	6	18	108	
2–4	1	3	3	9	$= \sqrt{25.59} = 5.06$
	$N = 66$		$\sum fX = 990$		$\sum fX^2 = 16{,}488$

multiplying fX times X. Since the latter requires one step less and is not as likely to lead to error, the authors prefer this method for arriving at fX^2 values.

To summarize the calculation of the standard deviation from a grouped frequency distribution:

Step 1. Go through all the steps involved in calculating the mean from a grouped frequency distribution.

Step 2. Add an additional column, headed fX^2.

Step 3. Multiply the entries in the fX column by the corresponding values in the X column and enter them in the fX^2 column.

Step 4. Sum the fX^2 column to obtain $\sum fX^2$.

Step 5. Substitute the values $\sum fX^2$, N, $\sum fX$, $N - 1$ into the formula and solve for the standard deviation.

7.5.4 Errors to Watch for

In calculating the standard deviation, using the raw score method, it is common for students to confuse the similar-appearing terms $\sum X^2$ and $(\sum X)^2$. It is important to remember that the former represents the sum of the squares of each of the individual scores, whereas the latter represents the square of the sum of the scores. By definition, it is impossible to obtain a negative sum of squares or a negative standard deviation. In the event that you obtain a negative value under the square root sign, you have probably confused these two terms.

We have noted that students tend to square the entries in the fX column in obtaining fX^2. If this is done, you will obtain f^2X^2 and will greatly over-estimate the standard deviation. A rule of thumb for estimating the standard deviation is that the ratio of the range to the standard deviation is rarely smaller than 2 or greater than 6. In our preceding example, the ratio is 26/5.06, or 5.14. If we obtain a standard deviation which yields a ratio greater than 6 or smaller than 2, we have almost certainly made an error. If it is greater than 6, check the fX^2 column and the placement of the decimal. If it is less than 2, check the placement of the decimal.

7.6 INTERPRETATION OF THE STANDARD DEVIATION

An understanding of the meaning of the standard deviation hinges on a knowledge of the relationship between the standard deviation and the normal distribution. Thus, in order to be able to interpret the standard deviations that are calculated in this chapter, it will be necessary to explore the relationship between the raw scores, the standard deviation, and the normal distribution. This material is presented in the following chapter.

CHAPTER SUMMARY

We have seen that to fully describe a distribution of scores, we require more than a measure of central tendency. We must be able to describe how these scores are dispersed about central tendency. In this connection we discussed five measures of dispersion: the range, the interquartile range, the mean deviation, the standard deviation, and the variance.

For normally distributed variables, the two measures based on the squaring of deviations about the mean (the variance and the standard deviation) are maximally useful: We discussed and demonstrated the procedures for calculating the standard deviation employing the mean deviation method, the raw score method with ungrouped frequency distributions, and the raw score method with grouped frequency distributions. We also pointed out several of the errors commonly made in calculating standard deviations. Procedures for calculating the standard deviation and variance are summarized in Tables 7.5 and 7.6.

Terms to Remember:

Dispersion	***Raw score method***
Range	***Standard deviation***
Interquartile range	***Variance***
Mean deviation	***Absolute value of a number***
Mean deviation method	***Sum of squares***

EXERCISES

7.1 Calculate s^2 and s for the following array of scores: 3, 4, 5, 6, 7.

a) Add a constant, say, 2, to each score. Recalculate s^2 and s. Would the results be any different if you had added a larger constant, say, 200? Generalize: What is the effect on s and s^2 of adding a constant to an array of scores? Does the variability increase as we increase the magnitude of the scores?

b) Subtract the same constant from each score. Recalculate s^2 and s. Would the results be any different if you had subtracted a larger constant, say, 200? Generalize: What is the effect on s and s^2 of subtracting a constant from an array of scores?

c) Alternately add and subtract the same constant from each score (i.e., $3 + 2$, $4 - 2$, $5 + 2$, etc). Recalculate s and s^2. Would the results be any different if you had added and subtracted a larger constant? Generalize: What is the effect on s and s^2 of adding and subtracting a constant from an array of scores? (*Note*: this generalization is extremely important with relation to subsequent chapters when we discuss the effect of random errors on measures of variability.)

d) Multiply each score by a constant, say 2. Recalculate s and s^2. Generalize: What is the effect on s and s^2 of multiplying each score by a constant?

e) Divide each score by the same constant. Recalculate s and s^2. Generalize: What is the effect on s and s^2 of dividing each score by a constant?

Table 7.5

Summary of procedures: calculating the variance and the standard deviation from an array of scores

X	X^2
7	49
6	36
6	36
5	25
5	25
5	25
4	16
4	16
3	9
2	4
1	1
0	0
$\sum 48$	242

Steps:

1. Count the number of scores to obtain N. $N = 12$.
2. Sum the scores in the X column to obtain $\sum X$. $\sum X = 48$.
3. Square each score and place it in the adjacent column.
4. Sum the X^2 column to obtain $\sum X^2 = 242$.
5. Substitute the values found in steps 2 and 4 in the formula for s and solve:

$$s = \sqrt{\frac{N\sum X^2 - (\sum X)^2}{N(N-1)}}$$

$$= \sqrt{\frac{12(242) - 48^2}{12(11)}}$$

$$= \sqrt{\frac{600}{132}} = \sqrt{4.55}.$$

Also,

$$s^2 = \frac{N\sum X^2 - (\sum X)^2}{N(N-1)}$$

$$= \frac{12(242) - 48^2}{12(11)}$$

$$= \frac{600}{132} = 4.55.$$

Therefore

$$s = \sqrt{4.55} = 2.13.$$

Table 7.6

Summary of procedures: calculating the variance and standard deviation from a grouped frequency distribution

1 Class interval	2 f	3 Midpoint of interval X	4 fX	5 fX^2
24–26	1	25	25	625
21–23	3	22	66	1452
18–20	8	19	152	2888
15–17	11	16	176	2816
12–14	16	13	208	2704
9–11	10	10	100	1000
6– 8	9	7	63	441
3– 5	4	4	16	64
0– 2	1	1	1	1
	$N = 63$		$\sum fX = 807$	$\sum fX^2 = 11{,}991$

Steps:

1. Sum the f column to obtain $N = 63$.
2. Prepare column 3, showing the midpoint of each interval.
3. Multiply the f in each interval by the score at the midpoint of its corresponding interval. Values in column 2 multiplied by corresponding values in column 3.
4. Sum column 4 to obtain $\sum fX$.
5. Multiply values in column 4 by corresponding values in column 3 to obtain fX^2. Sum this column to obtain $\sum fX^2 = 11{,}991$.
6. Substitute the values in steps 4 and 5 in the formula for s^2 and s:

$$s^2 = \frac{N\sum fX^2 - (\sum fX)^2}{N(N-1)}$$

$$= \frac{63(11{,}991) - 807^2}{63(62)}$$

$$= \frac{104{,}991}{3906} = 26.88.$$

Then

$$s = \sqrt{26.88} = 5.18.$$

7.2 Compare these above generalizations with those made in relation to the mean (see Exercise 6.10).

7.3 A rigorous definition of a measure of variation as a descriptive statistic would involve the following properties: (a) If a constant is added to or subtracted from each score or observation, the measure of variation remains unchanged. (b) If each score is multiplied or divided by a constant, the measure of variation is also multiplied or divided by that number. Check the following for the satisfaction of these conditions:

i) the mean, ii) the median, iii) the mode
iv) the mean deviation, v) the standard deviation, vi) the variance.

If the properties defining measures of dispersion were extended to include *powers* of the constant by which each score is multiplied, would the variance qualify as a measure of dispersion?

7.4 How would the standard deviation be affected by the situations described in Exercises 6.4 and 6.5?

7.5 What is the nature of the distribution if $s = 0$?

7.6 Calculate the standard deviations for the following sets of measurements.

a) 10, 8, 6, 0, 8, 3, 2, 2, 8, 0
b) 1, 3, 3, 5, 5, 5, 7, 7, 9
c) 20, 1, 2, 5, 4, 4, 4, 0
d) 5, 5, 5, 5, 5, 5, 5, 5, 5, 5

7.7 Why is the standard deviation in part (c) of Exercise 7.6 so large? Describe the effect of extreme deviations on s.

7.8 Determine the range for the sets of measurements in Exercise 7.6. For which of these is the range a misleading index of variability, and why?

7.9 Calculate the mean and standard deviation for the 250 stocks listed in Exercise 3.7, employing two different grouping intervals. The calculated standard deviation has a somewhat different value in each instance. To what do you attribute the difference?

7.10 Calculate the mean and standard deviation for the frequency distributions obtained in Exercises 3.10, 3.11, 3.12, and 3.13 of Chapter 3. Compare and discuss the results.

7.11 A comparison shopper compares prices of chopped chuck at a number of different supermarkets. He finds the following prices per pound (in cents): 56, 65, 48, 73, 59, 72, 63, 65, 60, 63, 44, 79, 63, 61, 66, 69, 64, 71, 58, 63.

a) Find the mean.
b) Find the range, interquartile range, and mean deviation.
c) Find the standard deviation and variance.

7.12 Table 7.7 presents the age at first inaugural and the number of years lived after the first inaugural of the deceased presidents of the United States. Calculate the mean and standard deviation of each of these measures.

7.13 Referring back to Exercise 4.16, find the mean and standard deviation of the number of quarts of milk sold at the supermarket.

7.14 List at least three specific instances in which a measure of variability was important in describing a group.

Table 7.7

Longevity of Presidents of the U.S.

	Age, first inauguration	Years lived after first inauguration
George Washington	57	10.6
John Adams	61	29.3
Thomas Jefferson	57	25.3
James Madison	57	27.3
James Monroe	58	14.3
John Quincy Adams	57	23.0
Andrew Jackson	61	16.3
Martin Van Buren	54	25.4
William H. Harrison	68	.1
John Tyler	51	20.8
James K. Polk	49	4.3
Zachary Taylor	64	1.3
Millard Fillmore	50	23.7
Franklin Pierce	48	16.6
James Buchanan	65	11.3
Abraham Lincoln	52	4.1
Andrew Johnson	56	10.3
Ulysses S. Grant	46	16.4
Rutherford B. Hayes	54	15.9
James A. Garfield	49	.5
Chester A. Arthur	50	5.2
Grover Cleveland	47	23.3
Benjamin Harrison	55	12.0
William McKinley	54	4.5
Theodore Roosevelt	42	17.3
William H. Taft	51	21.0
Woodrow Wilson	56	10.9
Warren G. Harding	55	2.4
Calvin Coolidge	51	9.4
Herbert C. Hoover	54	35.6
Franklin D. Roosevelt	51	12.1
Harry S. Truman	60	27.7
Dwight D. Eisenhower	62	16.2
John F. Kennedy	43	2.8
Lyndon B. Johnson	55	9.2

Source. Statistical Bulletin, Metropolitan Life.

7.15 List at least three specific instances in which a measure of variability was important in comparing groups of people.

7.16 Table 7.8 lists the maximum daily temperature recorded for New York City for the months of January and May, 1965 and 1966.*

Table 7.8

Date	Jan. 1965	Jan. 1966	May 1965	May 1966
1	35	62	71	66
2	29	52	77	60
3	32	46	71	68
4	39	45	90	59
5	43	53	62	70
6	44	47	78	83
7	44	47	52	66
8	49	44	60	61
9	55	30	78	52
10	40	48	92	57
11	35	42	88	63
12	36	27	78	57
13	41	32	80	67
14	37	42	74	64
15	19	41	82	74
16	17	26	87	71
17	16	39	83	68
18	25	38	73	66
19	25	35	70	64
20	38	44	80	84
21	36	42	72	85
22	50	39	80	75
23	47	37	79	85
24	32	38	70	80
25	38	32	79	76
26	44	26	94	84
27	42	36	92	82
28	38	19	83	73
29	27	23	68	82
30	20	38	67	73
31	30	28	74	68

For each month and year:

a) Find the mean.
b) Find the range and mean deviation.
c) Find the standard deviation and variance.

* The data for this problem were extracted from *The World Almanac*, pp. 558–559. New York: Newspaper Enterprise Association, Inc., 1967. Reprinted by permission.

7.17 In late 1973, a national speed limit of 55 mph was established in order to conserve energy. An unexpected side effect has been an apparent drop in the rate of highway fatalities. Shown in Table 7.9 are grouped frequency distributions of the number of fatalities per 100 million passenger miles during the years 1973 and 1974 for each of the 50 states and the District of Columbia.

a) Calculate the mean rate of traffic fatalities for each year.
b) Calculate the standard deviation and variance for each year.

Table 7.9

Fatalities per 100 million passenger miles

Class interval	1974	1973
6.6–6.8		1
6.3–6.5		3
6.0–6.2		3
5.7–5.9	2	1
5.4–5.6	1	2
5.1–5.3	2	2
4.8–5.0	0	4
4.5–4.7	4	8
4.2–4.4	6	2
3.9–4.1	6	8
3.6–3.8	4	7
3.3–3.5	16	4
3.0–3.2	6	1
2.7–2.9	0	3
2.4–2.6	2	2
2.1–2.3	1	0
1.8–2.0	1	0

Source. 1976 *Information Please Almanac*, p. 728.

7.18 Table 7.10 shows the murder and rape rates per 100,000 population for each of the 50 states. Calculate the mean, the standard deviation, and the variance for murder and rape.

7.19 Referring to Exercise 3.20, calculate

a) the mean,
b) the crude range and the interquartile range,
c) the variance and the standard deviation.

7.20 For Exercise 6.33, determine the crude range, the variance, and the standard deviation for each manufacturer's product.

7.21 In Chapter 4, three types of normal distributions were reported: leptokurtic, mesokurtic, and platykurtic. Which would yield the largest amount of dispersion? Referring to the numbers you assigned to these distributions in Exercise 4.28, determine the range, the mean deviation, the variance, and the standard deviation for each distribution.

Table 7.10

	Murder	Rape
Alabama	13.2	21.2
Alaska	10.0	44.5
Arizona	8.1	31.0
Arkansas	8.8	19.5
California	9.0	40.6
Colorado	7.9	38.7
Connecticut	3.3	11.1
Delaware	5.9	15.8
Florida	15.4	31.9
Georgia	17.4	25.8
Hawaii	5.3	20.2
Idaho	2.6	14.2
Illinois	10.4	24.0
Indiana	7.2	21.0
Iowa	2.2	11.3
Kansas	6.0	18.0
Kentucky	9.7	16.3
Louisiana	15.4	22.2
Maine	2.1	7.8
Maryland	11.3	27.8
Massachusetts	4.4	16.3
Michigan	12.1	35.1
Minnesota	2.7	14.9
Mississippi	16.1	17.1
Missouri	9.0	28.2
Montana	6.0	16.4

7.22 Suppose merchant A sold a quart of milk for $0.40 and the standard deviation of this price was 0 during the last month. What was the price on the third day of the month? On the fifteenth day?

7.23 Calculate the mean, the variance, and the standard deviation for Exercises 5.25 and 5.26.

7.24 Table 7.11 shows the monthly normal precipitation recorded by U.S. weather stations in four areas. Determine the variance and the standard deviation for each city per year.

7.25 Refer to Exercise 6.38, Chapter 6. Calculate the standard deviation of the frequency distribution of sample means. [*Hint.* Treat each mean as a score, and employ the raw-score method for obtaining the standard deviation from an ungrouped frequency distribution.]

Table 7.10 (*Continued*)

	Murder	Rape
Nebraska	4.3	16.5
Nevada	12.2	46.0
New Hampshire	2.1	9.5
New Jersey	7.4	18.8
New Mexico	11.4	32.1
New York	11.1	26.1
North Carolina	13.0	16.1
North Dakota	0.8	7.3
Ohio	7.3	21.4
Oklahoma	6.6	20.0
Oregon	4.9	29.3
Pennsylvania	6.3	14.9
Rhode Island	3.4	8.3
South Carolina	14.4	22.5
South Dakota	3.8	12.8
Tennessee	13.2	26.9
Texas	12.7	25.5
Utah	3.2	22.9
Vermont	2.2	11.2
Virginia	8.5	20.7
Washington	4.0	26.2
West Virginia	5.7	9.3
Wisconsin	2.6	10.8
Wyoming	6.8	15.6

Source. Federal Bureau of Investigation. Uniform Crime Reports—1973.

Table 7.11

Monthly normal precipitation (in inches)

Stations	Jan	Feb	Mar	Apr	May	June	July	Aug	Sept	Oct	Nov	Dec
Barrow, Alaska	0.2	0.2	0.1	0.1	0.1	0.4	0.8	0.9	0.6	0.5	0.2	0.2
Burlington, Vt.	2.0	1.8	2.1	2.6	3.0	3.5	3.9	3.4	3.3	3.0	2.6	2.1
Honolulu, Hawaii	3.8	3.3	2.9	1.3	1.0	0.3	0.4	0.9	1.0	1.8	2.2	3.0
Seattle-Tacoma, Washington	5.7	4.2	3.8	2.4	1.7	1.6	0.8	1.0	2.1	4.0	5.4	6.3

Source. National Climatic Center, NOAA, U.S. Department of Commerce.

8 Standard Deviation and Standard Normal Distribution

8.1 INTRODUCTION

We previously noted that scores derived from scales employed by behavioral scientists are generally meaningless by themselves. To take on meaning they must be compared to the distribution of scores from some reference group. Indeed, the scores derived from any scale, including those employed by the physical scientists, become more meaningful when they are compared to some reference group of objects or persons. Thus if we were to learn that Pierre Le Blanc caught a northern Canadian pike weighing 50 lb, we might or might not be impressed, depending upon the extent of our knowledge concerning the usual weight of this type of fish. However, once a reference group is established, the measurement becomes meaningful. Since most northern pike weigh under 10 lb and only rarely achieve weights as high as 20 lb, the achievement of our apocryphal fisherman must be considered Bunyanesque.

8.2 THE CONCEPT OF STANDARD SCORES

In interpreting a single score, we want to place it in some position with respect to a collection of scores from some reference group. In Chapter 5, you learned to place a score by determining its percentile rank. It will be recalled that the percentile rank of a score tells us the percentage of scores that are of lower scale value. Another approach for interpretation of a single score might be to view it with reference to some central point, such as the mean. Therefore a score of 20 in a distribution with a mean of 23 might be reported as -3. Finally, we might express this deviation score in terms of standard deviation units. Thus if our standard deviation is 1.5, the score of 20 would be two standard deviations below the mean (that is, $-3/1.5 = -2$). This process of dividing a deviation of a score from the mean by the standard deviation is known as the transformation to **z-scores**. Symbolically, z is defined as

$$z = \frac{X - \bar{X}}{s}. \tag{8.1}$$

Since we previously employed x to represent $(X - \bar{X})$, we may also state that

$$z = x/s. \tag{8.2}$$

It will be noted that every score in the distribution may be transformed into a z- (or **standard**) **score**, in which case each z will represent the *deviation of a specific score from the mean expressed in standard deviation units.*

What is the value of transforming to a z-score? Simply this: if the *population of scores* on a given variable is normal, we may express any score as a percentile rank by referring our z to the *standard normal distribution.* In addition, since z-scores represent abstract numbers as opposed to the concrete values of the original scores (inches, pounds, I.Q. scores, etc.), we may compare an individual's position on one variable with his position on a second. To understand these two important characteristics of z-scores, we must make reference to the standard normal distribution.

8.3 THE STANDARD NORMAL DISTRIBUTION

The **standard normal distribution** has a μ of 0, a σ of 1, and a total area equal to 1.00.*
There is a fixed proportion of cases between a vertical line, or ordinate, erected at any one point and an ordinate erected at any other point. Taking a few reference points along the normal curve, we can make the following statements:

1. Between the mean and 1 standard deviation above the mean are found 34.13% of all cases. Similarly, 34.13% of all cases fall between the mean and 1 standard deviation below the mean. Stated in another way, 34.13% of the *area* under the curve is found between the mean and 1 standard deviation above the mean, and 34.13% of the *area* falls between the mean and -1 standard deviation.

2. Between the mean and 2 standard deviations above the mean are found 47.72% of all cases. Since the normal curve is symmetrical, 47.72% of the area also falls between the mean and -2 standard deviations.

3. Finally, between the mean and 3 standard deviations above the mean are found 49.87% of all the cases. Similarly, 49.87% of the cases fall between the mean and -3 standard deviations. Thus 99.74% of all cases fall between ± 3 standard deviations. These relationships are shown in Fig. 8.1.

Now, by transforming the scores of a normally distributed variable to z-scores, we are, in effect, expressing these scores in units of the standard normal curve. For any given value of X with a certain proportion of area beyond it, there is a corresponding value of z with the same proportion of area beyond it. Thus, if we have a population in which $\mu = 30$ and $\sigma = 10$, the z of a score at the mean $(X = 30)$

* It will be recalled that the Greek letters μ and σ represent the population mean and the standard deviation, respectively.

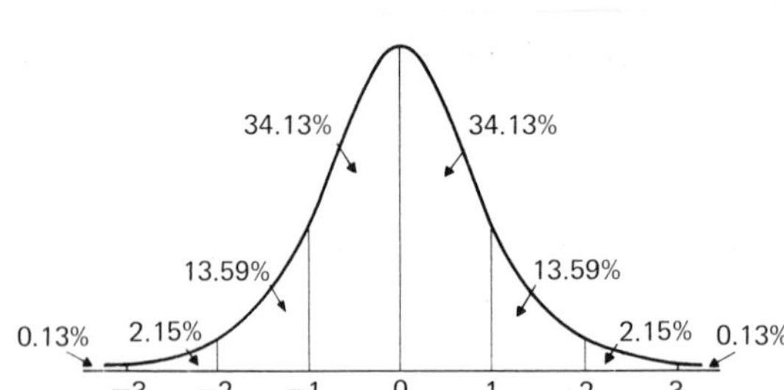

Fig. 8.1 Areas between selected points under the normal curve.

will equal zero, and the z of scores 1 standard deviation above and below the mean ($X = 40$ and $X = 20$) will be $+1.00$ and -1.00, respectively.

8.3.1 Finding Area Between Given Scores

For expositional purposes, we confined our preceding discussion of area under the standard normal curve to selected points. As a matter of actual fact, however, it is possible to determine the percent of areas between *any* two points by making use of the tabled values of the area under the normal curve (Table A). The left-hand column headed by z represents the deviation from the mean expressed in standard deviation units. *By referring to the body of the table, we can determine the proportion of total area between a given score and the mean* (Column B) *and the area beyond a given score* (Column C). Thus, if an individual obtained a score of 24.65 on a normally distributed variable with $\mu = 16$ and $\sigma = 5$, his z-score would be

$$z = \frac{24.65 - 16}{5} = 1.73.$$

Referring to Column B in Table A, we find that 0.4582 or 45.82%* of the area lies between his score and the mean. Since 50% of the area also falls below the mean in a symmetrical distribution, we may conclude that 95.82% of all the area falls below a score of 24.65. Note that we can now translate this score into a percentile rank of 95.82.

Let us suppose another individual obtained a score of 7.35 on the same normally distributed variable. Her z-score would be

$$z = \frac{7.35 - 16}{5} = -1.73.$$

Since the normal curve is symmetrical, only the areas corresponding to the positive z-values are given in Table A. Negative z-values will have precisely the

* The areas under the normal curve are expressed as proportions of area. To convert to percentage of area, multiply by 100 or merely move the decimal two places to the right.

same proportions as their positive counterparts. Thus the area between the mean and a z of -1.73 is also 45.82%. The percentile rank of a score below the mean may be obtained either by subtracting 45.82% from 50%, or directly from Column C. In either case, the percentile rank of a score of 7.35 is 4.18.

You should carefully note that these relationships apply *only to scores from normally distributed populations.* Transforming the raw scores to standard scores does not, in any way, alter the form of the original distribution. The only change is to convert the mean to zero and the standard deviation to one. Thus, if the original distribution of scores is nonnormal, *the distribution of* z-scores will be nonnormal. In other words, our transformation to z's will *not* convert a nonnormal distribution to a normal distribution.

Figure 8.2 further clarifies the relationships among raw scores, z-scores, and percentile ranks of a normally distributed variable. It assumes that $\mu = 50$ and $\sigma = 10$.

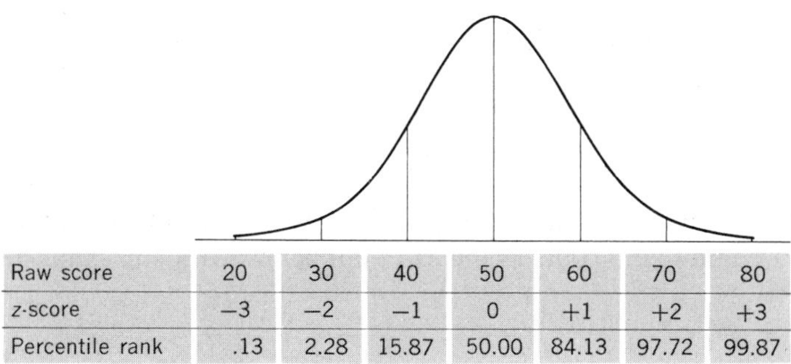

Raw score	20	30	40	50	60	70	80
z-score	−3	−2	−1	0	+1	+2	+3
Percentile rank	.13	2.28	15.87	50.00	84.13	97.72	99.87

Fig. 8.2 Relationships among raw scores, z-scores, and percentile ranks of a normally distributed variable in which $\mu = 50$ and $\sigma = 10$.

8.4 ILLUSTRATIVE PROBLEMS

Let us take several sample problems in which we assume that the mean of the general population, μ, is equal to 100 on a standard I.Q. test, and the standard deviation, σ, is 16. It is assumed that the variable is normally distributed.

Problem 1

John Doe obtains a score of 125 on an I.Q. test. What percent of cases fall between his score and the mean? What is his percentile rank in the general population?

At the outset, it is wise to construct a crude diagram representing the relationships in question. Thus in the present example, the diagram would appear as shown in Fig. 8.3. To find the value of z corresponding to $X = 125$, we subtract the population mean from 125 and divide by 16. Thus

$$z = \frac{125 - 100}{16} = 1.56.$$

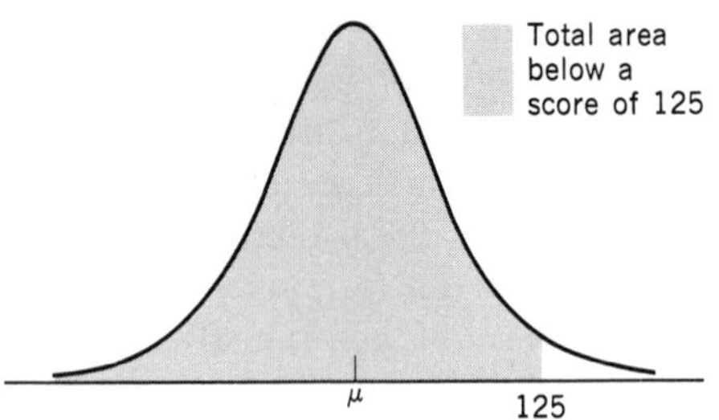

Fig. 8.3 Proportion of area below a score of 125 in a normal distribution with $\mu = 100$ and $\sigma = 16$.

Looking up 1.56 in Column B (Table A), we find that 44.06% of the area falls between the mean and 1.56 standard deviations above the mean. John Doe's percentile rank is therefore 50 + 44.06 or 94.06.

Problem 2

Mark Jones scores 93 on an I.Q. test. What is his percentile rank in the general population (Fig. 8.4)?

$$z = \frac{93 - 100}{16} = -0.44.$$

The minus sign indicates that the score is below the mean. Looking up 0.44 in Column C, we find that 33.00% of the cases fall below his score. Thus his percentile rank is 33.00.

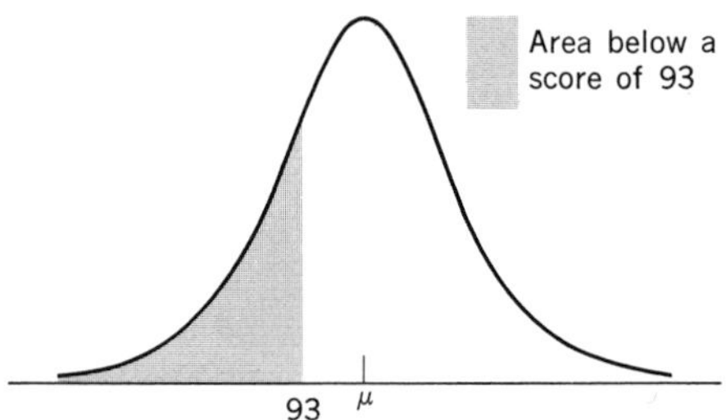

Fig. 8.4 Proportion of area below a score of 93 in a normal distribution with $\mu = 100$ and $\sigma = 16$.

Problem 3

What percent of cases fall between a score of 120 and a score of 88 (Fig. 8.5)?

Note that to answer this question we do *not* subtract 88 from 120 and divide by σ. The areas in the normal probability curve are designated in relation to the mean as a fixed point of reference. We must, therefore, separately calculate the area between the mean and a score of 120 and the area between the mean and a score of 88. We then add the two areas to answer our problem.

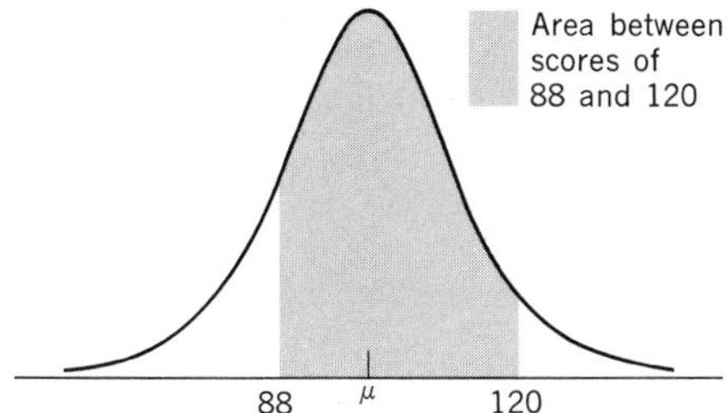

Fig. 8.5 Proportion of area between the scores 88 and 120 in a normal distribution with $\mu = 100$ and $\sigma = 16$

Procedure:

Step 1. Find the z corresponding to $X = 120$:

$$z = \frac{120 - 100}{16} = 1.25.$$

Step 2. Find the z corresponding to $X = 88$:

$$z = \frac{88 - 100}{16} = -0.75.$$

Step 3. Find the required areas by referring to Column B (Table A):

Area between the mean and $z = 1.25$ is 39.44%.
Area between the mean and $z = -0.75$ is 27.34%.

Step 4. Add the two areas together.

Thus, the area between 88 and 120 = 66.78%.

Problem 4

What percent of the area falls between a score of 123 and 135 (Fig. 8.6)?

Again, we cannot obtain the answer directly; we must find the area between the mean and a score of 123 and subtract this from the area between the mean and a score of 135.

Procedure:

Step 1. Find the z corresponding to $X = 135$.

$$z = \frac{135 - 100}{16} = 2.19.$$

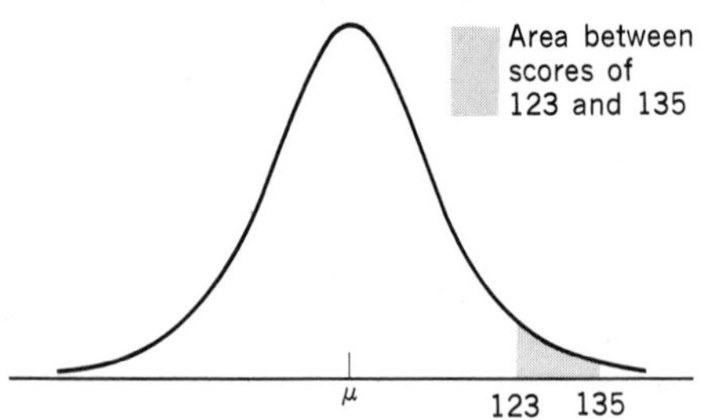

Fig. 8.6 Proportion of area between the scores 123 and 135 in a normal distribution with $\mu = 100$ and $\sigma = 16$.

Step 2. Find the z corresponding to $X = 123$.

$$z = \frac{123 - 100}{16} = 1.44.$$

Step 3. Find the required areas by referring to Column B.

Area between the mean and $z = 2.19$ is 48.57%.

Area between the mean and $z = 1.44$ is 42.51%.

Step 4. Subtract to obtain the area between 123 and 135. The result is

$$48.57 - 42.51 = 6.06\%$$

Problem 5

We stated earlier that our transformation to z-scores permits us to compare an individual's position on one variable with his position on another. Let us illustrate this important use of z-scores.

On a standard aptitude test, John G. obtained a score of 245 on the verbal scale and 175 on the mathematics scale. The means and the standard deviations of each of these normally distributed scales are as follows: Verbal, $\mu = 220$, $\sigma = 50$; Math, $\mu = 150$, $\sigma = 25$. On which scale did John score higher? All that we need to do is compare John's z-score on each variable. Thus:

$$\text{Verbal } z = \frac{245 - 220}{50}$$

$$= 0.50$$

$$\text{Math } z = \frac{175 - 150}{25}$$

$$= 1.00.$$

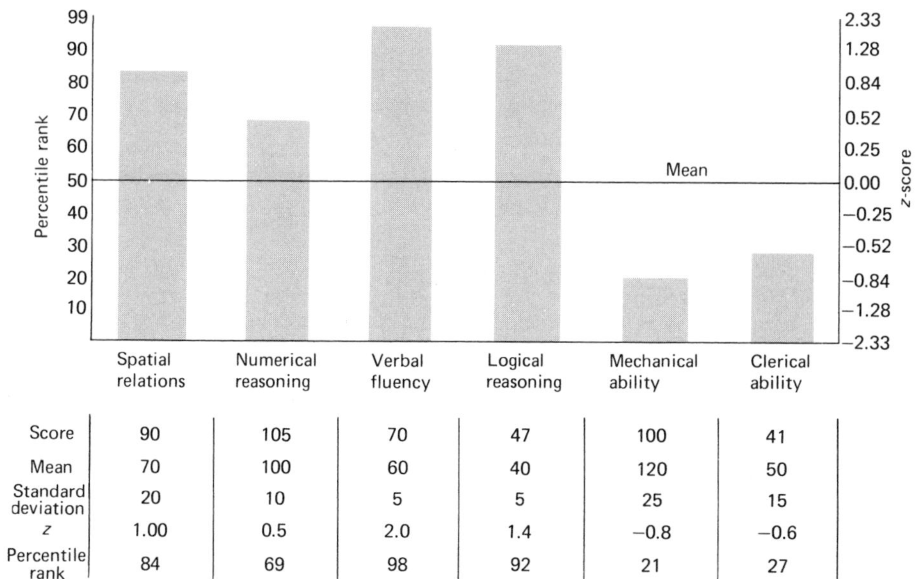

	Spatial relations	Numerical reasoning	Verbal fluency	Logical reasoning	Mechanical ability	Clerical ability
Score	90	105	70	47	100	41
Mean	70	100	60	40	120	50
Standard deviation	20	10	5	5	25	15
z	1.00	0.5	2.0	1.4	−0.8	−0.6
Percentile rank	84	69	98	92	21	27

Fig. 8.7 The use of the standard score conversion on six different aptitude scales, each with different means and standard deviations. Note that we may compare an individual's performance on all scales and construct a profile of his abilities.

We may conclude, therefore, that John scored higher on the math scale of the aptitude test. Of course, if we so desire, we may express these scores as percentile ranks. Thus John's percentile rank is 84.13 on the math scale and only 69.15 on the verbal scale. Figure 8.7 illustrates the use of the standard score conversion to permit the comparison of an individual's relative performance on six different aptitude scales.

8.5 THE STANDARD DEVIATION AS AN ESTIMATE OF ERROR

When discussing the mean deviation at the beginning of Chapter 7, we pointed out that the mean of a distribution can be considered our best single predictor of a score in the absence of any other information. The more compactly our scores are distributed about the mean, the smaller our errors will be in prediction, on the average. Since the standard deviation reflects the dispersion of scores, it becomes, in a sense, an estimate of error. Thus, if we have two distributions with identical means but with standard deviations of, say, 10 and 30 respectively, we will make larger errors, on the average, when we employ the mean as a basis for predicting scores in the latter distribution. This characteristic of the standard deviation is important in the understanding of certain derivatives of σ which we shall be discussing in subsequent chapters.

8.6 THE COEFFICIENT OF VARIATION

Our preceding discussion should make it clear that the standard deviation may be regarded as an estimate of precision in measurements. Imagine that a group of foresters were asked to estimate the height of two trees. Obviously, not all estimates of the height of each tree would be the same. In each case, we should obtain, instead, a distribution of estimates about some central value. Imagine further that the standard deviation of their estimates of the first tree was 20′ and that the second was 1′. On the surface, it would appear that the height of the second tree was estimated with greater precision than the first. But what if the first tree were 400′ tall and the second were only 20′ tall? Clearly, in comparing the relative precision of both estimates, we must take into account the magnitude of the original quantities which we are estimating. An error of $\frac{1}{16}''$ would be acceptable if we were joining two segments of an expansion bridge, but for a subminiature component in a space capsule, the same error would be intolerable.

The measure which expresses variation relative to magnitude is the **relative variation**. The most widely employed measure of relative variation divides the standard deviation by the mean. This measure, called the **coefficient of variation**, V, is defined as

$$V = \frac{100s}{\overline{X}}. \tag{8.3}$$

We multiply by 100 in order to express the coefficient of variation as a percentage. In the preceding examples, the coefficient of variation for the larger tree is

$$V = \frac{100(20)}{400} = 5\%.$$

For the smaller tree,

$$V = \frac{100(1)}{20} = 5\%.$$

In each case, then, the variation in estimations of the height of the two trees in relation to their means is the same. A useful characteristic of the coefficient of variation is that it is independent of the units in terms of which the measurements are made. Thus one can compare dollars to doughnuts if one pleases. Imagine that in recent weeks the English pound, expressed in dollars, has had a mean value of 2.035 dollars and a standard deviation of 0.013. The mean price of a dozen doughnuts is 1.80 dollars with a standard deviation of 0.20 dollars. Which shows greater variation?

For pound sterling, $$V = \frac{0.013(100)}{2.035} = 0.64\%$$

For doughnuts, $$V = \frac{0.20(100)}{1.80} = 11.11\%$$

Thus the value of the pound sterling varied less than the price of doughnuts in *relation to the mean.*

One word of caution in using the coefficient of variation: V may not be employed when the variable in question has no true zero point. Since many measures employed in education and psychology have no true zero point, the coefficient of variation has few applications in these fields.

CHAPTER SUMMARY

In this chapter, we demonstrated the value of the standard deviation for comparison of the dispersion of scores in different distributions of a variable, the interpretation of a score with respect to a single distribution, and the comparison of scores on two or more variables. We showed how to convert raw scores into units of the standard normal curve (transformation to z-scores) and the various characteristics of the standard normal curve were explained. A series of worked problems demonstrated the various applications of the conversion of normally distributed variables to z-scores.

Finally, we discussed the standard deviation as an estimate of error and as an estimate of precision. We demonstrated the use of the coefficient of variation, V, as a measure of variation relative to the mean.

Terms to Remember:

Standard normal distribution
Relative variation
Coefficient of variation (V)
Standard score (z-score)

EXERCISES

8.1 Given a normal distribution with a mean of 45.2 and a standard deviation of 10.4, find the standard score equivalents for the following scores.

a) 55 b) 41 c) 45.2
d) 31.5 e) 68.4 f) 18.9

8.2 Find the proportion of area under the normal curve between the mean and the following z-scores.

a) -2.05 b) -1.90 c) -0.25
d) $+0.40$ e) $+1.65$ f) $+1.96$
g) $+2.33$ h) $+2.58$ i) $+3.08$

8.3 Given a normal distribution based on 1000 cases with a mean of 50 and a standard deviation of 10, find

a) the proportion of area and the number of cases between the mean and the following scores:
   i) 60 ii) 70 iii) 45 iv) 25
b) the proportion of area and the number of cases *above* the following scores:
   i) 60 ii) 70 iii) 45 iv) 25 v) 50

c) the proportion of area and the number of cases between the following scores:
i) 60 – 70   ii) 25 – 60   iii) 45 – 70   iv) 25 – 45.

8.4 Below are given Student Jones' scores, the mean, and the standard deviation on each of three tests given to 3,000 students.

Test	μ	σ	Jones' score
Arithmetic	47.2	4.8	53
Verbal comprehension	64.6	8.3	71
Geography	75.4	11.7	72

a) Convert each of Jones' test scores to standard scores.
b) On which test did Jones stand highest? On which lowest?
c) Jones' score in arithmetic was surpassed by how many students? In verbal comprehension? In geography?
d) What assumption must be made in order to answer the preceding question?

8.5 On a normally distributed mathematics aptitude test, for females,

$$\mu = 60, \quad \sigma = 10,$$

and for males,

$$\mu = 64, \quad \sigma = 8.$$

a) Aaron obtained a score of 62. What is his percentile rank on both the male and the female norms?
b) Shelley's percentile rank is 73 on the female norms. What is her percentile rank on the male norms?

8.6 If frequency polygons were constructed for each of the following, which do you feel would approximate a normal curve?
a) Heights of a large representative sample of adult American males.
b) Means of a large number of samples with a fixed N (say, $N = 100$) drawn from a normally distributed population of scores.
c) Means of a large number of samples of a fixed N (say, $N = 100$), drawn from a moderately skewed distribution of scores.
d) Weights, in ounces, of ears of corn selected randomly from a cornfield.
e) Annual income, in dollars, of the "breadwinner" of a large number of American families selected at random.
f) Weight, in ounces, of all fish caught in a popular fishing resort.

8.7 In a normal distribution with $\mu = 72$ and $\sigma = 12$:
a) What is the score at the 25th percentile?
b) What is the score at the 75th percentile?
c) What is the score at the 90th percentile?
d) Find the percent of cases scoring above 80.
e) Find the percent of cases scoring below 66.
f) Between what scores do the middle 50 percent of the cases lie?

g) Beyond what scores do the most extreme 5 percent lie?
h) Beyond what scores do the most extreme 1 percent lie?

8.8 Answer the above questions (a through h) for

$$\mu = 72 \quad \text{and} \quad \sigma = 8;$$
$$\mu = 72 \quad \text{and} \quad \sigma = 4;$$
$$\mu = 72 \quad \text{and} \quad \sigma = 2.$$

8.9 In electronics, a resistor with 5% tolerance is one in which a reading within $\pm 5\%$ of the nominal value is considered acceptable. The daily output of 10,000 ohm 5% resistors at a given plant is 100,000 units. The mean resistance of a day's output is 10,000 ohms and $\sigma = 200$ ohms.

a) What percentage of resistors would be expected to give out-of-tolerance readings?
b) How many 10,000 ohm resistors on a given day would be expected to give out-of-tolerance readings?

8.10 Repeat Exercise 8.9 employing 1% tolerance resistors at 10,000 ohms, and assume the same mean and the same standard deviation.

8.11 A food price analyst obtains the following means and standard deviations of various food products being offered in a large metropolitan area:

	Mean	Standard deviation
Chopped chuck	80	7
Loin lamb chops	110	15
Sirloin steak	99	6
Pork loin	75	9
Lettuce (head)	25	3
Carrots (bunch)	15	3
String beans (pound)	35	6
Corn (frozen, pkg)	28	2

a) All food items considered, which shows the greatest relative variation? Which the least?
b) Among the meats, which shows the greatest relative variation? Which the least?
c) Among the produce, which shows the greatest relative variations? Which the least?

8.12 Shown in Table 8.1 are the number of accidental deaths resulting from falls, drowning, and fire or burns for each month of 1973.

a) Calculate the mean, the standard deviation and the variance for each type of accidental death.
b) Which shows the greatest relative variation and which shows the least? Account for these differences.

8.13 Referring back to Exercise 7.16, which of the four time periods shows the greatest relative variation? Which the least?

8.14 Employing the data in Exercise 7.16, convert each of the temperature readings for the four time periods to standard scores.

Table 8.1

Accidental deaths—falls, drowning, fire or burns

Month	Falls	Drowning	Fires, burns
January	1,586	270	903
February	1,258	265	829
March	1,328	430	583
April	1,282	620	576
May	1,354	950	444
June	1,334	1,670	357
July	1,414	1,710	309
August	1,392	1,160	335
September	1,306	690	294
October	1,420	420	489
November	1,358	260	539
December	1,474	280	845
Total	16,506	8,725	6,503

8.15 The following data lists the number of home runs made by the home-run leaders in the National and American Leagues from 1951 to 1974:

Year	National League	American League
1951	42	33
1952	37	32
1953	47	43
1954	49	32
1955	51	37
1956	43	52
1957	44	42
1958	47	42
1959	46	42
1960	41	40
1961	46	61
1962	49	48
1963	44	45
1964	47	49
1965	52	32
1966	44	49
1967	39	44
1968	36	44
1969	45	49
1970	45	44
1971	48	33
1972	40	37
1973	44	32
1974	36	32

a) Find the mean and the standard deviation for each league.
b) Convert the number of home runs for each league to standard scores.
c) Which shows greater relative variation, the National or American League?

8.16 Referring to Exercise 7.25, calculate the relative variation for all four areas.

8.17 Referring to Exercise 5.25, assume that the 25 employees comprise the total population of employees for that company.
a) Determine μ and σ.
b) Determine the z-score for each person.
c) What is the proportion of total area in the normal distribution curve below each score?
d) Calculate the percent of area falling between the scores of:
i) A and Y ii) M and V iii) G and S

8.18 Follow the procedures of Exercise 8.17, employing the data of Exercise 5.26.

8.19 Determine the coefficient of variation for:
a) Exercise 8.17
b) Exercise 8.18

8.20 A store owner found that the number of sodas sold for each date during September was as follows.

Date	Sodas	Date	Sodas	Date	Sodas
1	20	11	15	21	25
2	15	12	20	22	15
3	25	13	30	23	30
4	35	14	25	24	0
5	30	15	20	25	10
6	20	16	40	26	15
7	15	17	20	27	20
8	25	18	5	28	25
9	20	19	20	29	10
10	35	20	10	30	5

Determine the z-scores for the following dates.
a) 1 b) 10 c) 20
d) 30 e) 15 f) 16

8.21 Table 8.2 shows life expectancy by sex for selected countries. (Source: 1976 *Information Please Yearbook*, p. 738.)
a) Calculate the mean and the standard deviation for males and females.
b) Which shows greater relative variation, males or females?
c) Round the mean and the standard deviation to one decimal place and calculate the z-scores for males and females for the following countries:
i) U.S.A. ii) Mexico iii) Sweden iv) India

8.22 Table 8.3 shows the 1974 birth and death rates for the 50 states plus the District of Columbia ($N = 51$).
a) Calculate the mean and the standard deviation for birth and death rates.
b) Round the mean and the standard deviation to one decimal place and calculate the z-scores for birth and death rates for the following states:
i) Alaska ii) Connecticut iii) Florida
iv) Pennsylvania v) Utah

Table 8.2

Expectation of life by sex for selected countries

Country	Average lifetime in years	
	Males	Females
North America		
United States	64.7	75.2
Canada	68.7	75.1
Mexico	59.4	63.4
Puerto Rico	68.9	75.2
South America		
Argentina	64.1	70.2
Bolivia	49.7	49.7
Chile	60.5	66.0
Colombia	44.2	45.9
Europe		
Austria	66.8	74.1
Belgium	67.7	73.5
Czechoslovakia	66.2	73.2
Denmark	70.7	75.9
England and Wales	68.9	75.1
Finland	65.9	73.6
France	68.5	76.1
Germany (West)	67.2	73.4
Germany (East)	68.9	74.2
Greece	67.5	70.7
Hungary	66.3	72.1
Iceland	70.7	76.3
Ireland	68.6	72.9
Italy	67.9	73.4
Netherlands	70.8	76.8
Norway	71.1	76.8
Poland	66.8	73.8
Portugal	65.3	71.0
Scotland	67.2	73.5
Spain	67.3	71.9
Sweden	72.0	77.4
Switzerland	69.2	75.0
U.S.S.R.	65.0	74.0
Yugoslovia	65.3	70.1
Asia		
Burma	40.8	43.8
India	41.9	40.6
Israel	70.1	72.8
Japan	70.5	75.9
Jordan	52.6	52.0
Korea	59.7	64.1

(Continued)

Table 8.2 (*Continued*)

Country	Average lifetime in years	
	Males	Females
Africa		
Egypt	51.6	53.8
Nigeria	37.2	36.7
South Africa (white population)	64.7	71.7
Oceania		
Australia	67.9	74.2
New Zealand	68.4	73.8

Table 8.3

United States birth and death rates—1974

State	1974 Birth rates	1974 Death rates	State	1974 Birth rates	1974 Death rates
Alabama	16.4	9.6	Nebraska	15.4	9.8
Alaska	20.8	4.4	Nevada	15.1	8.3
Arizona	18.5	8.1	New Hampshire	14.0	9.4
Arkansas	16.4	10.6	New Jersey	12.4	8.9
California	14.5	8.0	New Mexico	18.8	7.2
Colorado	15.9	7.4	New York	13.3	9.7
Connecticut	11.8	8.5	North Carolina	15.8	8.7
Delaware	14.5	9.1	North Dakota	16.8	9.2
D.C.	28.1	14.3	Ohio	14.7	9.1
Florida	13.4	11.0	Oklahoma	15.0	9.8
Georgia	17.9	10.3	Oregon	14.8	9.0
Hawaii	18.4	5.4	Pennsylvania	12.5	10.3
Idaho	18.9	7.8	Rhode Island	12.5	9.8
Illinois	14.9	9.5	South Carolina	17.0	8.6
Indiana	15.7	9.0	South Dakota	15.8	9.6
Iowa	14.2	10.0	Tennessee	16.3	10.0
Kansas	13.8	9.5	Texas	18.4	8.3
Kentucky	16.2	9.8	Utah	26.1	6.6
Louisiana	17.5	9.0	Vermont	14.5	9.5
Maine	13.9	10.3	Virginia	13.9	8.1
Maryland	11.4	7.8	Washington	13.7	8.5
Massachusetts	12.3	9.5	West Virginia	15.5	10.9
Michigan	15.0	8.3	Wisconsin	14.3	9.0
Minnesota	14.1	8.7	Wyoming	17.4	8.6
Mississippi	19.5	9.7			
Missouri	15.1	10.5			
Montana	16.5	8.8			

Source. National Center for Health Statistics, Dept. of Health, Education, and Welfare, 1976 *Information Please*, p. 717, 723.

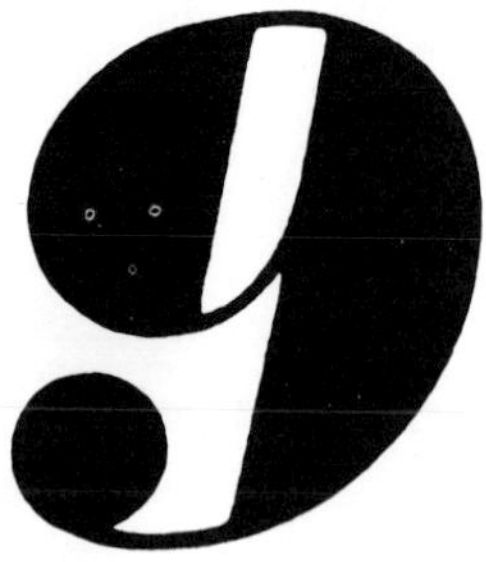

Correlation

BOX 9.1

Since the advent of the energy crisis, many energy experts are calling for conservation of energy. They frequently claim that smaller cars obtain better mileage than their heavier counterparts. What they are saying, in effect, is that there is a relationship between car weight and miles per gallon.

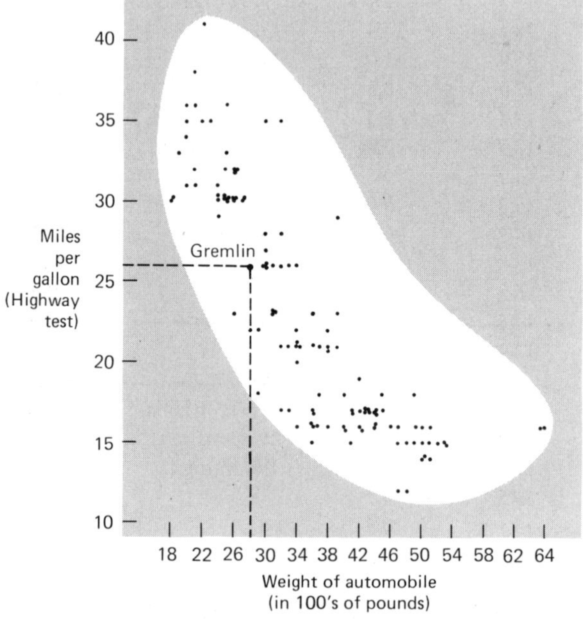

Fig. 9.1 Scatter diagram.

In Box 1.1 we presented the weight of 124 automobiles and the number of miles per gallon they attained in EPA highway tests. Let us call these variables X and Y, where X is the weight of the car and Y is the highway miles per gallon. Thus for each car, we are directing our attention to two scores—one for the X-variable and one for the Y-variable.

To help visualize these relationships, a **scatter diagram** may be drawn. It is traditional to represent the values of the X-variable along the horizontal axis (also called X-axis or *abscissa*) and the values of the Y-variable on the vertical axis (also called Y-axis or *ordinate*). The points shown on the scatter diagram, Fig. 9.1, represent the paired X and Y scores for each car. Let us illustrate the procedure for constructing a scatter diagram.

Note that the X-value for a Gremlin is 2831, and the corresponding Y-value is 26. Follow the X-axis until a value of 28 is located. Draw a line perpendicular (at a right angle) to the X-axis at this point. Now locate a value of 26 on the Y-axis. Draw a line perpendicular to the Y-axis. The point at which the two lines intersect represents the value of each variable, X and Y, for the Gremlin. Scatter diagrams are useful for depicting the relationship between two variables. Figure 9.1 shows that a negative relationship exists between X and Y (i.e., low scores on X are generally associated with high scores on Y; high scores on X are associated with low scores on Y). Stated another way, heavy cars show poor mileage performance whereas light cars obtain high mileage ratings.

9.1 THE CONCEPT OF CORRELATION

Up to this point in the course, we have been interested in calculating various statistics which permit us to thoroughly describe the distribution of the values of a single variable and to relate these statistics to the interpretation of individual scores. However, as you are well aware, many of the problems in the behavioral sciences go beyond the description of a single variable in its various and sundry ramifications. We are frequently called upon to determine the relationships among two or more variables. For example, college administration officers are vitally concerned with the relationship between high school averages or College Entrance Examination Board (CEEB) scores and performance at college. Do students who do well in high school or who score high on the CEEB also perform well in college? Conversely, do poor high school students or those who score low on the CEEB perform poorly at college? Do parents with high intelligence tend to have children of high intelligence? Is there a relationship between the declared dividend on stocks and their paper value in the exchange? Is there a relationship between the socio-economic class and recidivism in crime?

As soon as we raise questions concerning the relationships among variables, we are thrust into the fascinating area of **correlation**. In order to express quantitatively the extent to which two variables are related, it is necessary to calculate a **correlation coefficient**. There are many types of correlation coefficients. The decision of which one to employ with a specific set of data depends upon such factors as (1) the type of scale of measurement in which each variable is expressed; (2) the nature of the underlying distribution (continuous or discrete); and (3) the characteristics of the distribution of the scores (linear or nonlinear). We present two correlation coefficients in this text: the **Pearson** r, or the *Pearson* **product moment correlation coefficient**, employed with interval or ratio scaled variables, and r_{rho} or the **Spearman** r (*Spearman* **rank order correlation coefficient**) employed with ordered or ranked data.

No matter which correlational technique we use, all have certain characteristics in common.

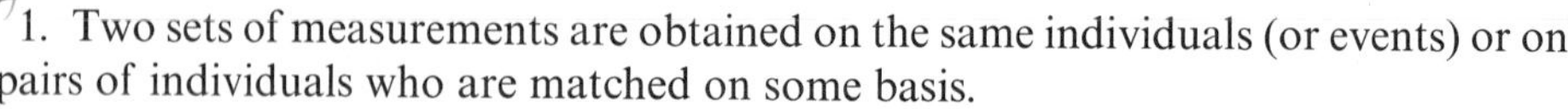

1. Two sets of measurements are obtained on the same individuals (or events) or on pairs of individuals who are matched on some basis.
2. The values of the correlation coefficients vary between +1.00 and −1.00. Both of these extremes represent perfect relationships between the variables, and 0.00 represents the absence of a relationship.
3. A **positive relationship** means that individuals obtaining high scores on one variable tend to obtain high scores on a second variable. The converse is also true, i.e., individuals scoring low on one variable tend to score low on a second variable.
4. A **negative relationship** means that individuals scoring low on one variable tend to score high on a second variable. Conversely, individuals scoring high on one variable tend to score low on a second variable.

Figure 9.2 shows a series of **scatter diagrams** illustrating various degrees of relationships between two variables, X and Y. In interpreting the figures it is important to remember that every circle represents two values: an individual's score on

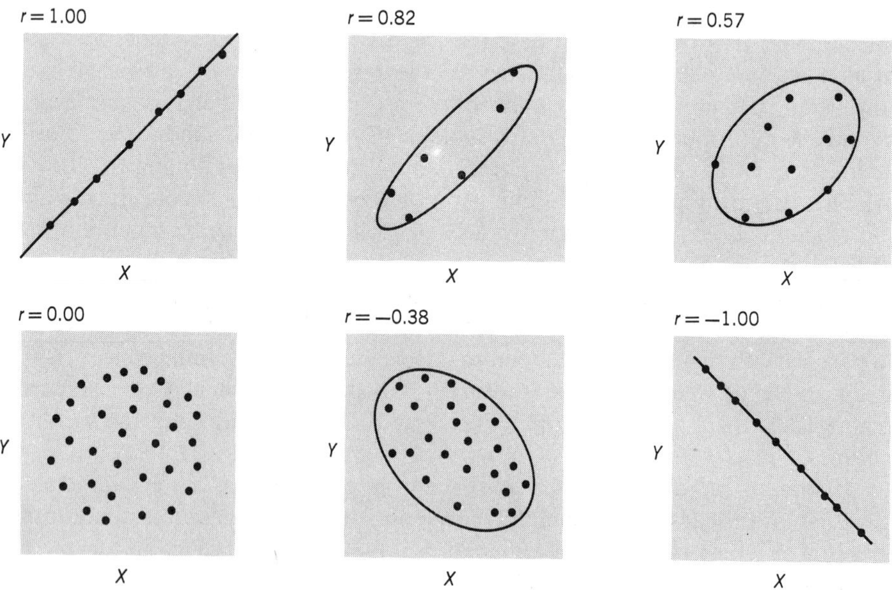

Fig. 9.2 Scatter diagrams showing various degrees of relationship between two variables.

the X-variable and the same person's score on the Y-variable. As indicated earlier (Section 4.2), the X-variable is represented along the abscissa and the Y-variable along the ordinate.

9.2 PEARSON r AND z-SCORES

A high positive Pearson r indicates that each individual obtains approximately the same z-score on both variables. In a perfect positive correlation, each individual obtains *exactly* the same z-score on both variables.

With a high negative r, each individual obtains approximately the same z-score on both variables, but the sigma scores are opposite in sign.

Remembering that the z-score represents a measure of relative position on a given variable (i.e., a high positive z represents a high score relative to the remainder of the distribution, and a high negative z represents a low score relative to the remainder of the distribution), we may now generalize the meaning of the Pearson r.

Pearson r represents the extent to which the same individuals or events occupy the same relative position on two variables.

In order to explore the fundamental characteristics of the Pearson r, let us examine a simplified example of a perfect positive correlation. In Table 9.1, we find the paired scores of 7 individuals on the two variables, X and Y.

Table 9.1

Raw scores and corresponding z-scores made by 7 subjects on two variables (hypothetical data)

Subject	X	x	x^2	z_x	Y	y	y^2	z_y	$z_x z_y$
A	1	−6	36	−1.39	4	−9	81	−1.39	1.93
B	3	−4	16	−0.93	7	−6	36	−0.93	0.86
C	5	−2	4	−0.46	10	−3	9	−0.46	0.21
D	7	0	0	0	13	0	0	0	0
E	9	2	4	0.46	16	3	9	0.46	0.21
F	11	4	16	0.93	19	6	36	0.93	0.86
G	13	6	36	1.39	22	9	81	1.39	1.93

$\sum X = 49$ $\quad \sum x^2 = 112$ $\quad \sum Y = 91$ $\quad \sum y^2 = 252$ $\quad \sum(z_x z_y) = 6.00$

$\bar{X} = 7.00$ $\quad s_x = \sqrt{\frac{112}{6}} = 4.32$ $\quad \bar{Y} = 13.00$ $\quad s_y = \sqrt{\frac{252}{6}} = 6.48$

It will be noted that the scale values of X and Y do not need to be the same for the calculation of a Pearson r. In the example, we see that X ranges from 1 through 13, whereas Y ranges from 4 through 22. This independence of r from specific scale values permits us to investigate the relationships among an unlimited variety of variables. We can even correlate the length of the big toe with the I.Q. if we feel so inclined!

Note, also, as we have already pointed out, that the z-scores of each subject on each variable are identical in the event of a perfect positive correlation. Had we reversed the order of either variable, i.e., paired 1 with 22, paired 3 with 19, etc., the z-scores would still be identical, but would be opposite in sign. In this latter case, our correlation would be a maximum negative ($r = -1.00$).

If we multiply our paired z-scores and then sum the results, we will obtain maximum values only when our correlation is 1.00. Indeed, as the correlation approaches zero, the sum of the products of the paired z-scores also approaches zero. Note that when the correlation is perfect, the sum of the products of the paired z-scores is equal to $n - 1$, where n equals the number of pairs. These facts lead to one of the many different but algebraically equivalent formulas for r.

$$r = \frac{\sum(z_x z_y)}{n - 1}. \tag{9.1}$$

It is suggested that you take the data in Table 9.1, rearrange them in a number of different ways, and calculate r, employing the above formula. You will arrive at a far more thorough understanding of r in this way than by reading the text (not that we are discouraging the latter).

It so happens that the formula is unwieldy in practice since it requires the calculation of separate z's for each score of each individual. Imagine the Herculean task of calculating r when n exceeds 50 cases, as it often does in research!

For this reason, a number of different computational formulas are employed. In this text, we shall illustrate the use of two: (1) the mean deviation formula, and (2) the raw score formula.

9.3 CALCULATION OF PEARSON r

9.3.1 Mean Deviation Method

The mean deviation method for calculating a Pearson r, like the z-score formula above, is not often employed by behavioral scientists because it involves more time and effort than other computational techniques. It is being presented here primarily because it sheds further light on the characteristics of the Pearson r. However, with small n's, it is as convenient a computational formula as any, unless an automatic calculator is available. The computational formula for the Pearson r, employing the mean deviation method is

$$r = \frac{\sum xy \text{ (cross products)}}{\sqrt{\sum x^2 \cdot \sum y^2}}. \tag{9.2}$$

Let us illustrate the mean deviation method employing the figures in Table 9.1 but arranged in a different sequence (Table 9.2).

The computational procedures, employing the mean deviation method, should be perfectly familiar to you. The $\sum x^2$ and the $\sum y^2$ have already been confronted when

Table 9.2

Computational procedures for Pearson r employing mean deviation method (hypothetical data)

Subject	X	x	x^2	Y	y	y^2	xy
A	1	−6	36	7	−6	36	36
B	3	−4	16	4	−9	81	36
C	5	−2	4	13	0	0	0
D	7	0	0	16	3	9	0
E	9	2	4	10	−3	9	−6
F	11	4	16	22	9	81	36
G	13	6	36	19	6	36	36

$\sum x^2 = 112$ $\qquad \sum y^2 = 252 \qquad \sum xy = 138$

$$r = \frac{\sum xy}{\sqrt{\sum x^2 \cdot \sum y^2}} = \frac{138}{\sqrt{(112)(252)}} = \frac{138}{168.00} = 0.82$$

we were studying the standard deviation. In fact, in calculating r only one step has been added, namely, the one to obtain the sum of the cross products ($\sum xy$). This is obtained easily enough by multiplying the deviation of each individual's score from the mean of the X-variable by his corresponding deviation on the Y-variable and then summing all of the cross products. Incidentally, you should notice the similarity of $\sum xy$ to $\sum(z_x z_y)$ which is discussed in Section 9.2. Everything that has been said with respect to the relationship between the variations in $\sum(z_x z_y)$ and r holds also for $\sum xy$ and r. Note that, if maximum deviations in X had lined up with maximum deviations in Y, and so on down through the array, $\sum xy$ would have been equal to 168.00, which is the same as the value of the denominator, and would have produced a correlation of 1.00.

9.3.2 Raw Score Method

We have already seen that the raw score formula for calculating the sum of squares is

$$\sum x^2 = \sum X^2 - \frac{(\sum X)^2}{N} \qquad \text{and} \qquad \sum y^2 = \sum Y^2 - \frac{(\sum Y)^2}{N}.$$

By analogy, the raw score formula for the sum of the cross products is

$$\sum xy = \sum XY - \frac{(\sum X)(\sum Y)}{n}. \qquad (9.3)$$

In calculating the Pearson r, by the raw score method, you have the option of calculating all the above quantities separately and substituting them into formula

(9.2) or defining r in terms of raw scores as in formula (9.4) as follows:

$$r = \frac{\sum XY - \frac{(\sum X)(\sum Y)}{n}}{\sqrt{\left[\sum X^2 - \frac{(\sum X)^2}{n}\right]\left[\sum Y^2 - \frac{(\sum Y)^2}{n}\right]}}. \tag{9.4}$$

The procedures for calculating r by the raw score method are summarized in Table 9.3. Here we find exactly the same coefficient as we did before. As with the mean deviation method, all the procedures, except those of obtaining the cross products, are familiar to you from our earlier use of the raw score formula to obtain the standard deviation. The quantity $\sum XY$ is obtained very simply by multiplying each X-value by its corresponding Y and then summing these products.

Table 9.3

Computational procedures for Pearson r employing raw score method (hypothetical data)

Subject	X	X^2	Y	Y^2	XY
A	1	1	7	49	7
B	3	9	4	16	12
C	5	25	13	169	65
D	7	49	16	256	112
E	9	81	10	100	90
F	11	121	22	484	242
G	13	169	19	361	247
	$\sum X = 49$	$\sum X^2 = 455$	$\sum Y = 91$	$\sum Y^2 = 1435$	$\sum XY = 775$
	$\bar{X} = 7$		$\bar{Y} = 13$		

$$r = \frac{\sum XY - \frac{(\sum X)(\sum Y)}{n}}{\sqrt{\left[\sum X^2 - \frac{(\sum X)^2}{n}\right]\left[\sum Y^2 - \frac{(\sum Y)^2}{n}\right]}}$$

$$= \frac{775 - \frac{(49)(91)}{7}}{\sqrt{\left(455 - \frac{49^2}{7}\right)\left(1435 - \frac{91^2}{7}\right)}}$$

$$= \frac{138}{\sqrt{(112)(252)}} = 0.82.$$

9.4 A WORD OF CAUTION

When low correlations are found, one is strongly tempted to conclude that there is little or no relationship between the two variables under study. However, it must be remembered that the Pearson r reflects only the *linear* relationship between two variables. The failure to find evidence of a relationship may be due to one of two possibilities: (1) the variables are, in fact, unrelated, or (2) the variables are related in a *nonlinear* fashion. In the latter instance, the Pearson r would not be an appropriate measure of the degree of relationship between the variables. To illustrate, if we were plotting the relationship between age and strength of grip, we might obtain a picture somewhat like Fig. 9.2.

It is usually possible to determine whether there is a substantial departure from linearity by examining the scatter diagram. If the distribution of points in the scatter diagram is elliptical, without the decided bending of the ellipse that occurs in Fig. 9.3, it may safely be assumed that the relationship is linear. Any small departures from linearity will not greatly influence the size of the correlation coefficient.

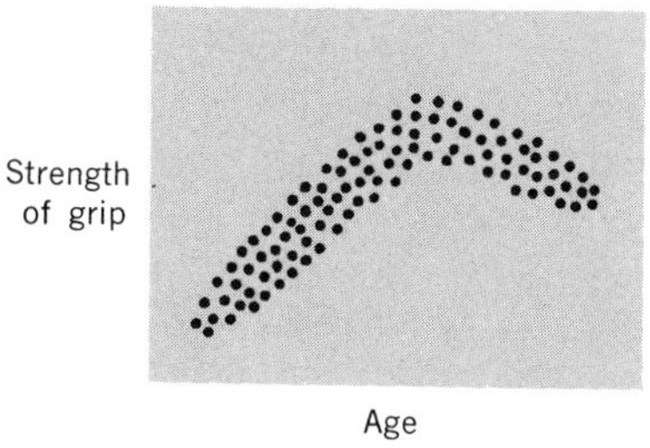

Fig. 9.3 Scatter diagram of two variables which are related in a nonlinear fashion (hypothetical data).

BOX 9.2*

A classic example of the faulty conclusions that can arise from culturally limited sampling is apparent in research on the effect of *age of weaning* on later *emotional disturbance*. In an effort to determine the nature of the relationship between these variables, Sears and Wise (1950) studied a sample of 80 children living in Kansas City. The investigators found a clear positive relationship between the age of weaning and the degree to which the infant gave indications of emotional disturbance. That is, the later a child was weaned, the more disturbance he showed; therefore, judging from these data, one might conclude that the earlier the child

* *Source*. Wiggins, J., E. Renner, G. Clore, and R. Rose, *The Principles of Personality*. Reading, Mass: Addison-Wesley, 1976.

was weaned, the better off he would be. However, the normal range of variation in American weaning practices is only from birth to one year of age, and we know from anthropologists that many societies delay weaning until considerably later. For example, the Kurtatchi of the Solomon Islands do not wean their children until they are over three years old. Information from the Kansas City sample would lead us to assume that Kurtatchi children would show severe emotional disturbance, but according to the anthropologists' report, they show virtually no evidence of such disturbance. Thus little can be concluded from these two studies by themselves, except that the relationship between age of weaning and later emotional disturbance is more complex than was initially suspected. Clearly more information is needed to fill the gap between the weaning ages of one and three years.

Fortunately, from a previous study that used the Human Relations Area files (Whiting and Child, 1953), information was available on the variables of age of weaning and emotional disturbance for 37 different societies. The range of weaning age in these cultures stretches from 12 months to 6 years. Contrary to the relationship found in the Kansas City sample, weaning and emotional disturbance across these 37 societies are in general negatively related, indicating that children are less rather than more disturbed when weaning is delayed. However, the results of these studies are not really in conflict since they represent different segments of the weaning-age continuum; and, as can be seen (in Fig. 9.4) the information fits nicely together. There is practically no overlap between ages of weaning in the primitive

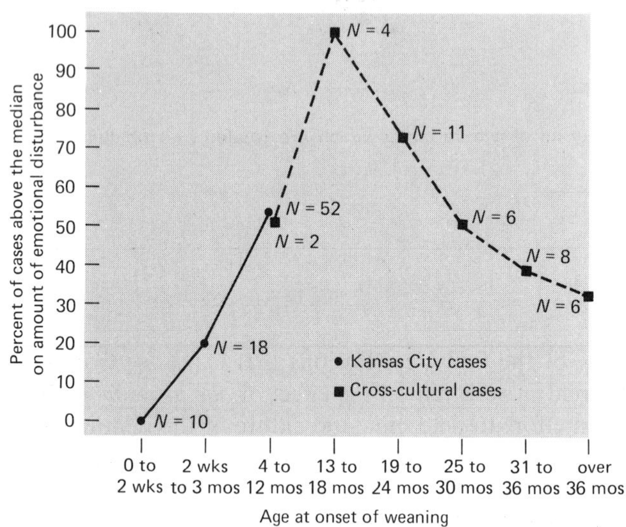

Fig. 9.4 Relation between age of weaning and amount of emotional disturbance shown by child. Comparable data for 80 individual children from Kansas City and 37 societies from Human Relations Area files (Whiting and Child, 1953) are presented. (From J. W. M. Whiting, "Methods and problems in cross-cultural research," in *The Handbook of Social Psychology,* Vol. II, Second edition, edited by Lindzey and Aronson, 1968, Addison-Wesley, Reading, Mass.)

societies sampled from the H. R. files and those in Kansas City, but the two age ranges in combination allow a conclusion to be reached which would not have emerged from either sample alone. Apparently age of weaning and the chances of emotional disturbance are curvilinearly related so that disturbance is likely to increase in the range of weaning age from 0 to 18 months but to decrease from that point as weaning is delayed beyond 18 months (Whiting, 1968). It is apparent, then, that investigators who limit the range of variability they study may emerge with only part of the picture, and in many cases (such as weaning age) the normal range varies considerably from culture to culture.

On the other hand, where there is marked curvilinearity, as in Fig. 9.3, a curvilinear coefficient of correlation would better reflect the relationship between the two variables under investigation. Although it is beyond the scope of this text to investigate nonlinear coefficients of correlation, you should be aware of this possibility and, as a matter of course, you should construct a scatter diagram prior to your calculation of the Pearson r.

The assumption of linearity of relationship is the most important requirement to justify the use of the Pearson r as a measure of relationship between two variables. It is not necessary that r be calculated only with normally distributed variables. So long as the distributions are unimodal and relatively symmetrical, a Pearson r may legitimately be computed.

Another situation giving rise to spuriously low correlation coefficients results from restricting the range of values of one of the variables. For example, if we were interested in the relationship between age and height for children from 3 years to 16 years of age, undoubtedly we would obtain a rather high coefficient of correlation between these two variables. However, suppose that we were to restrict the range of one of our variables? What effect would this have on the size of the coefficient? That is, let us look at the same relationship between age and height but only for those children between the ages of 9 and 10. We would probably end up with a rather low coefficient. Let us look at this graphically.

You will note that the overall relationship illustrated in Fig. 9.5 is rather high. The inset illustrates what happens when we restrict our range. Note also that the scatter diagram contained in the inset represents a very low correlation. This restriction of the range is frequently referred to as the **truncated range**. The problem of truncated range is not uncommon in behavioral research, since much of this research is conducted in the colleges and universities where subjects have been preselected for intelligence and related variables. Thus they represent a fairly homogeneous group with respect to these variables. Consequently, when an attempt is made to demonstrate the relationship between variables like CEEB scores and college grades, the resulting coefficient may be lowered because of the truncated range. Furthermore, the correlations would be expected to be lower for colleges which select their students from within a narrow range of CEEB scores.

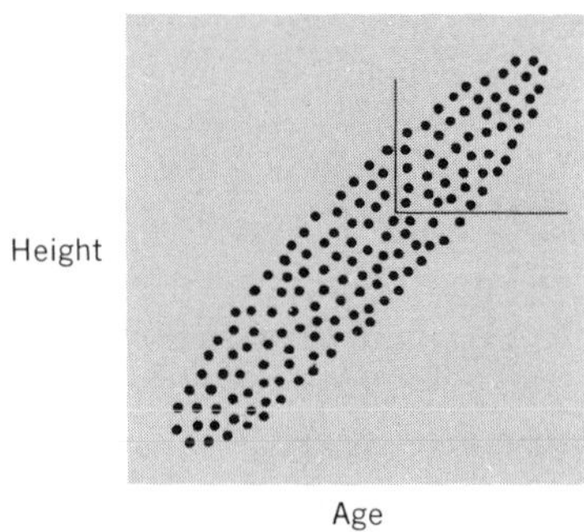

Fig. 9.5 Scatter diagram illustrating high correlation over entire range of X- and Y-values, but low correlation when range is truncated (hypothetical data).

9.5 ORDINALLY SCALED VARIABLES AND r_{rho}

There are occasions in which at least one of the variables under study is ordinal and we wish to determine its correlation with another variable that is either ordinal or interval in nature. Under these circumstances, we express both variables as ranks. Although we could obtain a Pearson r with ranked data, a variant of the product moment correlation coefficient, which is referred to as the Spearman r, r_{rho}, or the rank correlation coefficient, reduces the computational task involved in obtaining the correlation. The Spearman r is appropriate when one scale constitutes ordinal measurement and the remaining scale is either ordinal or higher. However, prior to applying the r_{rho} formula, both scales must be expressed as ranks.

The rank correlation coefficient requires that you obtain the differences in the ranks, square and sum the squared differences, and substitute the resulting values into the formula

$$r_{\text{rho}} = 1 - \frac{6\sum D^2}{n(n^2 - 1)} \tag{9.5}$$

in which $D = \text{rank } X - \text{rank } Y$.

In Table 9.4 we present data relating family size to I.Q. (Reed and Reed, 1965) and the procedures involved in calculating r_{rho}.

As a matter of course, $\sum D$ should be obtained even though it is not used in any of the calculations. It constitutes a useful check on the accuracy of your calculations up to this point since $\sum D$ must equal zero. If you obtain any value other than zero, you should rechec[illegible] your original ranks and the subsequent subtractions.

Occasionally, when it is necessary to convert scores to ranks, as in the present example, you will find two or more tied scores. In this event, assign the mean rank to each of the tied scores. The next score in the array receives the rank normally assigned to it. Thus the ranks of the scores 128, 122, 115, 115, 115, 107, 103 would be 1, 2, 4, 4, 4, 6, 7.

You may wonder how the formula for rho reflects the degree of relationship between two ordinally scaled variables. To engage in extensive mathematical proofs is, as you know, beyond the scope of this book. However, a few of the fundamentals

Table 9.4

Computational procedures for calculating r_{rho} from ranked variables

Mean I.Q. of children	I.Q. Rank	Family Size	D	D^2	
106	6.5	1	5.5	30.25	
110	10	2	8	64.00	$r_{rho} = 1 - \frac{6\sum D^2}{n(n^2-1)}$
107	8	3	5	25.00	
109	9	4	5	25.00	
106	6.5	5	1.5	2.25	$= 1 - \frac{6 \cdot 309.5}{10(99)}$
99	5	6	−1	1.00	
93	4	7	−3	9.00	
84	2	8	−6	36.00	$= 1 - \frac{1857}{990}$
90	3	9	−6	36.00	
62	1	10	−9	81.00	$= 1 - 1.88$
			$\sum D = 0$	$\sum D^2 = 309.50$	$= -0.88$

of the mathematics involved may permit you to grasp intuitively "how the formula works."

First, you may note that, when the correlation is maximum positive (+1.00), the difference between each rank is equal to zero and, therefore, $\sum D^2 = 0$. On the other hand, when the correlation is maximum negative (−1.00), the difference between ranks is maximum and $\sum D^2$ is maximum. Now, if we wanted to "invent" a formula which reflects these facts, we could define rho as follows:

$$r_{rho} = 1 - \frac{2\sum D^2}{\sum D_{max}{}^2}.$$

For a moment, let us concentrate on one term in our "invented" formula, that is, $\sum D^2/\sum D^2_{max}$. This term reflects, in any data we are analyzing, the *obtained proportion of the maximum possible sum of the squared differences* in ranks. If our correlation is maximum positive, in which $\sum D^2 = 0$, the term reduces to zero. If we were to multiply it by any constant, it would, of course, remain zero. On the other hand, if our correlation is maximum negative, the term reduces to unity, i.e., $\sum D^2_{max}/\sum D^2_{max} = 1$. In order that the correlation coefficient may vary between +1.00 and −1.00, it is necessary to multiply the numerator by the constant 2 and subtract the entire term from 1.00.

Now, it can be proved mathematically that $\sum D^2_{max} = n(n^2 - 1)/3$. You may check this empirically by setting up several sets of data with maximum negative correlations. You will find that the above formulation always holds.

Substituting this term in our "invented" formula, rho becomes

$$r_{rho} = 1 - \frac{3(2\sum D^2)}{n(n^2-1)} = 1 - \frac{6\sum D^2}{n(n^2-1)},$$

which is, of course, the formula for the Spearman rank coefficient.

CHAPTER SUMMARY

In this chapter we discussed the concept of correlation and demonstrated the calculation of two correlation coefficients, i.e., the Pearson r, employed with interval or ratio scaled data, and r_{rho}, used with ordinally scaled variables.

We saw that correlation is concerned with determining the extent to which two variables are related or tend to vary together. The quantitative expression of the extent of the relationship is given in terms of the magnitude of the correlation coefficient. Correlation coefficients vary between values of -1.00 to $+1.00$; both extremes represent perfect relationships. A coefficient of zero indicates the absence of a relationship between two variables.

We noted that the Pearson r is appropriate only for variables which are related in a linear fashion. With ranked data, the Spearman rank correlation coefficient is the exact counterpart of the Pearson r. The various computational formulas for the Pearson r may be employed in calculating r_{rho} from ranked data. However, a computational formula for r_{rho} was demonstrated which considerably simplifies the calculation of the rank correlation coefficient.

Terms to Remember:

Correlation
Correlation coefficient
Pearson r (product moment correlation coefficient)
Spearman r (r_{rho} or rank correlation coefficient)
Positive relationship
Negative relationship
Scatter diagram
Cross products
Truncated range

EXERCISES

9.1 The data below show the scores obtained by a group of 20 students on a college entrance examination and a verbal comprehension test. Prepare a scatter diagram and calculate a Pearson r for these data.

Student	College entrance exam (X)	Verbal comprehension test (Y)	Student	College entrance exam (X)	Verbal comprehension test (Y)
A	52	49	K	64	53
B	49	49	L	28	17
C	26	17	M	49	40
D	28	34	N	43	41
E	63	52	O	30	15
F	44	41	P	65	50
G	70	45	Q	35	28
H	32	32	R	60	55
I	49	29	S	49	37
J	51	49	T	66	50

9.2 The data below represent scores obtained by 10 students on a statistics examination and their final grade point average. Prepare a scatter diagram and calculate a Pearson r for these data.

Student	Statistics examination	Grade point average	Student	Statistics examination	Grade point average
A	90	2.50	F	70	1.00
B	85	2.00	G	70	1.00
C	80	2.50	H	60	0.50
D	75	2.00	I	60	0.50
E	70	1.50	J	50	0.50

9.3 A psychological study involved the rating of rats along a dominance-submissiveness continuum. In order to determine the reliability of the ratings, the ranks given by two different observers were tabulated:

Animal	Rank observer *A*	Rank observer *B*	Animal	Rank observer *A*	Rank observer *B*
A	12	15	I	6	5
B	2	1	J	9	9
C	3	7	K	7	6
D	1	4	L	10	12
E	4	2	M	15	13
F	5	3	N	8	8
G	14	11	O	13	14
H	11	10	P	16	16

Are the ratings reliable? Explain your answer.

9.4 Explain in *your own words* the meaning of correlation.

9.5 In each of the examples presented below, identify a possible source of contamination in the collection and/or interpretation of the results of a correlational analysis.

a) The relationship between age and reaction time for subjects from three months to 65 years of age.

b) The correlation between I.Q. and grades for honor students at a university.

c) The relationship between vocabulary and reading speed among children in a "culturally deprived" community.

9.6 For a group of 50 individuals $\sum z_x z_y$ is 41.3. What is the correlation between the two variables?

9.7 The following scores were made by five students on two tests. Calculate the Pearson r $\left(\text{using } r = \dfrac{\sum z_x z_y}{n-1}\right)$. Convert to ranks and calculate r_{rho}.

Student	Test X	Test Y
A	5	1
B	5	3
C	5	5
D	5	7
E	5	9

Generalize: What is the effect of tied ranks on r_{rho}?

9.8 Show algebraically that

$$\sum xy = \sum XY - \frac{\sum X \sum Y}{n}.$$

9.9 In Table 2.1 (Section 2.5.2), we presented data showing the number of wins at each post position on a circular track. What is the correlation between these two variables?

9.10 Exercise 3.8 presented data showing the prices of stock on January 1, 1967 and their net change over the course of the year.
a) Find the correlation between 1967 prices and the net change.
b) Find the correlation between 1967 prices and the percentage of change.

9.11 In Exercise 6.15, we showed the mean grades obtained by classes with varying numbers of students. What is the correlation between class size and mean grades on the final exam?

9.12 Table 9.6 shows scores on college entrance examinations and college grade-point averages following the first semester. What is the relationship between these two variables?

Table 9.6

Entrance examinations	Grade point averages	Entrance examinations	Grade point averages
440	1.57	528	2.08
448	1.83	550	2.15
455	2.05	582	3.44
460	1.14	569	3.05
473	2.73	585	3.19
485	1.65	593	3.42
489	2.02	620	3.87
500	2.98	650	3.00
512	1.79	690	3.12
518	2.63		

9.13 Following are data showing the latitude of 35 cities in the northern hemisphere and the mean high and mean low annual temperatures.
a) What is the correlation between latitude and mean high temperature?

b) What is the correlation between latitude and mean low temperature?
c) What is the correlation between mean high temperature and mean low temperature?

City	Latitude to nearest degree	Mean high temperature	Mean low temperature
Acapulco	17	88	73
Accra	6	86	74
Algiers	37	76	71
Amsterdam	52	54	46
Belgrade	45	62	45
Berlin	53	55	40
Bogota	5	66	50
Bombay	19	87	74
Bucharest	44	62	42
Calcutta	22	89	70
Casablanca	34	72	55
Copenhagen	56	52	41
Dakar	15	84	70
Dublin	53	56	42
Helsinki	60	46	35
Hong Kong	22	77	68
Istanbul	41	64	50
Jerusalem	32	74	54
Karachi	25	87	70
Leningrad	60	46	33
Lisbon	39	67	55
London	52	58	44
Madrid	40	66	47
Manila	15	89	73
Monrovia	6	84	73
Montreal	46	50	35
Oslo	60	50	36
Ottawa	45	51	32
Paris	49	59	43
Phnom Penh	12	89	74
Prague	50	54	42
Rangoon	17	89	73
Rome	42	71	51
Saigon	11	90	74
Shanghai	31	69	53

9.14 Obtain rank correlation coefficients for the data in Exercise 9.13. How closely do the r_{rho}'s approach the Pearson r's?

9.15 Refer back to Exercise 7.16, Chapter 7:

a) Find the correlation between the maximum daily temperature in January, 1965 and January, 1966.

b) Find the correlation between the maximum daily temperature in May, 1965 and May, 1966.

c) Is there a correlation between the day of the month and the maximum daily temperature in January, 1965? January, 1966? May, 1965? May, 1966?

9.16 Obtain rank correlation coefficients for Exercise 9.15 (a and b), using the data in Exercise 7.16. Compare the r_{rho}'s with the previously obtained Pearson r's.

9.17 Listed below are the 1967 earnings per share of 37 United States industries and the closing cost per share of stock for February 15, 1968.

a) Find the correlation between cost per share and 1967 earnings per share.

b) Express the earnings as a percentage of cost per share.

c) Is there a correlation between percentage of earnings and cost per share?

1967 Earnings per share (in cents)	Cost per share (to nearest dollar)	1967 Earnings per share (in cents)	Cost per share (to nearest dollar)
130	34	33	15
56	23	36	14
8	7	157	40
9	6	304	30
200	48	115	8
195	22	68	44
177	28	73	11
7	20	69	34
361	46	16	35
167	22	88	17
93	21	122	16
94	14	147	42
168	27	160	40
86	41	367	50
70	19	35	18
287	41	386	45
214	37	159	34
84	28	105	12
237	29		

9.18 Employ the data in Exercise 8.15:

a) Determine whether there is any relationship between the number of home runs obtained by the National and American League leaders over the 24-year period.

b) Assuming that the year constitutes an ordinal scale, determine the correlation between the year and the number of home runs hit by the leaders in each league.

9.19 Determine the correlation between the scores in Exercises 5.25 and 5.26.

9.20 Referring to Exercise 8.20, assume that the temperatures for the dates in September were as given in the following table.

Date	Temp.	Date	Temp.	Date	Temp.
1	70	11	60	21	70
2	60	12	65	22	65
3	75	13	70	23	80
4	80	14	65	24	60
5	80	15	70	25	60
6	65	16	85	26	60
7	65	17	70	27	65
8	75	18	50	28	75
9	75	19	55	29	50
10	85	20	50	30	45

a) Construct a scatter diagram relating the number of sodas sold and the temperature for each date. By inspection, determine if a correlation is suggested.
b) Determine the Pearson r.

9.21 The table below shows the consumer price index for all items and the per capita personal income from 1951–1974. Determine the extent of the relationship between these two variables (Source: Bureau of Labor Statistics). For ease of computation, round consumer price index to the nearest whole number.

Year	Consumer price index*	Weekly per capita personal income
1951	77.8	32
1955	80.2	36
1960	88.7	43
1965	94.5	54
1970	116.3	76
1971	121.3	81
1972	125.3	87
1973	133.1	97
1974	147.7	86

* (1967 = 100.0)

9.22 The following are wholesale food prices reported for March 7, 1972, and March 8, 1971. Determine the correlation between the prices on the two dates.

Food	3/7/72	3/8/71	Food	3/7/72	3/8/71
Flour	\$ 6.65	\$ 6.96	Eggs	$.33\frac{1}{2}$	$.31\frac{3}{4}$
Coffee	$.44\frac{1}{2}$	.45	Broilers	.29	.28
Cocoa	$.27\frac{7}{8}$	$.27\frac{1}{4}$	Pork bellies	.32	.22
Sugar	.1300	.1250	Hogs	23.80	17.25
Butter	$.68\frac{3}{4}$	$.70\frac{3}{4}$	Steers	35.25	31.50

(*The Wall Street Journal*, March 9, 1972, p. 26:4.)

9.23 On the finance page of your newspaper, find the over-the-counter quotations. Select ten issues and determine correlations between price asked and price bid.

9.24 Suppose you wanted to study the relation between the amount of time-saving machinery a manufacturer possessed and the mean price of the leather belts produced. Because it was difficult to order the amount of machinery on a ratio scale, you ranked the machines on an ordinal scale, with a rank of 15 indicating the most advanced machinery. You found the relation of price and type of the machines to be as follows:

15	$3.50	10	$4.25	5	$4.95
14	$3.75	9	$4.50	4	$5.50
13	$3.50	8	$4.45	3	$5.75
12	$4.00	7	$4.75	2	$5.45
11	$3.95	6	$5.00	1	$6.00

Determine the r_{rho} between the amount of machinery and the price.

9.25 A store owner recorded the number of times consumers bought or asked for a given item. He called this amount the demand. Each month he had 15 of the items to sell. In addition, the owner recorded the price of the item each month.

Month	Demand	Price	Month	Demand	Price
Jan.	25	$0.50	July	13	$0.80
Feb.	10	.90	Aug.	19	.70
Mar.	12	.80	Sept.	18	.72
April	18	.75	Oct.	16	.74
May	11	.85	Nov.	15	.75
June	20	.70	Dec.	15	.75

a) Determine the relation between demand and price, using Pearson r.
b) Determine the relation between demand and price, using r_{rho}.

9.26 Referring to Exercise 9.13, determine the correlation between latitude and mean high temperature for those areas with a latitude on or below 25 degrees. Compare this correlation with that obtained in Exercise 9.13(a). Why are the correlations different?

9.27 Again referring to Exercise 9.13, determine the correlation between latitude and mean high temperature for those areas with a latitude on or above 45 degrees. Compare this correlation with those obtained in Exercises 9.13(a) and 9.26. Why are they different?

9.28 In discussing the formula for r_{rho}, it was stated that $\sum D^2_{\text{max}} = n(n^2 - 1)/3$. Show that this is true for the following ranks:

1	8
2	7
3	6
4	5
5	4
6	3
7	2
8	1

9.29 Demonstrate that $\sum D^2 = 0$ for the following paired ranks:

1	1
2	2
3	3
4	4
5	5
6	6
7	7
8	8

9.30 a) Calculate r_{rho} for Exercise 9.28.
b) Calculate r_{rho} for Exercise 9.29.

9.31 Construct a scatter diagram for each of the following sets of data:

a) *X*	*Y*	b) *X*	*Y*	c) *X*	*Y*	d) *X*	*Y*
1.5	0.5	0.5	5.0	0.5	0.5	0.5	1.0
1.0	0.5	0.5	4.5	1.0	1.0	0.5	2.5
1.0	2.0	1.0	3.5	1.0	1.5	0.5	4.5
1.5	1.5	1.5	4.0	1.5	2.5	1.0	3.5
1.5	2.0	1.5	2.5	1.5	3.5	1.5	1.0
2.0	2.0	2.0	3.0	2.0	2.5	1.5	2.5
2.5	2.5	2.5	2.0	2.0	3.5	1.5	4.0
2.5	3.2	2.5	3.5	2.5	4.5	2.0	1.0
3.0	2.5	3.0	2.5	3.0	3.5	3.0	2.0
3.0	3.5	3.0	2.0	3.5	3.0	3.0	3.5
3.5	3.5	3.5	2.0	3.5	2.5	3.0	4.5
3.5	4.5	3.5	2.5	3.5	2.0	3.5	1.0
4.0	3.5	4.0	1.5	4.0	2.5	3.5	1.0
4.0	4.5	4.0	0.7	4.0	2.0	3.5	3.5
4.5	4.5	5.0	0.5	4.5	1.0	4.0	3.5
5.0	5.0			5.0	1.0	4.0	4.5
				5.0	0.5	4.5	2.5
						4.5	1.0

9.32 By inspecting the scatter diagrams for the data in Exercise 9.31, determine which one represents:
a) a curvilinear relation between *X* and *Y*.
b) a positive correlation between *X* and *Y*.
c) little or no relation between *X* and *Y*.
d) a negative correlation between *X* and *Y*.

9.33 Employing the data in Exercise 8.21, determine the correlation between males and females for the countries listed.

9.34 Following are the marriage, divorce, and birth rates for selected years from 1910–1974.
a) Calculate the correlation between marriage and divorce rates.
b) Calculate the correlation between marriage and birth rates.
c) Calculate the correlation between divorce and birth rates.

Year	Marriage rate	Divorce rate	Birth rate
1910	10.3	0.9	30.1
1919	11.0	1.3	26.1
1937	11.3	1.9	18.7
1939	10.7	1.9	18.8
1941	12.7	2.2	20.3
1943	11.7	2.6	22.7
1944	10.9	2.9	21.2
1945	12.2	3.5	20.4
1946	16.4	4.3	24.1
1947	13.9	3.4	26.6
1948	12.4	2.8	24.9
1949	10.6	2.7	24.5
1950	11.1	2.6	24.1
1951	10.4	2.5	24.9
1952	9.9	2.5	25.1
1953	9.8	2.5	25.1
1954	9.2	2.4	25.3
1955	9.3	2.3	25.0
1956	9.5	2.3	25.2
1957	8.9	2.2	25.3
1958	8.4	2.1	24.5
1959	8.5	2.2	24.3
1960	8.5	2.2	23.7
1961	8.5	2.3	23.3
1962	8.5	2.2	22.4
1963	8.8	2.3	21.7
1964	9.0	2.4	21.0
1065	9.3	2.5	19.4
1966	9.5	2.5	18.4
1967	9.7	2.6	17.8
1968	10.4	2.9	17.5
1969	10.6	3.2	17.8
1970	10.6	3.5	18.4
1971	10.6	3.7	17.2
1972	11.0	4.1	15.6
1973	10.9	4.4	14.9
1974	10.5	4.6	15.0

Source. National Center for Health, Dept. of Health Educational and Welfare, 1976 *Information Please*, p. 711, 718.

9.35 In ongoing research involving the development of premature infants, behavioral tests developed by Arnold Gesell were administered at the ages of 4, 9, and 24 months. Shown below are the test scores of 50 randomly selected subjects. (*Source.* Infant Studies Project, Dept. of Pediatrics, UCLA.)

Subject	4 months	9 months	24 months	Subject	4 months	9 months	24 months
1	100	110	133	26	94	95	88
2	89	110	104	27	76	92	88
3	100	98	92	28	89	100	83
4	106	93	92	29	106	93	83
5	129	98	96	30	100	87	104
6	84	75	54	31	117	105	117
7	100	102	83	32	112	108	104
8	100	97	83	33	111	100	125
9	94	95	106	34	106	93	83
10	124	103	92	35	106	105	104
11	110	108	108	36	94	95	104
12	112	98	108	37	88	90	83
13	125	107	88	38	112	126	100
14	106	105	104	39	116	102	92
15	95	95	88	40	106	92	92
16	100	100	95	41	105	102	96
17	111	102	88	42	95	100	92
18	107	102	108	43	117	113	98
19	106	108	87	44	94	95	88
20	100	102	92	45	112	110	117
21	106	110	104	46	106	103	83
22	106	98	100	47	94	102	96
23	111	98	104	48	118	108	92
24	111	100	92	49	100	105	96
25	94	98	79	50	72	72	79

Determine the correlations:
a) between 4 and 9 months.
b) between 4 and 24 months.
c) between 9 and 24 months.

9.36 Shown in the following table are the highway miles per gallon and the cubic inches of engine displacement among twenty 1976 automobiles with manual transmission.
a) Prepare a scatter diagram in which the X-variable is engine displacement and the Y-variable is miles per gallon.
b) Calculate the correlation between the two variables.

Engine displacement (cubic inches)	Highway mpg	Engine displacement (cubic inches)	Highway mpg
236	26	171	27
304	16	250	20
97	36	231	26
231	26	225	23
140	30	231	26
85	41	260	21
168	25	97	35
107	30	156	22
171	24	240	27
250	20	111	32

9.37 The table below shows highway miles per gallon and cubic inches of engine displacement for twenty 1976 automobiles with automatic transmission.

a) Prepare a scatter diagram in which the Y-variable is engine displacement and the Y-variable is miles per gallon.

b) Calculate the correlation between the two variables.

Engine displacement (cubic inches)	Highway mpg
236	19
304	20
97	32
231	24
140	27
85	35
168	22
107	25
171	20
250	21
171	23
250	21
231	24
225	20
231	24
260	18
97	26
156	18
240	23
111	28

Regression and Prediction

10.1 INTRODUCTION TO PREDICTION

Knowing a person's I.Q., what can we say about his prospects of satisfactorily completing a college curriculum? Knowing his prior voting record, can we make any informed guesses concerning his vote in the coming election? Knowing his mathematics aptitude score, can we estimate the quality of his performance in a course in statistics?

Let us look at an example. Suppose we are trying to predict Student Jones' score on the final exam. If the only information available was that the class mean on the final was 75 ($\bar{Y} = 75$), the best guess we could make is that he would obtain a score of 75 on the final.* However, far more information is usually available, e.g., Mr. Jones obtained a score of 62 on the midterm examination. How can we use this information to make a better prediction about his performance on the final exam? If we know that the class mean on the midterm examination was 70 ($\bar{X} = 70$), we could reason that since he scored below the mean on the midterm, he would probably score below the mean on the final. At this point, we appear to be closing in on a more accurate prediction of his performance. How might we further improve the accuracy of our prediction? Simply knowing that he scored below the mean on the midterm does not give us a clear picture of his relative standing on this exam. If, however, we know the standard deviation on the midterm, we could express his score in terms of his relative position, i.e., his z-score. Let us imagine that the standard deviation on the midterm was 4 ($s_x = 4$). Since he scored 2 standard deviations below the mean ($z_x = -2$), would we be justified in guessing that he would score 2 standard deviations below the mean on the final ($z_y = -2$)? That is, if $s_y = 8$, would you predict a score of 59 on the final? No! You will note that an important piece of information is missing, i.e., the correlation between the midterm and the final. You may recall from our discussion of correlation† that the Pearson r represents

* See Section 6.2.2, in which we demonstrated that the sum of the deviations from the mean is zero and that the sum of squares of deviations from the arithmetic mean is less than the sum of squares of deviations about any other score or potential score.

† See Section 9.2.

the extent to which the same individuals or events occupy the same relative position on two variables. Thus we are only justified in predicting a score of exactly 59 on the final when the correlation is perfect (that is, when $r = +1.00$). Suppose that the correlation is equal to zero. Then it should certainly be obvious that we are not justified in predicting a score of 59; rather, we are once again back to our original prediction of 75 (that is, $\bar{Y}$).

In summary, when $r = 0$, our best prediction is 75 ($\bar{Y}$); when $r = +1.00$, our best prediction is 59 ($z_y = z_x$). It should be clear that predictions from intermediate values of r will fall somewhere between 59 and 75.*

An outstanding advantage, then, of a correlational analysis stems from its application to problems involving predictions from one variable to another. Psychologists, educators, biologists, sociologists, and economists are constantly being called upon to perform this function. To provide an adequate explanation of r and to illustrate its specific applications, it is necessary to digress into an analysis of linear regression.

10.2 LINEAR REGRESSION

To simplify our discussion, let us start with an example of two variables which are usually perfectly or almost perfectly related: monthly salary and yearly income. In Table 10.1 we have listed the monthly income of eight wage earners in a small electronics firm. These data are shown graphically in Fig. 10.1. It is customary to refer to the horizontal axis as the X-axis, or abscissa, and to the vertical axis as the

Table 10.1

Monthly salaries and annual income of eight wage earners in an electronics firm (hypothetical data)

Employee	Monthly salary	Annual income
A	400	4800
B	450	5400
C	500	6000
D	575	6900
E	600	7200
F	625	7500
G	650	7800
H	675	8100

* We are assuming that the correlation is positive. If the correlation were -1.00, our best prediction would be a score of 91, that is, $z_y = -z_x$.

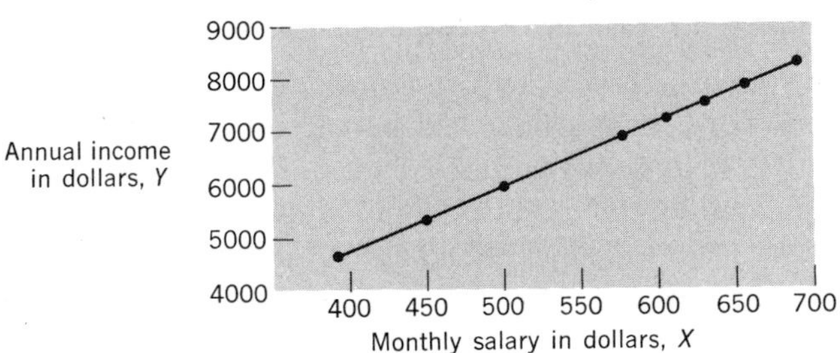

Fig. 10.1 Relation of monthly salaries to annual income for eight employees in an electronics firm.

ordinate, or Y-axis. If the variables are temporally related, the prior one is represented on the X-axis. It will be noted that all salaries are represented on a straight line extending diagonally from the lower left-hand corner to the upper right-hand corner.

10.2.1 Formula for Linear Relationships

The formula relating monthly salary to annual salary may be represented as

$$Y = 12X.$$

You may substitute any value of X into the formula and obtain directly the value of Y. For example, if another employee's monthly salary were $700, his annual income would be

$$Y = 12 \cdot 700 = 8400.$$

Let us add one more factor to this linear relationship. Let us suppose that the electronics firm had an exceptionally good year and that it decided to give each of its employees a Christmas bonus of $500. The equation would now read

$$Y = 500 + 12X.$$

Perhaps, thinking back to your high school days of algebra, you will recognize the above formula as a special case of the general formula for a straight line, that is,

$$Y = a + b_y X, \tag{10.1}$$

in which Y and X represent variables which change from individual to individual, and a and b_y represent constants for a particular set of data. More specifically, b_y represents the slope of a line relating values of Y to values of X. This is referred to as the regression of Y on X. In the present example, the slope of the line is 12 which means that Y changes by a factor of 12 for each change in X. The letter a represents the value of Y when $X = 0$.

You may also note that the above formula may be regarded as a method for predicting Y from known values of X. When the correlation is 1.00 (as in the present case), the predictions are perfect.

10.2.2 Predicting *X* and *Y* from Data on Two Variables

In behavioral research, however, the correlations we obtain are almost never perfect. Therefore we must find a straight line which best fits our data and we make predictions from that line. But what do we mean by "best fit?" You will recall that, when discussing the mean and the standard deviation, we defined the mean as that point in a distribution that makes the sum of squares of deviations from it minimal (least sum squares). When applying the least sum square method to correlation and regression, the **line of best fit** is defined as that line which minimizes the squared deviations around it. This straight line is referred to as a **regression line**.

We might note at this time that the term **prediction**, as employed in statistics, does not carry with it any necessary implication of futurity. The term "predict" simply refers to the fact that we are using information about one variable to obtain information about another. Thus, if we know a student's grade point average in college, we may use this information to "predict" his or her intelligence (which in our more generous moods, we assume preceded entrance into college).

At this point, we shall introduce two new symbols: X' and Y'. These may be read as "X prime and Y prime," "X and Y predicted," or "estimated X and Y." We use these symbols whenever we employ the regression line or the **regression equation** to estimate or predict a score on one variable from a known score on another variable.

Returning to the formula for a straight line, we are faced with the problem of determining b and a for a particular set of data so that Y' may be obtained.

The formula for obtaining the slope of the line relating Y to X, which is known as the line of regression of Y on X, is

$$b_y = \frac{\sum xy}{\sum x^2}. \qquad (10.2)$$

From formula (10.2) we may derive another useful formula for determining the slope of the line of Y on X:

$$b_y = \frac{\sum xy}{\sum x^2}.$$

But

$$\sum xy = r\sqrt{\sum x^2 \cdot \sum y^2}$$

and

$$\sum x^2 = (N - 1)s_x^2,$$

$$\sum y^2 = (N - 1)s_y^2.$$

Thus

$$b_y = \frac{r\sqrt{(N-1)^2 s_x^2 \cdot s_y^2}}{(N-1)s_x^2}$$

$$= r\frac{(N-1)s_x s_y}{(N-1)s_x^2}$$

$$= r\frac{s_y}{s_x}. \qquad (10.3)$$

Similarly, the slope of the regression line of X on Y may be expressed as:

$$b_x = \frac{\sum xy}{\sum y^2} \tag{10.4}$$

or

$$b_x = r\frac{s_x}{s_y}. \tag{10.5}$$

It will be noted that all the quantities shown in formulas (10.2) through (10.5) may be readily obtained in the course of calculating Pearson r.

$$a = \bar{Y} - b_y\bar{X}. \tag{10.6}$$

In the computation of Y', it is unwieldy to obtain each of these values separately and substitute them into the formula for a straight line. However, by algebraically combining formulas (10.3) and (10.6) and relating the result to formula (10.1), we obtain a much more useful formula for Y':

$$Y' = \bar{Y} + r\frac{s_y}{s_x}(X - \bar{X}).^* \tag{10.7}$$

Since there is also a separate regression equation for predicting scores on the X-variable from values of the Y-variable, the formula for X' is

$$X' = \bar{X} + r\frac{s_x}{s_y}(Y - \bar{Y}). \tag{10.8}$$

Concentrating our attention upon the second term on the right of each equation, we can see that the larger the r, the greater the magnitude of the entire term. This term also represents the *predicted deviation from the sample mean resulting from the regression of Y on X or X on Y.* Thus we may conclude that the greater the correlation, the greater the predicted deviation from the sample mean. In the event of a perfect correlation, the entire predicted deviation is maximal. On the other hand, when $r = 0$, the predicted deviation is also zero. Thus when $r = 0$, we have $X' = \bar{X}$, and $Y' = \bar{Y}$. All of this is another way of saying that in the absence of a correlation between two variables, our best prediction of any given score on a specified variable is the mean of the distribution of that variable.

10.2.3 Illustrative Regression Problems

Let us solve two sample problems employing the data introduced in Section 10.1.

---

* Since $Y' = a + b_yX$, $a = \bar{Y} - b_y\bar{X}$, and $b_y = r(s_y/s_x)$, then

$$Y' = \bar{Y} - r\left(\frac{s_y}{s_x}\right)\bar{X} + r\left(\frac{s_y}{s_x}\right)X = \bar{Y} + r\left(\frac{s_y}{s_x}\right)(X - \bar{X}).$$

Problem 1

Mr. Jones, you will recall, scored 62 on the midterm examination. What is our prediction concerning his score on the final examination? The relevant statistics are reproduced below.

X Midterm	Y Final
$\bar{X} = 70$	$\bar{Y} = 75$
$s_x = 4$	$s_y = 8$

$$r = 0.60$$

Employing formula (10.7), we find

$$\begin{aligned} Y' &= 75 + 0.60\left(\tfrac{8}{4}\right)(62 - 70) \\ &= 75 - 9.60 = 65.40. \end{aligned}$$

Problem 2

Ms. Smith, on the same midterm test, scored 76. What is our prediction concerning her score on the final examination? Employing the data appearing in Problem 1, we obtain the following results:

$$\begin{aligned} Y' &= 75 + 0.60\left(\tfrac{8}{4}\right)(76 - 70) \\ &= 75 + 7.20 = 82.20. \end{aligned}$$

Had our problem been to "predict" X-scores from known values of Y, the procedures would have been precisely the same as above except that formula (10.8) would have been employed.

A reasonable question at this point is, "Since we know $\bar{Y}$ and s_y in the above problems, we presumably have all the observed data at hand. Therefore why do we wish to predict Y from X?" It should be pointed out that the purpose of these examples was to acquaint you with the prediction formulas. In actual practice, however, correlational techniques are most commonly employed in making predictions about future samples where Y is unknown.

For example, let us suppose that the admissions officer of a college has constructed an entrance examination which she has administered to all the applicants over a period of years. During this time she has accumulated much information concerning the relationship between entrance scores and subsequent quality point averages in school. She finds that it is now possible to use scores on the entrance examination (X-variable) to predict subsequent quality point averages (Y-variable), and then use this information to establish an entrance policy for future applicants.

Since we have repeatedly stressed the relationship between Pearson r and z-scores, it should be apparent that the prediction formulas may be expressed in terms of z-scores. Mathematically, it can be shown that

$$z_{y'} = rz_x,^{*} \tag{10.9}$$

where $z_{y'} = Y'$ expressed in terms of a z-score.

Returning to Problem 1, Mr. Jones' score of 62 on the midterm can be expressed as a $z = -2.00$. Thus $z_{y'} = 0.60(-2.00) = -1.20$.

To assure yourself of the comparability of the two prediction formulas, that is, Eqs. (10.7) and (10.9), you should translate the $z_{y'}$-score into a raw score, Y'.

10.2.4 Constructing Lines of Regression

Let us return to the problem of constructing regression lines for predicting scores on the variables X and Y. As we have already pointed out, the regression line will not pass through all the paired scores. It will, in fact, pass among the paired scores in such a way as to minimize the squared deviations between the regression line (predicted scores) and the obtained scores. Earlier we pointed out that the mean is the point in a distribution which makes the squared deviations around it minimal. In discussing regression, the regression line is analogous to the mean, since, as we shall demonstrate, the sum of deviations of scores around the regression line is zero and the sum squares of these deviations are minimal.

It will be recalled that all the values required to calculate predicted scores are readily found during the course of calculating r, that is, $\overline{X}$, $\overline{Y}$, s_x, s_y. Now to construct our regression line for predicting Y from X, all we need to do is take two extreme values of X, predict Y from each of these values, and then join these two points on the scatter diagram. The line joining these points represents the regression line for predicting Y from X, which is also referred to as the line of regression of Y on X. Similarly, to construct the regression line for predicting X from Y, we take two extreme values of Y, predict X for each of these values, and then join these two points on the scatter diagram. This is precisely what was done in Fig. 10.2 to construct the two regression lines from the data in Table 9.3.

You will note that both regression lines intersect at the means of X and Y. In conceptualizing the relationship between the regression lines and the magnitude of r, it is helpful to think of the regression lines as rotating about the joint means of X and Y. When $r = 1.00$, both regression lines will have identical slopes and will be superimposed upon one another since they pass directly through all the paired scores. However, as r becomes increasingly small, the regression lines rotate

---

* $Y' = \overline{Y} + r(s_y/s_x)(X - \overline{X})$. By transposing terms:

$$\frac{Y' - \overline{Y}}{s_y} = r\frac{(X - \overline{X})}{s_x}, \quad \text{but} \quad \frac{X - \overline{X}}{s_x} = z_x \quad \text{and} \quad \frac{Y' - \overline{Y}}{s_y} = z_{y'}.$$

Therefore $z_{y'} = rz_x$.

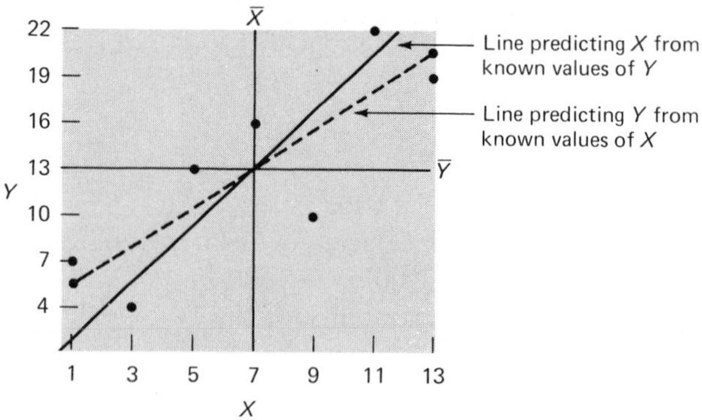

Fig. 10.2 Scatter diagram representing paired scores on two variables and regression lines for predicting X from Y and Y from X.

away from each other so that in the limiting case when $r = 0$, they are perpendicular to each other. At this point the regression line for predicting X from known values of Y is $\bar{X}$, and the regression line for predicting Y from known values of X is $\bar{Y}$.

10.3 RESIDUAL VARIANCE AND STANDARD ERROR OF ESTIMATE

Figure 10.3 shows a series of scatter diagrams, each reproduced from Fig. 10.2, showing only one regression line: the line for predicting Y from known values of X. Although our present discussion will be directed only to this regression line, all the conclusions we draw will be equally applicable to the line predicting X from known values of Y.

The regression line represents our best basis for predicting Y scores from known values of X. As we can see, not all the obtained scores fall on the regression line. However, if the correlation had been 1.00, all the scores would have fallen right on the regression line. The deviations $(Y - Y')$ in Fig. 10.3 represent our errors in prediction.

You will note the similarity of $Y - Y'$ (the deviation of scores from the regression line) to $Y - \bar{Y}$ (the deviation of scores from the mean). The algebraic sum of these deviations around the regression line is equal to zero. Earlier, we saw that the algebraic sum of the deviations around the mean is also equal to zero. In a sense, then, the regression line is a sort of "floating mean:" one that takes on different values depending on the values of X that are employed in prediction.

You will also recall that in calculating the variance, s^2, we squared the deviations from the mean, summed, and divided by $N - 1$. Finally, the square root of the variance provided our standard deviation. Now, if we were to square and sum the deviations of the scores from the regression line, $\sum(Y - Y')^2$, we would have a basis for calculating another variance and standard deviation. The variance around

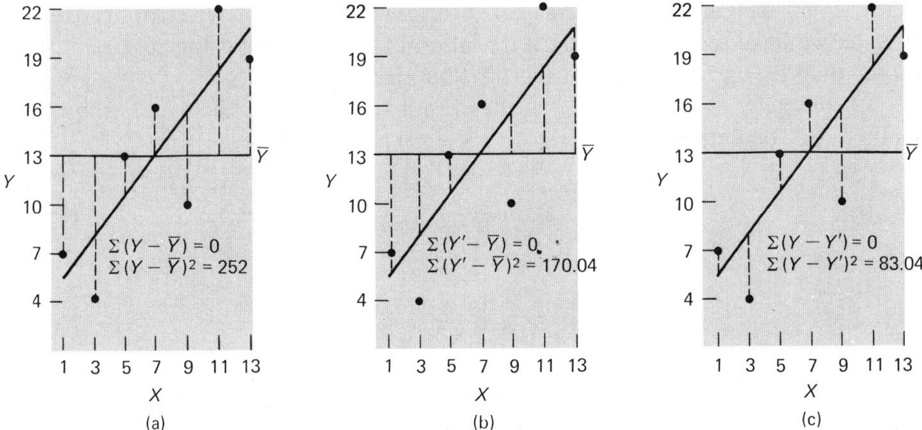

Fig. 10.3 Scatter diagram of paired scores on two variables, regression line for predicting *Y*-values from known values of *X*, and the mean of the distribution of *Y*-scores ($\bar{Y}$); $r = 082$ (from data in Table 9.3. (a) Deviations of scores from the mean of *Y* ($Y - \bar{Y}$). (b) Deviations of predicted scores from the mean of *Y* ($Y' - \bar{Y}$); explained variation. (c) Deviations of scores from the regression line ($Y - Y'$); unexplained variation.

the regression line is known as the **residual variance** and is defined as follows:

$$s^2_{\text{est } y} = \frac{\sum(Y - Y')^2}{n - 1}. \tag{10.10}$$

When predictions are made from *Y* to *X*, the residual variance of *X* is

$$s^2_{\text{est } x} = \frac{\sum(X - X')^2}{n - 1}. \tag{10.11}$$

The standard deviation around the regression line (referred to as the **standard error of estimate**) is, of course, the square root of the residual variance. Thus

$$s_{\text{est } y} = \sqrt{\frac{\sum(Y - Y')^2}{n - 1}}. \tag{10.12}$$

When predictions are made from *Y* to *X*, the standard error of estimate of *X* is

$$s_{\text{est } x} = \sqrt{\frac{\sum(X - X')^2}{n - 1}}. \tag{10.13}$$

You may be justifiably aghast at the amount of computation that is implied in the above formulas for calculating the standard error of estimate. However, as has been our practice throughout this text, we have shown the basic formulas so that you may have a conceptual grasp of the meaning of the standard error of estimate. It is, as we have seen, the standard deviation of scores around the regression line rather than around the mean of the distribution.

Fortunately, as in all previous illustrations in the text, there is a simplified method for calculating $s_{\text{est } y}$ and $s_{\text{est } x}$:

$$s_{\text{est } y} = s_y\sqrt{1 - r^2},^* \tag{10.14}$$

and

$$s_{\text{est } x} = s_x\sqrt{1 - r^2}.^* \tag{10.15}$$

You will note that when $r = \pm 1.00$, $s_{\text{est } y} = 0$, which means there are no deviations from the regression line and therefore no errors in prediction. On the other hand, when $r = 0$, the errors of prediction are maximal for that given distribution, that is, $s_{\text{est } y} = s_y$.

With the data in Exercise 9.1 the following statistics were calculated:

College entrance exam X	Verbal comprehension exam Y
$\bar{X} = 47.65$	$\bar{Y} = 39.15$
$s_x = 14.18$	$s_y = 12.67$

$$r = 0.85.$$

Thus

$$\begin{aligned} s_{\text{est } y} &= 12.67\sqrt{1 - 0.85^2} \\ &= 12.67(0.5268) = 6.67. \end{aligned}$$

As already indicated, the standard error of estimate has properties that are similar to those of the standard deviation. For example, if we were to construct lines parallel to the regression line for predicting Y from X at distances $1s_{\text{est } y}$, $2s_{\text{est } y}$, and $3s_{\text{est } y}$, we would find that approximately 68% of the cases fall between $\pm 1s_{\text{est } y}$, 95% between $\pm 2s_{\text{est}}\ y$, and 99% between $\pm 3s_{\text{est } y}$. These relationships between standard error of estimate and percentage of area will be more closely approximated when the variability within the columns and the rows is homogeneous. This condition is known as **homoscedasticity**.

Using the above data, we have drawn two lines parallel to the regression line for predicting Y from known values of X in Fig. 10.4. These lines are both 1 standard error of estimate (± 6.67) from the regression line. Each circle represents an

* The values of $\sqrt{1 - r^2}$ may be obtained directly from Table H. Thus for $r = 0.82$, we have $\sqrt{1 - r^2} = (0.5724)$.

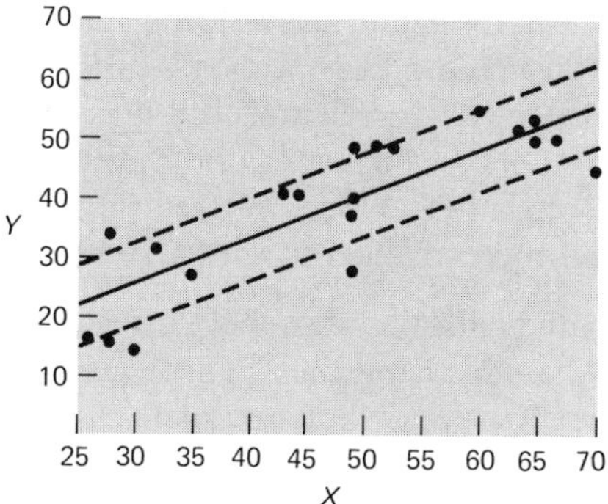

Fig. 10.4 Line of regression for predicting Y from X with parallel lines $1s_{est\,y}$ above and below the regression line from data in Exercise 9.1. Circles indicate individuals' scores on X and Y.

individual's scores on the X- and the Y-variables. It can be seen that 13 of the 20 scores, or 65% of the cases, fall between $\pm 1s_{est\,y}$. This figure is in fairly good agreement with the expected percentage of 68. With a larger n, the approximation would be better.

10.4 EXPLAINED AND UNEXPLAINED VARIATION*

If we look again at Fig. 10.3, we can see that there are three separate sum squares that can be calculated from the data. These are:

1. Variation of scores around the sample mean. This variation is given by $(Y - \bar{Y})^2$ and is, of course, basic to the determination of the variance and the standard deviation of the sample.

2. Variation of scores around the regression line (or predicted scores). This variation is given by $(Y - Y')^2$ and is referred to as *unexplained variation*. The reason for this choice of terminology should be clear. If the correlation between two variables is ± 1.00, all the scores fall on the regression line. Consequently, we have, in effect, explained *all* the variation in Y in terms of the variation in X and, conversely, all the variation of X in terms of the variation in Y. In other words, in the event of a perfect relationship, there is no unexplained variation. However, when the correlation is less than perfect, many of the scores will not fall right on the regression line, as we have seen. The deviations of these scores from the regression line represent variation which is not accounted for in terms of the correlation between two variables. Hence, the term **unexplained variation** is employed.

3. Variation of predicted scores about the mean of the distribution. This variation is given by $(Y' - \bar{Y})^2$ and is referred to as **explained variation**. The reason for this

* Although analysis of variance is not covered until Chapter 16, much of the material in this section will serve as an introduction to some of the basic concepts of analysis of variance.

terminology should be clear from our discussion in the preceding paragraph and our prior reference to predicted deviation (Section 10.3). You will recall our previous observation that the greater the correlation, the greater the predicted deviation from the sample mean. It follows further that the greater the predicted deviation, the greater the explained variation. When the predicted deviation is maximum, the correlation is perfect, and the explained variation is 100%.

It can be shown mathematically that the total sum squares consists of two components which may be added together. These two components represent explained variation and unexplained variation respectively. Thus

$$\sum(Y - \bar{Y})^2 = \sum(Y - Y')^2 + (Y' - \bar{Y})^2. \tag{10.16}$$

Total variation = unexplained variation + explained variation.

These calculations are shown in Fig. 10.3. You will note that the sum of the explained variation (170.04) and the unexplained variation (83.04) is equal to the total variation. The slight discrepancy found in this example is due to rounding r to 0.82 prior to calculating the predicted scores.

Now, when $r = 0.00$, then $\sum(Y' - \bar{Y})^2 = 0.00$. (Why? See Section 10.2.) Consequently, the total variation is equal to the unexplained variation. Stated another way, when $r = 0$, all the variation is unexplained. On the other hand, when $r = 1.00$, then $\sum(Y - Y')^2 = 0.00$, since all the scores are on the regression line. Under these circumstances, total variation is the same as explained variation. In other words, all the variation is explained when $r = 1.00$.

The ratio of the explained variation to the total variation is referred to as the **coefficient of determination** and is symbolized by r^2. The formula for the coefficient of determination is

$$r^2 = \frac{\text{explained variation}}{\text{total variation}} = \frac{\sum(Y' - \bar{Y})^2}{\sum(Y - \bar{Y})^2}.^* \tag{10.17}$$

It can be seen that the coefficient of determination indicates the proportion of total variation which is explained in terms of the magnitude of the correlation coefficient. When $r = 0$, the coefficient of determination, r^2, equals 0. When $r = 0.5$, the coefficient of determination is 0.25. In other words, 25% of the total variation is accounted for. Finally, when $r = 1.00$, then $r^2 = 1.00$ and all variation is accounted for.

Figure 10.5 depicts graphically the proportion of variation in one variable that is accounted for by the variation in another variable when r takes on different values.

You have undoubtedly noted that the square root of the coefficient of determination provides another definition of r. Thus

$$r = \pm\sqrt{\frac{\text{explained variation}}{\text{total variation}}} = \pm\sqrt{\frac{\sum(Y' - \bar{Y})^2}{\sum(Y - \bar{Y})^2}}. \tag{10.18}$$

* Table H in Appendix IV presents a number of functions of r for various values of r, including such useful functions as r^2, $1 - r^2$, and $\sqrt{1 - r^2}$. You should familiarize yourself with this table.

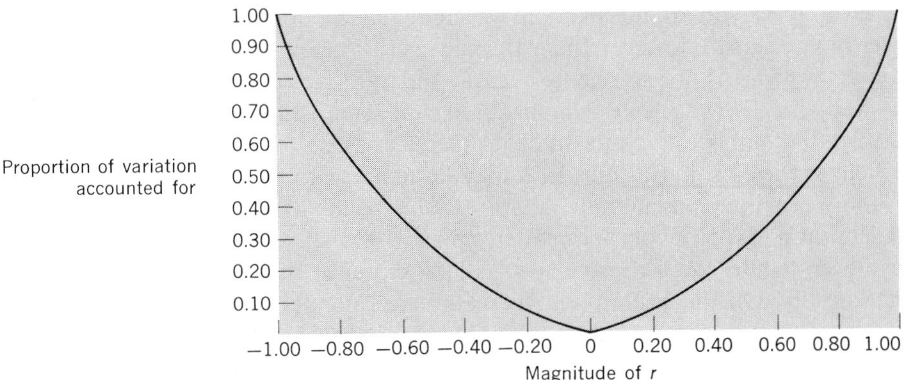

Fig. 10.5 The proportion of the variation on one variable accounted for in terms of variations of a correlated variable at varying values of r.

10.5 CORRELATION AND CAUSATION

You have seen that when two variables are related it is possible to predict one from your knowledge of the other. This relationship between correlation and prediction often leads to a serious error in reasoning, i.e., the relationship between two variables frequently carries with it the implication that one has caused the other. This is especially true when there is a temporal relationship between the variables in question, i.e., when one precedes the other in time. What is often overlooked is the fact that the variables may not be causally connected in any way, but that they may vary together by virtue of a common link with a third variable. Thus, if you are a bird watcher, you may note that as the number of birds increases in the spring, the grass becomes progressively greener. However, recognizing that the extended number of hours and the increasing warmth of the sun is a third factor influencing both of

Fig. 10.6 It is hot because the thermometer is up? What is wrong with this conclusion?

these variables, you are not likely to conclude that the birds cause the grass to turn green or vice versa. However, there are many occasions, particularly in the behavioral sciences, when it is not so easy to identify the third factor (Fig. 10.6).

Suppose that you have demonstrated that there is a high positive correlation between the number of hours students spend studying for an exam and their subsequent grades on that exam. You may be tempted to conclude that the number of hours of study causes grades to vary. This seems to be a perfectly reasonable conclusion, and is probably in close agreement with what your parents and instructors have been telling you for years. Let us look closer at the implications of a causal relationship. On the assumption that a greater number of hours of study causes grades to increase, we would be led to expect that *any* student who devotes more time to study is guaranteed a high grade and that one who spends less time with his books is going to receive a low grade. This is not necessarily the case. We have overlooked the fact that it might be that the better student (by virtue of higher intelligence, stronger motivation, better study habits, etc.) who devotes more time to study, performs better simply because of a greater capacity to do so.

What we are saying is that correlational studies simply do not permit inferences of causation. Correlation is a necessary but not a sufficient condition to establish a causal relationship between two variables. In short, to establish a causal relationship it is necessary to conduct an experiment in which an independent variable is manipulated by the experimenter, and the effects of these manipulations are reflected in the dependent, or criterion, variable. A correlational study lacks the requirement of independent manipulation.

Table 10.2

Murder victims, by weapons used: 1963–1973

Year	Murder victims, total	% Guns	% Cuttings or stabbings	% Blunt objects	% Strangulations and beatings	% Drownings, arson, etc.	% All other
1963	7,549	56.0	23	6	9	3	3
1964	7,990	55.0	24	5	10	3	3
1965	8,773	57.2	23	6	10	3	1
1966	9,552	59.3	22	5	9	2	1
1967	11,114	63.0	20	5	9	2	1
1968	12,503	64.8	19	6	7	2	1
1969	13,575	65.4	19	5	8	2	1
1970	13,649	66.2	18	4	8	3	1
1971	16,183	66.2	19	4	8	2	1
1972	15,832	65.6	19	4	8	2	1
1973	17,123	65.7	17	5	8	1	2

Adapted from *The American Almanac: The Statistical Abstract of the U.S.* New York: Grosset & Dunlap, 1976.

There are many ways we can be deceived by correlational data. Consider the following* fact and the conclusions drawn from it.

Fact: There is a positive relationship between the number of handguns produced each year and the number of murders committed by the use of this instrument.

Conclusion: We should reduce the annual production and importation of handguns since they are the cause of about two-thirds of all murders. Since fewer murders by handguns are committed when fewer weapons are available, we can lower the murder rate by restricting the production of handguns.

Let us examine the evidence. Table 10.2 shows the number and percentage of murders committed with various weapons between 1963 and 1973. There is no doubt about it. Aside from the automobile, the gun is our favorite weapon for abruptly terminating the lives of others. What is more, it is becoming increasingly popular with each passing year.

"But wait," you say. "Murder is on the increase, but so also is the population." Perhaps the two events are merely keeping pace with one another. So let's compare the rates of growth of both population and murder . . . . We'll calculate fixed base index numbers between 1964 and 1969, using 1964 as our base year. The time index numbers are shown in Table 10.3.

When these index numbers are placed on a graph, the picture is abundantly clear. The rate of increase in murder is far outstripping the rate of increase in population [Fig. 10.7]. Although not shown here, it is also far outstripping the growth of population in that segment of the population that commits most of the murders (17-year-olds and older).

So far we have established the fact that murder is on the rise and that guns are number one on the hit list. Now let's relate this increase to the number of firearms that are produced and

Table 10.3

Fixed-base index numbers between 1964 and 1973, using 1964 as base year

Year	Murder by gun	Fixed base time index 1964	Population (in millions)	Fixed base time index 1964
1964	4,393	100	191.9	100
1965	5,015	114	194.3	101
1966	5,660	129	196.6	102
1967	6,998	159	198.7	104
1968	8,105	184	200.7	105
1969	8,876	202	202.7	106
1970	9,039	207	204.9	107
1971	10,712	244	207.1	108
1972	10,379	236	208.8	109
1973	11,249	256	210.4	110

Source. The American Almanac: *The Statistical Abstract of the U.S.* New York: Grossett & Dunlap, 1976.

* *Winning with Statistics* by Richard P. Runyon. Reading, Mass: Addison-Wesley, 1977.

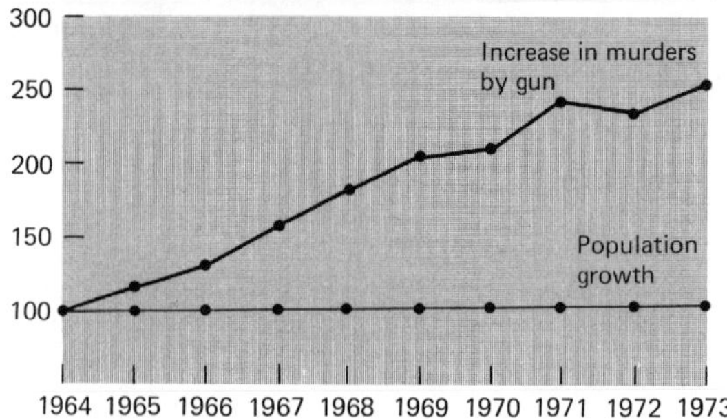

Fig. 10.7 Time indexes showing rates of increase both in murders by gun and in population.

imported each year. Presumably, greater production means greater availability and greater availability means greater opportunity and temptation. The graph is shown in Fig. 10.7. Pretty convincing evidence, huh?

'Fraid not. Please don't get me wrong. I am not trying to add ammunition to the arsenal of the "Have guns, will travel" crowd. Quite frankly, I think our insistence on the right to bear arms is a racial memory from the Neanderthal days. But my own feelings are irrelevant to the issue. If the weight of evidence is the high correlation between availability of guns and murder by gun, I must admit to being a skeptic. Why?

Just about everything—marriage, divorce, banana production, abortions—has been either growing or decreasing since 1964. Any of these events will correlate with the increased availability of guns. Take a look at Fig. 10.8. It shows that the number of murders by *cutting* or *stabbing* increases as the number of available *firearms* increases. Does anyone seriously entertain the view that the greater availability of firearms increases the risk of death by stabbing or that we can reduce the number of gunshot murders by cutting down on the availability of knives? And how about Fig. 10.9? We see a beautiful negative relationship between the number of farm workers and the number of murders committed by guns. If we take this sort of evidence seriously, we had better figure out how we can get them back on the farm, even after they have seen Paree.

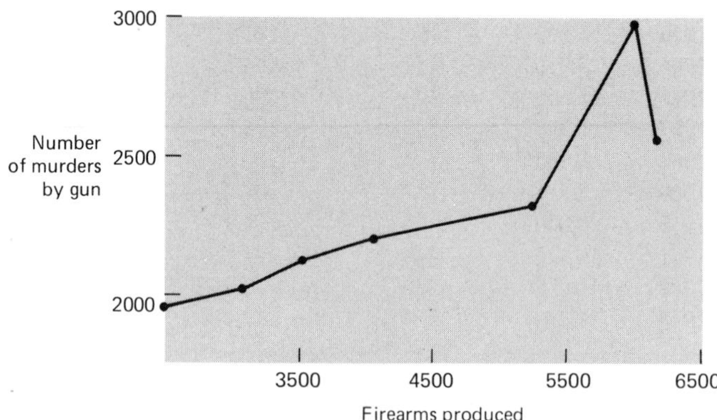

Fig. 10.8 Relationship between firearms produced and murders by gun.

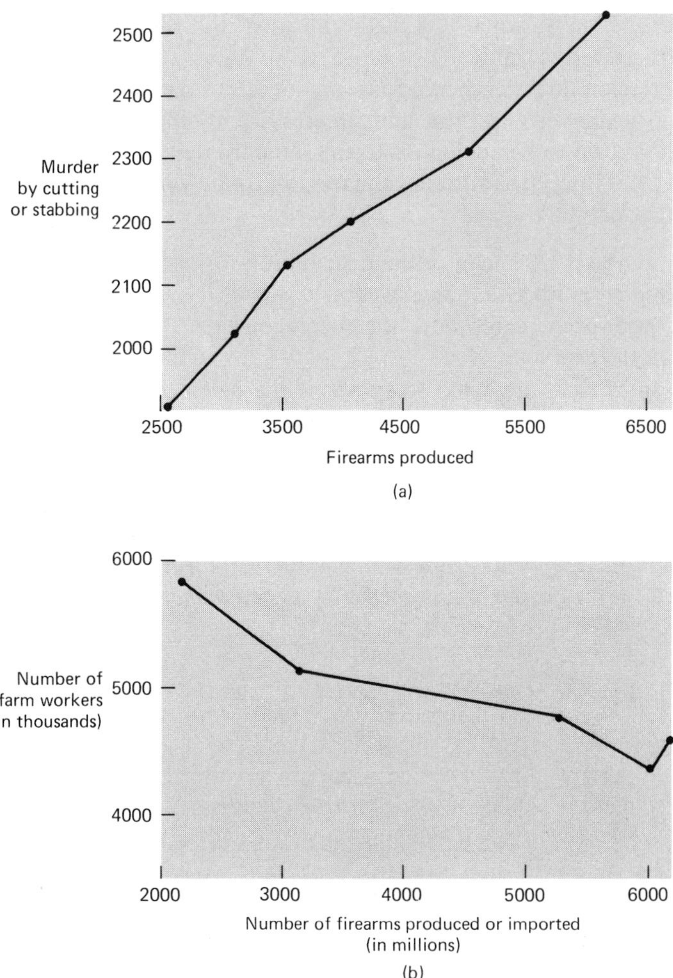

Fig. 10.9 (a) Relationship between firearms produced and murders by cutting and stabbing; (b) relationship between firearms produced and number of farm workers.

CHAPTER SUMMARY

Let us briefly review what we have learned in this chapter. We have seen that it is possible to "fit" two straight lines to a bivariate distribution of scores, one for predicting Y-scores from known X-values and one for predicting X-scores from known Y-values.

We saw that, when the correlation is perfect, all the scores fall upon the regression line. There is, therefore, no error in prediction. The lower the relationship, the greater the dispersion of scores around the regression line, and the greater the errors of prediction. Finally, when $r = 0$, the mean of the sample provides our "best" predictor for a given variable.

The regression line was shown to be analogous to the mean: the summed deviations around it are zero and the sum squares are minimal. The standard error of estimate was shown to be analogous to the standard deviation.

We saw that three separate sum squares, reflecting variability, may be calculated from correlational data.

1. Variation about the mean of the distribution for each variable. This variation is referred to as the *total sum squares.*
2. Variation of each score about the regression line. This variation is known as *unexplained variation.*
3. Variation of each predicted score about the mean of the distribution for each variable. This variation is known as *explained variation.*

We saw that the sum of the explained variation and the unexplained variation is equal to the total variation.

Finally, we saw that the ratio of the explained variation to the total variation provides us with the proportion of the total variation which is explained. The term applied to this proportion is *coefficient of determination.*

EXERCISES

10.1 Find the equation of the regression line for the following data.

X	1	2	3	4	5
Y	5	3	4	2	1

10.2 In a study concerned with the relationship between two variables, X and Y, the following was obtained,

$$\bar{X} = 119 \qquad \bar{Y} = 1.30$$
$$s_x = 10 \qquad s_y = 0.55$$
$$r = 0.70$$
$$n = 100$$

a) Sally B. obtained a score of 130 on the X-variable. Predict her score on the Y-variable.
b) A score of 1.28 on the Y-variable was predicted for Bill B. What was his score on the X-variable?
c) Determine the standard error of estimate of Y.

10.3 A study was undertaken to find the relationship between "emotional stability" and performance in college. The following results were obtained.

Emotional stability	College average
$\bar{X} = 49$	$\bar{Y} = 1.35$
$s_x = 12$	$s_y = 0.50$

$$r = 0.36$$
$$n = 60$$

a) Norma obtained a score of 65 on the X-variable. What is our prediction of her score on the Y-variable?
b) Determine the standard error of estimate of X and Y.
c) What proportion of total variation is accounted for by explained variation?

10.4 Assume that $\bar{X} = 30$, $s_x = 5$; $\bar{Y} = 45$, $s_y = 8$. Draw a separate graph for each pair of regression lines for the following values of r.
a) 0.00 b) 0.20 c) 0.40
d) 0.60 e) 0.80 f) 1.00
Generalize: What is the relationship between the size of r and the angle formed by the regression lines? If the values of r given above, (b) through (f), were all negative, what is the relationship?

10.5 Given: The standard deviation of scores on a standardized vocabulary test is 15. The correlation of this test with I.Q. is 0.80. What would you expect the standard deviation on the vocabulary test to be for a large group of students with the same I.Q.? Explain your answer.

10.6 A student obtains a score on test X which is 1.5 standard deviations above the mean. What standard score would you predict for her on test Y if r equals
a) 0.00, b) 0.40, c) 0.80, d) 1.00, e) −0.50, f) −0.80?

10.7 A personnel manager has made a study of employees involved in one aspect of a manufacturing process. He finds that after they have been on the job for a year he is able to obtain a performance measure which accurately reflects their proficiency. He designs a selection test aimed at predicting their eventual proficiency and obtains a correlation of 0.65 with the proficiency measure (Y). The mean of the test is 50, $s_x = 6$; $\bar{Y} = 100$, $s_y = 10$. Answer the following questions based on these facts:
a) George H. obtained a score of 40 on the selection test. What is his predicted proficiency score?
b) How likely is it that he will score as high as 110 on the proficiency scale?
c) A score of 80 on the Y-variable is considered satisfactory for the job; below 80 is unsatisfactory. If the X-test is to be used as a selection device, which score should be used as a cutoff point? (*Hint*: Find the value of X which leads to a prediction of 80 on Y. Be sure to employ the appropriate prediction formula.)
d) Harry H. obtained a score of 30 on X. How likely is it that he will achieve an acceptable score on Y?
e) Joan G. obtained a score of 60 on X. How likely is it that she will *fail* to achieve an acceptable score on Y?
f) For a person to have prospects for a supervisory position, a score of 120 or higher on Y is deemed essential. What value of X should be employed for the selection of potential supervisory personnel?
g) If 1000 persons achieve a score on X which predicts a $Y = 120$, approximately how many of them will obtain Y scores below 120? Above 130? Below 110? Above 110?

10.8 An owner of a mail-order house advertises that all orders are shipped within 24 hours of receipt. Since personnel in the shipping department are hired on a day-to-day basis, it is important for her to be able to predict the number of orders contained in each batch of daily mail so that she can hire sufficient personnel for the following day. She hit on the idea of weighing each day's mail and correlating the weight with the actual number of orders. Over a successive 30-day period, she obtained the following results:

Weight in pounds	Number of orders	Weight in pounds	Number of orders
20	5400	26	5400
15	4200	21	5000
23	5800	24	5400
17	5000	16	4300
12	3500	34	6700
35	6400	28	6100
29	6000	15	3600
21	5200	11	3200
10	4000	18	5300
13	3800	27	5800
25	5700	30	5900
14	4000	22	5500
18	4800	20	5200
30	6200	24	5000
33	6600	13	3700

a) Find the correlation between the weight of mail and the number of orders. (*Hint*: In calculating the correlation, consider dropping the final two digits on the Y-variable).
b) If 10 persons are required to handle 1000 orders per day, how many should be hired to handle 22 pounds of mail? 15? 30? 38? (*Note*: Assume that employees are hired only in groups of 10.)

10.9 Peruse the magazine section of your Sunday newspaper, monthly magazines, television, and radio advertisements for examples of inferences of causation based on correlational data.

10.10 Referring to Exercise 9.1:
a) Morris obtained a score of 40 on his college entrance examination. Predict his score on the verbal comprehension test.
b) How likely is it that he will score at least 40 on the verbal comprehension test?
c) Nancy obtained a score of 40 on the verbal comprehension test. Predict her score on the college entrance examination.
d) How likely is it that she will score at least 40 on the college entrance examination?
e) REC University finds that students who score at least 45 on the verbal comprehension test are most successful. What score on the college entrance examination should be used as the cutoff point for selection?
f) Arthur obtained a score of 50 on the college entrance examination. Would he be selected by REC University? What are his chances of achieving an acceptable score on the verbal comprehension test?
g) Dianne obtained a score of 60 on the college entrance examination. Would she be accepted by REC University? How likely is it that she will *not* achieve an acceptable score on the verbal comprehension test?

10.11 On the basis of the obtained data (below) an experimenter asserts that the older a child is the fewer irrelevant responses made in an experimental situation.
a) Determine whether this conclusion is valid.
b) Johnny, age 13, enters the experimental situation:

i) What is the most probable number of irrelevant responses the experimenter would predict for Johnny?
ii) What is the likelihood that he will make no irrelevant responses?

Age	Number irrelevant responses	Age	Number irrelevant responses
2	11	7	12
3	12	9	8
4	10	9	7
4	13	10	3
5	11	11	6
5	9	11	5
6	10	12	5
7	7		

10.12 The per capita gross national product (GNP) is widely recognized as an estimate of the living standard of a nation. It has been claimed that per capita energy consumption is, in turn, a good predictor of per capita GNP. Shown below are the GNP (expressed in dollars per capita) and the per capita energy consumption (expressed in Btu's per capita) of various nations.

Country	Energy consumption (in millions of Btu's)	GNP
India	3.4	55
Ghana	3.0	270
Portugal	7.7	240
Colombia	12.0	290
Greece	12.0	390
Mexico	23.0	310
Japan	30.3	550
USSR	69.0	800
Netherlands	75.0	1100
France	58.0	1390
Norway	67.0	1330
West Germany	90.0	1410
Australia	88.0	1525
United Kingdom	113.0	1400
Canada	131.0	1900
United States	180.0	2900

a) Construct a scatter diagram from the data shown above.
b) Determine the correlation between GNP and energy expenditures.

c) Construct the regression line for predicting per capita GNP from per capita energy consumption.

d) The following nations were not represented in the original sample. Their per capita energy expenditures were: Chile, 21; Ireland, 49; Belgium, 88. Calculate the predicted per capita GNP for each country. Compare the predicted values with the actual values which are, respectively, 400; 630; 1400.

10.13 Referring to Exercise 9.21, predict the number of runs batted in by Shamsky, who scored five home runs. (For the actual number of rbi's, see *The 1972 World Almanac*, p. 911.)

10.14 Referring to Exercise 9.22, predict the cash price on March 8, 1971, for pepper. The cash price for this product was $0.45 on March 7, 1972. (See *The Wall Street Journal*, March 9, 1972, p. 26: 4.)

10.15 Referring to Exercise 9.23, predict the price bid for the following prices asked:
a) Mobile G Sv: 12.75 b) Fifth Dimension: 3.75 c) Zenith Labs: 13.0
(See *The New York Times*, March 2, 1972, p. 64: 1–4.)
d) What is the error of estimate for predicting the price asked?

10.16 Referring to Exercises 5.25 and 5.26 (and Exercise 9.19), suppose the manager hires three more people. Predict the total number of units they will complete per day if their scores on the math test were as follows:
a) 65 b) 15 c) 30

10.17 Referring to Exercise 9.20 and Exercise 8.20, predict the number of sodas sold if the temperature were:
a) 85 degrees on August 31.
b) 40 degrees on October 1.
c) 50 degrees on October 2.
d) Assuming the Y-variable to be the number of sodas sold, calculate the error of estimate for y.

10.18 Referring to Exercise 10.17, predict the temperature if the number of sodas sold were:
a) 40 on August 29.
b) 45 on August 28.
c) 10 on January 3.
d) Calculate the error of estimate for x.

10.19

X	3	4	5	6	7	8	9	10	11
Y	4	3	5	6	8	7	9	9	11

a) Determine the correlation for the above scores.
b) Given the above correlation, and the above mean for Y, predict the value of y for each X.
c) Calculate $\sum(Y - \bar{Y})^2$.
d) Calculate the unexplained variance.
e) Calculate the explained variance.
f) Calculate the coefficient of determination. Show that the square root of that value equals r.

10.20 A manager of a catering service found a correlation of 0.70 between the number of people at a party and the loaves of bread consumed. (See table on page 197.)
a) For a party of 60 people, calculate the predicted number of loaves needed.
b) For a party of 35, calculate the predicted number of loaves needed.
c) What is the $s_{\text{est } y}$?

Number of people	Number of loaves
$\bar{X} = 50$	$\bar{Y} = 5$
$s_x = 15$	$s_y = 1.2$

10.21 We have randomly selected 19 automobiles from the list in Box 1.1. Shown below are their weights and city and highway mileage.

a) Determine the correlation between weight and city mileage.

b) Using the regression formula for predicting Y from X, predict the city mpg from the weights of the following automobiles: Hornet, 2975; Monaco Wagon, 5035; Corolla, 2225. Compare the predicted values with the actual values which are, respectively, 16, 10, and 20.

c) Determine the correlation between weight and highway mileage and predict highway mpg for the cars listed in part (b). The actual values are, respectively, 26, 14, and 35.

	X Weight	Y_1 City	Y_2 Highway
Gremlin	2831	16	26
Fox	2086	24	36
Century Regal	3837	16	22
Cadillac	5025	11	15
Nova	3188	15	21
Corvette	3445	13	16
Datsun 710 Wagon	2600	21	30
Coronet Wagon	4455	11	17
Fiat 124 Sport	2370	19	31
Mercury	4637	11	16
Cutlass Wagon	4449	12	16
Valiant/Duster	3142	16	23
Astre	2439	19	30
Pontiac	4266	13	17
Subaru	1985	24	34
Corona Mark II	2845	16	22
Volvo 245 Wagon	3180	18	28
Ford Pickup	3587	16	23
Jeep	2945	13	18

Review of Section I: Descriptive Statistics

DESCRIPTIVE STATISTICS

In the preceding 10 chapters you have seen that there are many different ways to describe data. The application of the various descriptive techniques on the following two sets of data will give you an opportunity to see how these techniques can be integrated to describe data.

Review Problem 1

Presented in the following table are the number of hits and home runs obtained during the 1975 season by players from the Cincinnati Reds, winners of the 1975 World Series.

a) Draw a *scatter diagram* to illustrate the relationship between hits and home runs.
b) Calculate:
   i) the *mean* number of home runs.
   ii) the *median* number of hits.
   iii) the *standard deviation* of the distribution of hits.

Player	Batting average rank	Hits	Home runs
Morgan	1	163	17
Rose	2	210	7
Griffey	3	141	4
Foster	4	139	23
Bench	5	150	28
Perez	6	144	20
Driessen	7	59	7
Concepcion	8	139	5
Flynn	9	34	1
Geronimo	10	129	6
Rettenmund	11	45	2
Chaney	12	35	2
Plummer	13	29	1

c) Determine the relationship between:
   i) the number of hits and home runs.
   ii) the number of home runs and batting average rank.

Review Problem 2*

a) Draw a graph to illustrate the changes from 1946 to 1963 in average daily cost per hospital patient.
b) Draw a frequency distribution of the average length of hospital stay from the data presented below. Describe the distribution.
c) For the 18-year period from 1946 to 1963, calculate the mean and standard deviation of the average length of hospital stay.
d) Determine the relationship between the average cost per patient per hospital day and the average length of hospital stay.

	Average cost per patient per hospital day	Average length of hospital stay (in days)		Average cost per patient per hospital day	Average length of hospital stay (in days)
1946	$ 9.39	9.1	1955	$23.12	7.8
1947	11.09	8.0	1956	24.15	7.7
1948	13.09	8.7	1957	26.02	7.6
1949	14.33	8.3	1958	28.27	7.6
1950	15.62	8.1	1959	30.19	7.8
1951	16.77	8.3	1960	32.23	7.6
1952	18.35	8.1	1961	34.98	7.6
1953	19.95	7.9	1962	36.83	7.6
1954	21.76	7.8	1963	38.91	7.7

i) A person who spends one week (7 days) in a hospital can expect his stay to average how much daily?
ii) On the basis of these calculations, patient X concludes that the longer he stays in the hospital the less it will cost him per day. He decides that if he can prolong his hospital stay for 10 days he can expect an average daily cost of ______? What is the fallacy in his conclusion? [*Hint*: Criticize this inference in the light of what you know about correlation and causation.]

* The data for this problem were adapted from *Reader's Digest Almanac*, p. 489. New York: Reader's Digest Association, 1966, with permission.

Section II

Inferential Statistics

Part A

PARAMETRIC TESTS OF SIGNIFICANCE

11 Probability

11.1 AN INTRODUCTION TO PROBABILITY THEORY

In the past few chapters, we have been primarily concerned with the exposition of techniques employed by statisticians to describe and present data in the most economical and meaningful form. However, we pointed out in Chapter 1 that the interests of most disciplines employing statistics go beyond the mere description of data. Fundamental to the strategy of any rigorous discipline is the formulation of general statements about populations or the effects of experimental conditions on criterion variables. Thus, as we have already pointed out, the scientist is not usually satisfied to report merely that the arithmetic mean of the drug group tested on variable X is higher or lower than the mean of the placebo group tested on this variable. He or she also wants to make general statements such as: "The difference between the two groups is of such magnitude that we cannot reasonably ascribe it to chance variation. We may therefore conclude that the drug had an effect on the variable studied. More specifically, this effect was . . . etc., etc."

The problem of chance variation is an important one. We all know that the variability of our data in the behavioral and social sciences engenders the risk of drawing an incorrect conclusion. Take a look at the following example: From casual observations, Experimenter A hypothesizes that first grade girls have higher I.Q. scores than first grade boys. In an I.Q. test administered to four boys and four girls in a first grade class, the mean of the girls was found to be higher: 110 to 103. Is Experimenter A justified in concluding that the hypothesis has been confirmed? The answer is obviously in the negative. But why? After all, there is a difference between the sample means, isn't there? (See Fig. 11.1.) Intuitively, we might argue that the variability of intelligence among first graders is so great, and the N in the study so small, that *some* differences in the means are inevitable as a result of our selection procedures. The critical questions that must be answered by inferential statistics then become: "Is the apparent difference in intelligence among first graders reliable? That is, will it appear regularly in repetitions of the study? Or is the difference the result of unsystematic factors which will vary from study to study, and thereby

© 1957 United Feature Syndicate, Inc.

Fig. 11.1 Is Lucy justified in her conclusion?

produce sets of differences without consistency?" A prime function of inferential statistics is to provide rigorous and logically sound procedures for answering these questions. As we shall see in this chapter and the next, probability theory provides the logical basis for deciding among all the various alternative interpretations of research data.

Probability theory is not as unfamiliar as many would think. Indeed, in everyday life we are constantly called upon to make probability judgments although we may not recognize them as such.

For example, let us suppose that, for various reasons, you are unprepared for today's class. You seriously consider not attending class. What are the factors that will influence your decision? Obviously, one consideration would be the likelihood that the instructor will detect your lack of preparation. If the risk is high, you decide not to attend class; if low, then you will attend.

Let us look at this example in slightly different terms. There are two alternative possibilities

event A: Your lack of preparation *will* be detected.

event B: Your lack of preparation *will not* be detected.

There is uncertainty in this situation because more than one alternative is possible. Your decision whether or not to attend class will depend upon the degree of assurance you associate with each of these alternatives. Thus, if you are fairly certain that the first alternative will prevail, you will decide not to attend class.

Suppose that your instructor frequently calls upon students to participate in class discussion. In fact, you have noted that most of the students are called upon in any given class session. This is an example of a situation in which a high degree of assurance is associated with the first alternative. Stated another way, the probability of event A is higher than the probability of event B. Thus you decide not to attend class.

Although you have not used any formal probability laws in this example, you have actually made a judgment based upon an intuitive use of probability.

You may be aware of the fact that many of the questions raised in the exercises began with, "What is the likelihood that . . .?" These questions were in preparation for the formal discussion of probability occurring in the present and subsequent chapters. However, before discussing the elements of probability theory, it is desirable to understand one of the most important concepts in inferential statistics, that of *randomness.*

11.2 THE CONCEPT OF RANDOMNESS

You will recall that, when discussing the role of inferential statistics, we pointed out the fact that population parameters are rarely known or knowable. It is for this reason that we are usually forced to draw samples from a given population and estimate the parameters from the sample statistics. Obviously, we want to select these samples in such a way that they are representative of the populations from which they are drawn. One way to achieve representativeness is to employ simple

random sampling: *selecting samples in such a way that each sample of a given size has precisely the same probability of being selected* or, alternatively, *selecting the events in the sample such that each event is equally likely to be selected in a sample of a given size.*

Consider selecting samples of $N = 2$ from a population of five numbers: 0, 1, 2, 3, 4. If, for any reason, any number is more likely to be drawn than any other number, each sample would *not* have an equal probability of being drawn. For example, if for any reason the number 3 were twice as likely to be drawn as any other number, there would be a preponderance of samples containing the value 3. Such sampling procedures are referred to as being **biased**. In the naturalistic type of study alluded to earlier, in which our purpose is to describe certain characteristics of a population, the problem of bias is an ever-present danger. When we are interested in learning the characteristics of the general population on a given variable, we dare not select our sample from automobile registration lists or "at random" on a street corner in New York City. The dangers of generalizing to the general population from such biased samples should be obvious to you. Unless the condition of randomness is met, we may never know to what population we should generalize our results. Furthermore, with nonrandom samples, we find that many of the rules of probability do not hold.

Moreover, the statistical tests presented in this text require *independent random sampling.* Two events are said to be **independent** if the selection of one has no effect upon the probability of selecting the other event. We can most readily grasp independence in terms of games of chance, assuming they are played honestly. Knowledge of the results of one toss of a coin, one throw of a die, one outcome of the roulette wheel, or one selection of a card from a well-shuffled deck (assuming replacement of the card after each selection) will not aid us one iota in our predictions of future outcomes.

It is beyond the scope of this text to delve deeply into sampling procedures, since that topic is a full course by itself. However, let us look at an illustration of the procedures by which we may achieve randomness in assigning subjects to experimental conditions.

Let us suppose that you are interested in comparing three different methods of teaching reading readiness to preschool children. There are 87 subjects, who are to be divided into three equal groups. The assignment of these subjects must be made in a random manner.

One method to achieve randomness would be to place each subject's name on a slip of paper. We shuffle these slips and then place them into three equal piles.

An alternative method would be to use the Table of Random Digits (Table R in Appendix III). Since the digits in this table have already been randomized, the effect of shuffling has been achieved. We assign to each of the 87 students a numeral from 1 to 87. We may start with any row or column of digits in Table R.

After we have selected 29 numerals that correspond to 29 different subjects, we have formed our first group. We continue until we have three groups, each consisting of 29 different subjects. For example, if we start with the fifteenth row and then choose consecutive pairs of digits, we obtain the following subjects: 65, 48,

11, 76, 74, 17, etc. If any numeral over 87 or any repeated numeral appears, we disregard it.

The reason for the paramount importance of random procedures will become clear in this chapter and the next. Fundamentally, it is based on a fascinating fact of inferential statistics; i.e., *each event may not be predictable when taken alone, but collections of random events can take on predictable forms.* The binomial distribution, which we shall discuss at greater length in Section 12.3, illustrates this fact. If we were to take, say, 20 unbiased coins and toss them into the air, we could not predict accurately the proportion that would land "heads." However, if we were to toss these 20 coins a great many times, record the number turning up heads on each trial, and construct a frequency distribution of outcomes, in which the horizontal axis varies between no heads and all heads, the plot would take on a characteristic and predictable form known as the *binomial distribution.* By employing the binomial model, we would be able to predict with considerable accuracy, over a large number of trials, the percentage of the time various outcomes will occur. The same is true with respect to the normal curve model. In the absence of any specific information, we might not be able to predict a person's status with respect to a given trait (intelligence, height, weight, etc.). However, as we already know, frequency distributions of scores on these traits commonly take the form of the normal curve. Thus we may predict the proportion of individuals scoring between specified score limits.

What is perhaps of more importance, from the point of view of inferential statistics, is the fact that distributions of sample statistics ($\bar{X}$, s, median, etc.), based on random sampling from a population, also take on highly predictable forms. Chapter 12 deals with the concept of **sampling distributions**, which are theoretical probability distributions of a statistic which would result from drawing all possible samples of a given size from some population.

With this brief introduction to the concept of randomness, we are prepared to look at probability theory.

11.3 APPROACHES TO PROBABILITY

Probability may be regarded as a theory that is concerned with the possible outcomes of experiments. The experiments must be potentially repetitive; i.e., we must be able to repeat them under similar conditions. It must be possible to enumerate every outcome that can occur, and we must be able to state the expected relative frequencies of these outcomes.

It is the method of assigning relative frequencies to each of the possible outcomes that distinguishes the classical from the empirical approach to probability theory.

11.3.1 Classical Approach to Probability

The theory of probability has always been closely associated with games of chance. For example, suppose that we want to know the probability that a coin will turn up heads. Since there are only two possible outcomes (heads or tails) we assume an ideal situation in which we expect that each outcome is equally likely to occur. Thus the probability that heads, $p(H)$, will occur is $\frac{1}{2}$. This kind of reasoning has led to the following classical definition of probability:

$$p(A) = \frac{\text{number of outcomes favoring event } A}{\text{total number of events (those favoring } A + \text{ those not favoring } A)}. \qquad (11.1)$$

It should be noted that probability is defined as a proportion (p). The most important point in the classical definition of probability is the assumption of an *ideal* situation in which the structure of the population is known; i.e., the total number of possible outcomes (N) is known. The expected relative frequency of each of these outcomes is arrived at by deductive reasoning. Thus the probability of an event is interpreted as a theoretical (or an idealized) relative frequency of the event. In the above example, the total number of possible outcomes was 2 (heads or tails), and the relative frequency of each outcome was assumed to have an equal likelihood of occurrence.* Thus, $p(H) = \frac{1}{2}$ and $p(T) = \frac{1}{2}$.

11.3.2 Empirical Approach to Probability

Although it is usually easy to assign expected relative frequencies to the possible outcomes of games of chance, we cannot do this for most real-life experiments. In actual situations, expected relative frequencies are assigned on the basis of empirical findings. Thus we may not know the exact proportion of students in a university who have blue eyes, but we may study a random sample of students and estimate the proportion who will have blue eyes. Once we have arrived at an estimate, we may employ classical probability theory to answer questions such as: What is the probability that in a sample of 10 students, drawn at random from the student body, three or more will be blue-eyed? Or, what is the probability that student John, drawn at random from the student body, will have blue eyes?

If, in a random sample of 100 students, we found that 30 had blue eyes, we could estimate that the proportion of blue-eyed students in the university was 0.30 by employing formula (11.1):

$$p(\text{blue-eyed}) = \tfrac{30}{100} = 0.30.$$

Thus the probability is 0.30 that student John will have blue eyes. *Note*: This represents an *empirical* probability; i.e., the expected relative frequency was assigned on the basis of empirical findings.

Although we employ an idealized model in our forthcoming discussion about the properties of probability, we can apply the same principles to divers practical problems.

* Expected relative frequencies need not be assumed to be equal. In Chapter 18 we will deal with situations in which the expected relative frequencies are not assumed to be equal.

11.4 FORMAL PROPERTIES OF PROBABILITY

11.4.1 Probabilities Vary between 0 and 1.00

From the classical definition of probability, p is always a number between 0 and 1, inclusively. If an event is certain to occur, its probability is 1; if it is certain not to occur, its probability is 0. For example, the probability of drawing the ace of spades from an ordinary deck of 52 playing cards is $\frac{1}{52}$. The probability of drawing a red ace of spades is zero since there are no events favoring this result. If all events favor a result (for example, drawing a card with *some* marking on it), $p = 1$. Thus for any given event, say A, $0 \leq p(A) \leq 1.00$, in which the symbol $\leq$ means "less than or equal to."*

11.4.2 Expressing Probability

Probability can be expressed in several other ways besides as a proportion. It is sometimes convenient to express probability as a *percentage* or as the *number of chances in* 100.

To illustrate: If the probability of an event is 0.05, we expect this event to occur 5% of the time, or *the chances that this event will occur* are 5 in 100. This same probability may be expressed by saying that the odds are 95 to 5 *against* the event occurring or 19 to 1 against it.

Note that, when expressing probability as the *odds against* the occurrence of an event, we use the following formula:

$$\text{Odds against event } A = (\text{total number of outcomes} - \text{number favoring event } A) \text{ \textit{to} number favoring event } A. \quad (11.2)$$

Thus if $p(A) = 0.01$, the *odds against* the occurrence of event A are 99 to 1.

11.4.3 The Addition Rule

In the end of chapter exercises for Chapters 4, 5, 6 and 7, we presented a sampling experiment in which we drew all possible samples of $n = 2$ from a population of seven numbers. The purpose of these exercises was to lay the groundwork for a discussion of probability theory and, later, hypothesis testing.

Let us briefly review this experiment. Recall that we started with a population of the following scores: 0, 1, 2, 3, 4, 5, 6. We selected a score from a hat, recorded it, returned it to the hat, and randomly selected another score, added it to the first, and obtained a mean by dividing the sum by 2. After constructing a table of all possible draws of $n = 2$ from this population, we obtained the frequency distribution of 49 possible means shown in Table 11.1.

* The symbol $\geq$ means "greater than or equal to."

Table 11.1

Frequency and probability distribution of means of samples of size $n = 2$, drawn from a population of seven scores (0, 1, 2, 3, 4, 5, 6)

$\bar{X}$	f	$p(\bar{X})$	$\bar{X}$	f	$p(\bar{X})$
6.0	1	0.0204	2.5	6	0.1224
5.5	2	0.0408	2.0	5	0.1020
5.0	3	0.0612	1.5	4	0.0816
4.5	4	0.0816	1.0	3	0.0612
4.0	5	0.1020	0.5	2	0.0408
3.5	6	0.1224	0.0	1	0.0204
3.0	7	0.1429			
				$N_x = 49$*	$\sum p(\bar{X}) = 0.9997$†

* We use $N_{\bar{X}}$ to refer to the total number of means obtained in our sampling experiment and to distinguish it from n, which is the number of scores upon which each mean is based.
† $\sum p(\bar{X})$ should equal 1.000. The discrepancy of 0.0003 represents rounding error.

By dividing each frequency by $N_{\bar{X}}$, we obtain a probability distribution of means of sample size $n = 2$. Note that we are now in a position to answer such questions as: In a single draw of a mean when $n = 2$, what is the probability of obtaining a mean equal to zero, equal to 2, or equal to 5.5? The answers are, respectively, 0.0204, 0.1020, and 0.0408.

In inferential statistics, we often want to determine the probability of one of several different events. For example, we may be interested in answering such questions as: If we draw a single sample of $n = 2$, what is the probability of obtaining a mean of 5 or greater? To answer this question, we simply add together the separate probabilities of the events that are included in the statement "5 or greater," *as long as the separate events are* **mutually exclusive** (i.e., two or more cannot occur simultaneously). In the present example, these events are means of 5.0, 5.5, and 6.0. Note that these events are mutually exclusive. If you obtain a mean of 5.5 on a single draw, it cannot have any other value.

The probability of obtaining a mean equal to or greater than five becomes $p(\bar{X} = 5) + p(\bar{X} = 5.5) + p(\bar{X} = 6.0) = 0.0612 + 0.0408 + 0.0204 = 0.1224$.

Note that we can raise and answer such additional questions as: In a single sample of $n = 2$, what is the probability of obtaining a mean equal to or less than 2.0? Shown symbolically:

$$\begin{aligned} p(\bar{X} \leq 2.0) &= p(\bar{X} = 2) + p(\bar{X} = 1.5) + p(\bar{X} = 1.0) + p(\bar{X} = 0.5) + p(\bar{X} = 0) \\ &= 0.1020 + 0.0816 + 0.0612 + 0.0408 + 0.0204 \\ &= 0.3060 \end{aligned}$$

We can also ask, "What is the probability of selecting a mean between 3.5 and 5.5, inclusive?" We simply add together the separate probabilities of the following events:

$$p(\bar{X} = 3.5) + p(\bar{X} = 4.0) + p(\bar{X} = 4.5) + p(\bar{X} = 5.0) + p(\bar{X} = 5.5)$$
$$= 0.1224 + 0.1020 + 0.0816 + 0.0612 + 0.0408$$
$$= 0.4080.$$

These considerations lead us to the formulation of the **addition rule** for mutually exclusive events.

If A and B are mutually exclusive events, the probability of obtaining either of them is equal to the probability of A plus the probability of B.

In symbolic form, this reads:

$$p(A \text{ or } B) = p(A) + p(B). \tag{11.3}$$

This formula can be extended to include any number of mutually exclusive events. Thus

$$p(A \text{ or } B \text{ or} \ldots Z) = p(A) + p(B) + \cdots + p(Z).$$

In Chapters 12 and 18, we shall be dealing with problems based on *dichotomous, Yes–No,* or *two-category,* populations, in which the events in question are not only mutually exclusive, but are **exhaustive**. For example, in the sampling experiment with which we have been working, what is the probability of drawing either a mean equal to or greater than 5 or a mean less than 5? Not only is it impossible to obtain both events simultaneously (i.e., they are mutually exclusive) but there is *no possible outcome other than* a mean equal to or greater than 5 or a mean less than 5. In the case of *mutually exclusive* and *exhaustive events,* we arrive at the very useful formulation:

$$p(A) + p(B) = 1.00. \tag{11.4}$$

In treating dichotomous populations, it is common practice to employ the two symbols P and Q to represent, respectively, the probability of the occurrence of an event and the probability of the nonoccurrence of an event. Thus if we are flipping a single coin, we can let P represent the probability of occurrence of a head and Q the probability of the nonoccurrence of a head (i.e., the occurrence of a tail). These considerations lead to three useful formulations:

$$P + Q = 1.00, \tag{11.5}$$

$$P = 1.00 - Q, \tag{11.6}$$

$$Q = 1.00 - P, \tag{11.7}$$

when the events are *mutually exclusive and exhaustive.*

11.4.4 The Multiplication Rule

In the preceding section, we were concerned with determining the probability of obtaining one event or another based upon a *single* draw (or trial) from a set of 49 means. In statistical inference, we are often faced with the problem of ascertaining the probability of the joint or successive occurrence of two or more events when more than one draw or trial is involved.

Sampling with replacement: Let us return to the sampling experiment. Recall that, in obtaining the sample means based upon $n = 2$, we selected a score from the hat, recorded it, returned it to the hat, and randomly selected a second score. Imagine that, on each draw, you obtain a score equal to zero. You ask, "What is the probability that, by chance, you would obtain a mean equal to zero on two successive draws from the population of seven scores?" Actually, we have previously obtained this answer when we constructed Table 11.1 (Frequency and probability distribution of means of samples of size $n = 2$, drawn from a population of seven scores). We found a probability equal to 0.0204.

We may obtain the same result by an alternative method—the application of the **multiplication rule**. This rule states:

The probability of the simultaneous or successive occurrence of two events is the product of the separate probabilities of each event.

In symbolic form:

$$p(A \text{ and } B) = p(A)p(B). \tag{11.8}$$

Since there are seven scores, the probability that you will obtain a 0 on the first draw is $p(A) = \frac{1}{7}$.

The probability that you will obtain a 0 on the second draw, $p(B)$, is also $\frac{1}{7}$. Thus

$$\begin{aligned} p(A \text{ and } B) &= (\tfrac{1}{7})(\tfrac{1}{7}) \\ &= \tfrac{1}{49} = 0.0204. \end{aligned}$$

In the above example, the occurrence of A is not dependent upon the occurrence of B, and vice versa. The events are said to be *independent*. It is only when events are independent that the simple multiplication rule shown in formula (11.8) applies. The reason that the events are independent is that we sampled *with replacement*. Recall that we returned each score to the hat after each draw. As long as we thoroughly mix the scores after each draw, the outcome of the first draw can have no effect upon the outcome of the second draw.

Before proceeding to the case where the events are nonindependent, let us look at two additional examples.

Example 1 What is the probability of obtaining a 0 on the first draw (event A) and a 6 on the second draw (event B)?

$$p(A) = \tfrac{1}{7}; \qquad p(B) = \tfrac{1}{7};$$

$$\begin{aligned} p(A \text{ and } B) &= (\tfrac{1}{7})(\tfrac{1}{7}) \\ &= \tfrac{1}{49} = 0.0204. \end{aligned}$$

Example 2 What is the probability of obtaining a mean equal to 3 on two successive draws from the population of seven numbers? The answer to this question is not as straightforward, because there are a number of different ways to obtain a mean of 3. Listed below are the various ways in which a mean of 3 may be obtained.

First draw		Second draw	$p(A \text{ and } B)$
0	and	6	0.0204
1	and	5	0.0204
2	and	4	0.0204
3	and	3	0.0204
4	and	2	0.0204
5	and	1	0.0204
6	and	0	0.0204

To answer this question, we must combine the multiplication and addition rules. We ask the question, "What is the probability of obtaining 0 and 6 *or* 1 and 5 *or* 2 and 4, etc.?"

$$\begin{aligned} p(\bar{X} = 3) &= p(0 + 6) + p(1 + 5) + p(2 + 4) + p(3 + 3) + p(4 + 2) \\ &\quad + p(5 + 1) + p(6 + 0) \\ &= (\tfrac{1}{7})(\tfrac{1}{7}) + (\tfrac{1}{7})(\tfrac{1}{7}) + (\tfrac{1}{7})(\tfrac{1}{7}) + (\tfrac{1}{7})(\tfrac{1}{7}) + (\tfrac{1}{7})(\tfrac{1}{7}) + (\tfrac{1}{7})(\tfrac{1}{7}) + (\tfrac{1}{7})(\tfrac{1}{7}) \\ &= \tfrac{1}{49} + \tfrac{1}{49} + \tfrac{1}{49} + \tfrac{1}{49} + \tfrac{1}{49} + \tfrac{1}{49} + \tfrac{1}{49} \\ &= \tfrac{7}{49} = 0.1429. \end{aligned}$$

Note that this answer agrees with the probability of obtaining a mean equal to 3 shown in Table 11.1.

Sampling without replacement: Let's contrast sampling with replacement with a technique known as sampling *without* replacement. Imagine that, after selecting the first score, we do not return it to the hat. It would then be impossible to obtain the same score in the second draw. Indeed, it would now be impossible to obtain a mean equal to 6 or a mean equal to 0. Why?

When sampling without replacement, then, the results of the first draw influence the possible outcomes of the second draw. The events are said to be *nonindependent*.

For related or nonindependent events the multiplication rule becomes:

Given two events A and B, the probability of obtaining both A and B jointly is the product of the probability of obtaining one of these events times the conditional probability of obtaining one event, given that the other event has occurred.

Stated symbolically,

$$p(A \text{ and } B) = p(A)p(B|A) = p(B)p(A|B). \tag{11.9}$$

The symbols $p(B|A)$ and $p(A|B)$ are referred to as *conditional probabilities.* The symbol $p(B|A)$ means *the probability of B given that A has occurred.* The term *conditional probability* takes into account that the probability of B may depend on whether or not A has occurred, and conversely for $p(A|B)$. [*Note*: Either of the two events may be designated A or B since the symbols merely represent a convenient language for discussion, and there is no time order implied in the way they occur.]

If we were to sample, without replacement, from the population of seven scores, we would obtain the means shown in Table 11.2 for $n = 2$.

Table 11.2

All possible means that may be obtained by drawing samples of $n = 2$, without replacement, from a population of 7 scores

Second draw	First draw						
	0	1	2	3	4	5	6
0	—*	0.5	1.0	1.5	2.0	2.5	3.0
1	0.5	—	1.5	2.0	2.5	3.0	3.5
2	1.0	1.5	—	2.5	3.0	3.5	4.0
3	1.5	2.0	2.5	—	3.5	4.0	4.5
4	2.0	2.5	3.0	3.5	—	4.5	5.0
5	2.5	3.0	3.5	4.0	4.5	—	5.5
6	3.0	3.5	4.5	4.5	5.0	5.5	—

* The dashes in the diagonal cells indicate that these events cannot occur when sampling without replacement.

From Table 11.2, we may construct a frequency and probability distribution of means for sample size $n = 2$ (see Table 11.3).

Note that a mean of 0 has an associated probability of zero. This is due to the fact that, given the selection of 0 on the first draw, it cannot be selected again on the second draw. Stated symbolically $p(B|A) = 0$. Thus the probability of obtaining a mean of 0 is:

$$\begin{aligned} p(\bar{X} = 0) &= p(A)p(B|A) \\ &= (\tfrac{1}{7})(0) = 0. \end{aligned}$$

Table 11.3

Frequency and probability distribution of means of sample size $n = 2$, drawn, without replacement, from a population of seven scores.

$\bar{X}$	f	$p(\bar{X})$
6	0	0.0000
5.5	2	0.0476
5.0	2	0.0476
4.5	4	0.0952
4.0	4	0.0952
3.5	6	0.1429
3.0	6	0.1429
2.5	6	0.1429
2.0	4	0.0952
1.5	4	0.0952
1.0	2	0.0476
0.5	2	0.0476
0.0	0	0.0000
	$N_x = 42$	$\sum p(\bar{X}) = 0.9999$

The same is true concerning the probability of obtaining a mean of 6 when drawing, without replacement, from the population of seven scores. Why?

There is a second important feature of sampling without replacement: once one score is selected on the first draw, the probability becomes greater that each of the remaining scores will be selected on the second draw. In the population of seven numbers, the probability is $\frac{1}{7}$ that any given number will be selected on the first draw. However, once a number is selected, the probability becomes $\frac{1}{6}$ that each of the remaining scores will be selected on the second draw.

Let us look at two examples involving the multiplication rule when sampling without replacement from the population of seven scores.

Example 1 What is the probability of selecting in succession, scores of 0 and 2?

Let event A be the selection of a 0 and event B the selection of a 2:

$$p(A) = \tfrac{1}{7};$$

$$p(B) = p(B|A) = \tfrac{1}{6};$$

$$\begin{aligned} p(A \text{ and } B) &= (\tfrac{1}{7})(\tfrac{1}{6}) \\ &= \tfrac{1}{42} \\ &= 0.0238. \end{aligned}$$

BOX 11.1

The following article from *Time* shows how the distinction between independent events played a critical role in a legal decision.

Trial by Mathematics*

After an elderly woman was mugged in an alley in San Pedro, Calif., a witness saw a blonde girl with a ponytail run from the alley and jump into a yellow car driven by a bearded Negro. Eventually tried for the crime, Janet and Malcolm Collins were faced with the circumstantial evidence that she was white, blonde and wore a ponytail while her Negro husband owned a yellow car and wore a beard. The prosecution, impressed by the unusual nature and number of matching details, sought to persuade the jury by invoking a law rarely used in a courtroom—the mathematical law of statistical probability.

The jury was indeed persuaded, and ultimately convicted the Collinses (TIME, Jan. 8, 1965). Small wonder. With the help of an expert witness from the mathematics department of a nearby college, the prosecutor explained that the probability of a set of events actually occurring is determined by multiplying together the probabilities of each of the events. Using what he considered "conservative" estimates (for example, that the chances of a car's being yellow were 1 in 10, the chances of a couple in a car being interracial 1 in 1,000), the prosecutor multiplied all the factors together and concluded that the odds were 1 in 12 million that any other couple shared the characteristics of the defendants.

Only One Couple. The logic of it all seemed overwhelming, and few disciplines pay as much homage to logic as do the law and math. But neither works right with the wrong premises. Hearing an appeal of Malcolm Collins' conviction, the California Supreme Court recently turned up some serious defects, including the fact that not even the odds were all they seemed.

To begin with, the prosecution failed to supply evidence that "any of the individual probability factors listed were even roughly accurate." Moreover, the factors were not shown to be fully independent of one another as they must be to satisfy the mathematical law; the factor of a Negro with a beard, for instance, overlaps the possibility that the bearded Negro may be part of an interracial couple. The 12 million to 1 figure, therefore, was just "wild conjecture." In addition, there was not complete agreement among the witnesses about the characteristics in question. "No mathematical equation," added the court, "can prove beyond a reasonable doubt (1) that the guilty couple *in fact* possessed the characteristics described by the witnesses, or even (2) that only *one* couple possessing those distinctive characteristics could be found in the entire Los Angeles area."

* *Time*, April 26, 1968, p. 41. Reprinted by permission from *Time, The Weekly Newsmagazine*; © Time Inc., 1968.

Improbable Probability. To explain why, Judge Rayond Sullivan attached a four-page appendix to his opinion that carried the necessary math far beyond the relatively simple formula of probability. Judge Sullivan was willing to assume it was unlikely that such a couple as the one described existed. But since such a couple did exist—and the Collinses demonstrably did exist—there was a perfectly acceptable mathematical formula for determining the probability that another such couple existed. Using the formula and the prosecution's figure of 12 million, the judge demonstrated to his own satisfaction and that of five concurring justices that there was a 41% chance that at least one other couple in the area might satisfy the requirements.†

"Undoubtedly," said Sullivan, "the jurors were unduly impressed by the mystique of the mathematical demonstration but were unable to assess its relevancy or values." Neither could the defense attorney have been expected to know of the sophisticated rebuttal available to them. Janet Collins is already out of jail, has broken parole and lit out for parts unknown. But Judge Sullivan concluded that Malcolm Collins, who is still in prison at the California Conservation Center, had been subjected to "trial by mathematics" and was entitled to a reversal of his conviction. He could be tried again, but the odds are against it.

† The proof involved is essentially the same as that behind the common parlor trick of betting that in a group of 30 people, at least two will have the same birthday; in that case, the probability is 70%.

Example 2 What is the probability of selecting a mean equal to 1.00?

To answer this question, we combine the multiplication and the addition rules. There are three possible combinations of scores that can theoretically yield a mean equal to 1.00. These are shown below:

First draw		Second draw	$p(A \text{ and } B)$
0	and	2	$(\frac{1}{7})(\frac{1}{6}) = 0.0238$
2	and	0	$(\frac{1}{7})(\frac{1}{6}) = 0.0238$
1	and	1	$(\frac{1}{7})(0) = 0.0000$

Thus

$$\begin{aligned} p(\bar{X}) &= p(0 \text{ and } 2) + p(2 \text{ and } 0) + p(1 \text{ and } 1) \\ &= (\tfrac{1}{7})(\tfrac{1}{6}) + (\tfrac{1}{7})(\tfrac{1}{6}) + (\tfrac{1}{7})(0) \\ &= \tfrac{1}{42} + \tfrac{1}{42} + 0 \\ &= \tfrac{2}{42} = 0.0476. \end{aligned}$$

Note that this value agrees with the one shown in Table 11.3 for a mean equal to 1.00.

It is important to realize that, when selecting from a large population of scores, the difference between sampling with and without replacement becomes trivial. For example, if selecting samples of $n = 2$ from one million scores, the probability of obtaining the same score twice is 0, without replacement, and 0.000000000001 with replacement. Indeed, public opinion polls employ sampling without replacement (it doesn't make much sense to interview the same person twice except for follow-up purposes). Nevertheless, a probability model based upon sampling with replacement may safely be employed, since vast numbers of people usually make up the population of interest.

11.5 PROBABILITY AND CONTINUOUS VARIABLES

Up to this point we have considered probability in terms of the expected relative frequency of an event. In fact, as you will recall, probability was defined in terms of frequency and expressed as the following proportion [formula (11.1)]:

$$p(A) = \frac{\text{number of outcomes favoring event } A}{\text{total number of outcomes}}.$$

However, this definition presents a problem when we are dealing with continuous variables. As we pointed out in Section 4.5.1, it is generally advisable to represent frequency in terms of areas under a curve when we are dealing with continuous variables. Thus, for continuous variables, we may express probability as the following proportion:

$$p = \frac{\text{area under portions of a curve}}{\text{total area under the curve}}. \qquad (11.10)$$

Since the total area in a probability distribution is equal to 1.00, we define p as the proportion of total area under portions of a curve.

Chapters 12 through 16 employ the standard normal curve as the probability model. Let us examine the probability-area relationship in terms of this model.

11.6 PROBABILITY AND THE NORMAL CURVE MODEL

In Section 8.3 we stated that the standard normal distribution has a μ of 0, a σ of 1, and a total area that is equal to 1.00. We saw that when scores on a normally distributed variable are transformed into z-scores, we are, in effect, expressing these scores in units of the standard normal curve. This permits us to express the difference between any two scores as proportions of total area under the curve. Thus we may establish probability values in terms of these proportions as in formula (11.10).

Let us look at several examples which illustrate the application of probability concepts to the normal curve model.

11.6.1 Illustrative Problems*

For all problems, assume $\mu = 100$ and $\sigma = 16$.

Problem 1

What is the probability of selecting at *random*, from the general population, a person with an I.Q. score of at least 132? The answer to this question is given by the proportion of area under the curve above a score of 132 (refer to Fig. 11.2).

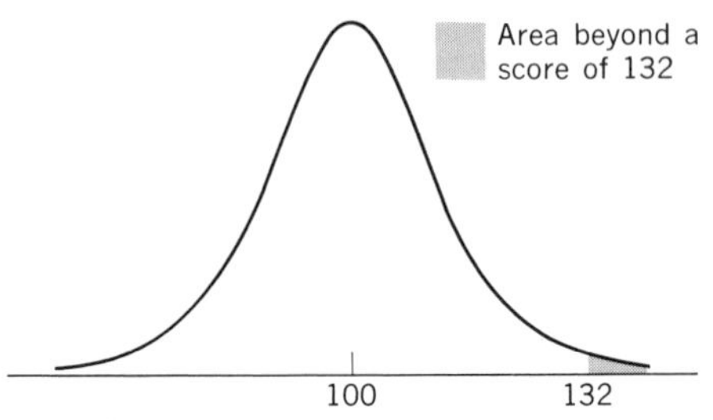

Fig. 11.2 Proportion of area above a score of 132 in a normal distribution with $\mu = 100$ and $\sigma = 16$.

First, we must find the z-score corresponding to $X = 132$:

$$z = \frac{132 - 100}{16} = 2.00.$$

In Column C (Table A), we find that 0.0228 of the area lies at or beyond a z of 2.00. Therefore the probability of selecting, at random, a score of at least 132 is 0.0228.

Problem 2

What is the probability of selecting, at random, an individual with an I.Q. score of at least 92?

We are dealing with two mutually exclusive and exhaustive areas under the curve. The area under the curve above a score of 92 is P; the area below a score of 92 is Q. In solving our problem, we may therefore employ formula (11.6):

$$P = 1.00 - Q.$$

By expressing a score of 92 in terms of its corresponding z, we may obtain the proportion of area below $X = 92$ (that is Q) directly from Column C (Table A).

The z-score corresponding to $X = 92$ is

$$z = \frac{92 - 100}{16} = -0.50.$$

* See Section 8.4.

The proportion of area below a z of -0.50 is 0.3085. Therefore the probability of selecting, at random, a score of at least 92 becomes

$$P = 1.00 - 0.3085 = 0.6915.$$

Figure 11.3 illustrates this relationship.

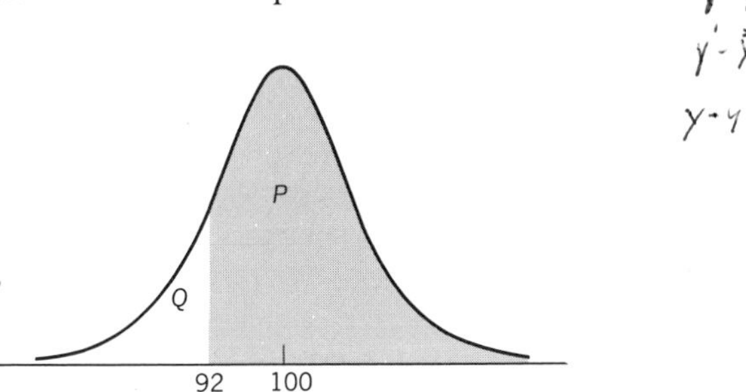

Fig. 11.3 Proportion of area above (P) and below (Q) a score of 92 in a normal distribution with $\mu = 100$ and $\sigma = 16$.

Problem 3

Let us look at an example involving the multiplication law. Given that sampling with replacement is employed, what is the probability of drawing, at random, three individuals with I.Q.'s equaling or exceeding 124? For this problem, we will again assume that $\mu = 100$ and $\sigma = 16$.

The z-score corresponding to $X = 124$ is

$$z = \frac{124 - 100}{16} = 1.5,$$

In Column C (Table A), we find that 0.0668 of the area lies at or beyond $X = 124$. Therefore

$$p(A, B, C) = (0.0668)^3 = 0.0003.$$

11.7 ONE- AND TWO-TAILED *p*-VALUES

In Problem 1 we posed the question: What is the probability of selecting a person with a score as high as 132? We answered the question by examining only one tail of the distribution, namely, scores as high as or higher than 132. For this reason, we refer to the probability value that we obtained as being a **one-tailed *p*-value**.

In statistics and research, the following question is more commonly asked: "What is the probability of obtaining a score (or statistic) this **deviant** from the mean? . . . or a score (or statistic) this **rare** . . . or a result this **unusual**?" Clearly, when the frequency distribution of scores is symmetrical, a score of 68 or lower is every bit as deviant from a mean of 100 as is a score of 132. That is, both are two standard deviation units away from the mean. When we express the probability value, taking into account both tails of the distribution, we call it a **two-tailed *p*-value**. In symmetrical distributions, two-tailed *p*-values may be obtained merely by doubling the one-tailed probability value. Thus in the preceding problem, the probability of selecting a person with a score as *rare* or *unusual* as 132 is $2 \times 0.0228 = 0.0456$.

We may illustrate the distinction between one- and two-tailed *p*-values by referring to the sampling experiment we have used throughout the text. Figure 11.4 shows a probability histogram based upon Table 11.1. Recall that this distribution was obtained by selecting, with replacement, all possible samples of $n = 2$ from a population of seven scores.

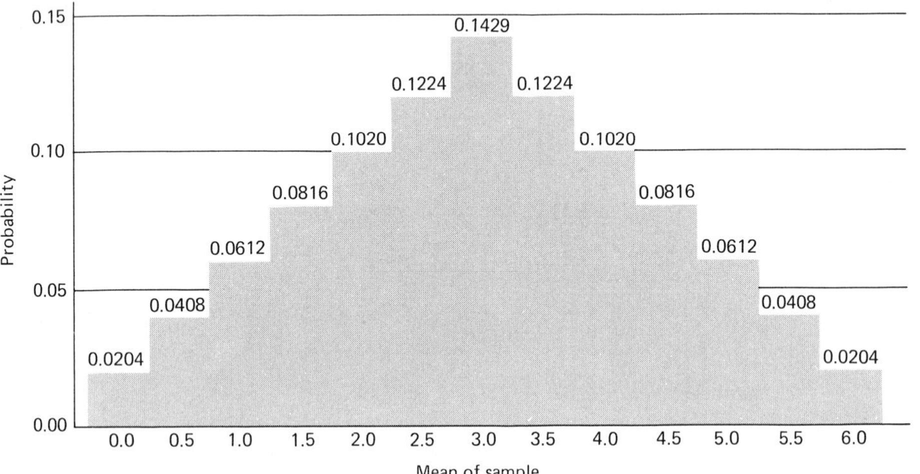

Fig. 11.4 Probability histogram of means based upon selecting, with replacement, samples of $n = 2$ from a population of seven numbers.

If we asked "What is the probability of obtaining a mean equal to zero?", we need refer only to the left tail of the probability histogram in order to find the answer of 0.0204. However, if we asked, "What is the probability of obtaining as deviant an outcome as a sample mean equal to zero?", we would have to look at both the left and the right tails. Since a mean of 6 is equally deviant from the mean of the distribution, we use the addition rule to obtain the two-tailed value:

$$P(\overline{X} = 0 \text{ or } \overline{X} = 6) = p(\overline{X} = 0) + p(\overline{X} = 6)$$
$$= 0.0204 + 0.0204 = 0.0408.$$

Let's look at two additional examples.

Example 1 What is the probability of obtaining a mean as low as 0.5 (i.e., a mean of 0.5 or lower)?

We find the probability of obtaining a mean of 0.5 and a mean of 0.0 (the only mean lower than 0.5) and add these probabilities together to obtain the one-tailed p-value:

$$\begin{aligned} p(\bar{X} \leq 0.5) &= p(\bar{X} = 0.5) + p(\bar{X} = 0.0) \\ &= 0.0408 + 0.0204 = 0.0612. \end{aligned}$$

Example 2 What is the probability of obtaining a mean as deviant from the distribution mean as a sample mean of 0.5?

The answer to this question calls for a two-tailed p-value obtained by applying the addition rule to

$$p(\bar{X} = 0.5) + p(\bar{X} = 0.0) + p(\bar{X} = 5.5) + p(\bar{X} = 6.0).$$

However, since the distribution is symmetrical, and means of 5.5 and 6.0 are equally as deviant as 0.5 and 0.0, respectively, we need only double the p-value obtained in Example 1:

$$\begin{aligned} p(0.5 \geq \bar{X} \geq 5.5) &= 2(0.0612) \\ &= 0.1224. \end{aligned}$$

Incidentally, the left-hand member of the above expression is read: *The probability of a mean equal to or less than* 0.5 *or equal to or greater than* 5.5.

The distinction between one- and two-tailed probability values takes on added significance as we progress into inferential statistics.

CHAPTER SUMMARY

In this chapter we discussed:

1. The importance of the concept of *randomness* in inferential statistics. Randomness refers to selecting the events in the sample such that each event is equally likely to be selected in a sample of a given size. Independent random sampling refers to the fact that the selection of one event has no effect upon the probability of selecting another event. Although the individual events are unpredictable, collections of random events take on characteristic and predictable forms. The binomial distribution and the normal curve were cited in this regard.

2. The theory of probability which is concerned with the outcomes of experiments. We can distinguish between probabilities established by assuming *idealized* relative frequencies, and those established empirically by determining relative frequencies. Probability was defined as:

$$p = \frac{\text{number of outcomes favoring event}}{\text{total number of outcomes}}.$$

3. The formal properties of probability:

   a) *Probabilities vary between* 0 *and* 1.00.

b) *The addition rule*:
If A and B are mutually exclusive events, the probability of obtaining either of them is equal to the probability of A plus the probability of B. Thus

$$p(A \text{ or } B) = p(A) + p(B).$$

If the two events are *mutually exclusive and exhaustive*, we obtain:

$$p(A) + p(B) = 1.00.$$

Allowing P to represent the probability of occurrence and Q to represent the probability of nonoccurrence, we find that three additional useful formulations for mutually exclusive and exhaustive events are:

$$P + Q = 1.00, \qquad P = 1.00 - Q, \qquad Q = 1.00 - P.$$

c) *The multiplication rule*:
When events are independent: When sampling with replacement, the selection on one trial is independent of the selection on another trial. Given two events, A and B, the probability of obtaining both A and B in successive trials is the product of the probability of obtaining one of these events times the probability of obtaining the second of these events:

$$p(A \text{ and } B) = p(A)p(B).$$

When events are nonindependent: When sampling without replacement, the selection of one event affects the probability of selecting each remaining event. Thus, given two events, A and B, the probability of obtaining both A and B jointly or successively is the product of the probability of obtaining one of the events times the conditional probability of obtaining one event, *given that* the other event has occurred. Symbolically:

$$p(A \text{ and } B) = p(A)p(B|A) = p(B)p(A|B).$$

4. In the application of probability theory to continuously distributed variables, probability is expressed in terms of the proportion of area under a curve. Hence

$$p = \frac{\text{area under portions of a curve}}{\text{total area under a curve}}.$$

We saw how we may employ z-scores and the standard normal curve to establish various probabilities for normally distributed variables.

5. Finally, we distinguished between one- and two-tailed probability values.

Terms to Remember:

Random	***Bias***
Independence	***Binomial model***

Sampling distribution
Probability
Mutually exclusive
Exhaustive
Joint occurrence
Conditional probability
Sampling with replacement
Sampling without replacement
Addition rule
Multiplication rule
One-tailed p-values
Two-tailed p-values
Deviant, rare, unusual

EXERCISES

11.1 List all the possible outcomes of a coin that is tossed three times. Calculate the probability of
   a) 3 heads.
   b) 3 tails.
   c) 2 heads and 1 tail.
   d) at least 2 heads.

11.2 A card is drawn at random from a deck of 52 playing cards. What is the probability that
   a) it will be the ace of spades?
   b) it will be an ace?
   c) it will be an ace or a face card?
   d) it will be a spade or a face card?

11.3 Express the probabilities, in Exercises 11.1 and 11.2, in terms of *odds against*.

11.4 In a single throw of two dice, what is the probability that
   a) a 7 will appear?
   b) a doublet (two of the same number) will appear?
   c) a doublet or an 8 will appear?
   d) an even number will appear?

11.5 On a slot machine (commonly referred to as a "one-armed bandit"), there are three reels with five different fruits plus a star on each reel. After inserting a coin and pulling the handle, the player sees that the three reels revolve independently several times before stopping. What is the probability that
   a) three lemons will appear?
   b) any three of a kind will appear?
   c) two lemons and a star will appear?
   d) two lemons and any other fruit will appear?
   e) no star will appear?

11.6 Three cards are drawn at random (without replacement) from a deck of 52 cards. What is the probability that
   a) all three will be hearts?
   b) none of the three cards will be hearts?
   c) all three will be face cards?

11.7 Calculate the probabilities in Exercise 11.6 if each card is replaced after it is drawn.

11.8 A well-known test of intelligence has a mean of 100 and a standard deviation of 16.
   a) What is the probability that someone picked at random will have an I.Q. of 122 or higher?
   b) There are I.Q.'s so *high* that the probability is 0.05 that such I.Q.'s would occur in a random sample of people. Those I.Q.'s are beyond what value?
   c) There are I.Q.'s so *extreme* that the probability is 0.05 that such I.Q.'s would occur in a random sample of people. Those I.Q.'s are beyond what values?
   d) The next time you shop you will undoubtedly see someone who is a complete stranger to you. What is the probability that his I.Q. will be between 90 and 110?

e) What is the probability of selecting two people, at random,
  i) with I.Q.'s of 122 or higher?
  ii) with I.Q.'s between 90 and 110?
  iii) one with an I.Q. of 122 or higher, the other with an I.Q. between 90 and 110?

f) What is the probability that on leaving your class, the first student you meet will have an I.Q. below 120? Can you answer this question on the basis of the information provided above? If not, why not?

11.9 Which of the following selection techniques will result in random samples? Explain your answers.

a) Population: Viewers of a given television program. Sampling technique: On a given night, interviewing every fifth person in the studio audience.

b) Population: A home-made pie. Sampling technique: A wedge selected from any portion of the pie.

c) Population: All the children in a suburban high school. Sampling technique: Selecting one child sent to you by each homeroom teacher.

11.10 In a study involving a test of visual acuity, four different hues varying slightly in brightness are presented to the subject. What is the probability that he will arrange them in order, from greatest brightness to least, by chance?

11.11 The proportion of people with type A blood in a particular city is 0.20. What is the probability that:

a) a given individual, selected at random, will have type A blood?
b) two out of two individuals will have type A blood?
c) a given individual will *not* have type A blood?
d) two out of two individuals will *not* have type A blood?

11.12 In a manufacturing process, the proportion of items that are defective is 0.10. What is the probability that:

a) in a sample of four items, none will be defective?
b) in a sample of four items, all will be defective?
c) one or more but less than four will be defective?

11.13 In the manufacture of machine screws for the space industry, millions of screws measuring 0.010 cm are produced daily. The standard deviation is 0.001. A screw is considered defective if it deviates from 0.010 by as much as 0.002. What is the probability that:

a) one screw, selected at random, will be defective?
b) two out of two screws will be defective?
c) one screw, selected at random, will *not* be defective?
d) two out of two screws will *not* be defective?
e) one screw, selected at random, will be too large?
f) two out of two screws will be too small?

11.14 A bag contains 6 blue marbles, 4 red marbles, and 2 green marbles. If you select a single marble at random from the bag, what is the probability that it will be:

a) red? b) blue? c) green? d) white?

11.15 Selecting *without* replacement from the bag described in Exercise 11.14, what is the probability that:

a) three out of three will be blue?
b) two out of two will be green?
c) none out of four will be red?

11.16 Selecting *with* replacement from the bag described in Exercise 11.14, what is the probability that:
a) three out of three will be blue?
b) two out of two will be green?
c) none out of four will be red?

11.17 Forty percent of the students at a given college major in business administration. Seventy percent of these are male and thirty percent female. Sixty percent of the students in the school are male. What is the probability that:
a) one student, selected at random, will be a BA major.
b) one female, selected at random, will be a BA major.
c) two students, selected at random, will both be BA majors.
d) one male and one female, selected at random, will both be BA majors.

11.18 What is the probability that a score chosen at random from a normally distributed population with a mean of 66 and a standard deviation of 8 will be:
a) greater than 70?
b) less than 60?
c) between 60 and 70?
d) in the 70's?
e) either less than 55 or greater than 72?
f) either less than 52 or between 78 and 84?
g) either between 56 and 64 or between 80 and 86?

Exercises 11.19 to 11.27 will provide insights that will be valuable in understanding concepts appearing in subsequent chapters.

11.19 Imagine that you have placed ten cardboard tabs in a receptable. Each tab has one of five numbers written on it (column 1). The frequency with which each number appears is shown in column 2.

Number	Frequency	p
0	1	
1	2	
2	4	
3	2	
4	1	

a) Calculate the mean and variance of the population.
b) Determine the probability that each number will be selected on a single draw. (*Hint*: what is the relative frequency with which each number occurs?)
c) On a separate piece of paper, fill in these probabilities in the column headed by p.

11.20 Assume you draw random samples, with replacement, of $n = 2$, from the population described in Exercise 11.19. The probability of drawing a sample with both scores equal to zero is (0.10)(0.10) = 0.01. What is the probability of drawing samples, in order, of the following scores?
a) 0 and 3 b) 3 and 0 c) 0 and 4 d) 2 and 2
e) 4 and 0 f) 3 and 3 g) 0 and 0 h) 4 and 4

11.21 A table can be constructed to summarize the probability of drawing all possible samples of $n = 2$ from the population shown in Exercise 11.19.

			Value of second draw				
			4	3	2	1	0
	Probability		0.1	0.2	0.4	0.2	0.1
	4	0.1	0.01				
Value of	3	0.2					0.02
first	2	0.4		0.08			
draw	1	0.2				0.04	
	0	0.1					

a) Copy the table and fill in the probabilities associated with all the missing cells.
b) What is the sum of all the probabilities shown in this table?
c) Are the probabilities shown in this table exhaustive?

11.22 A table can be constructed which shows the means of all samples of $n = 2$ drawn from the populations shown in Exercise 11.19.

Score	4	3	2	1	0
4	4.0				
3	3.5				
2		2.5	2.0		
1				1.0	
0					0.0

Copy the table and fill in the means of all the missing cells.

11.23 Construct a frequency distribution of the means calculated in Exercise 11.22.

Mean	f	p
4.0	1	0.04
3.5		
3.0	3	0.12
2.5		
2.0	5	0.20
1.5		
1.0		
0.5	2	
0.0		
	$N = 25$	$p = 1.00$

a) Copy the table and fill in all the missing values in column f.
b) Calculate the probability that a random sample with $n = 2$ will equal 3.5; 2.5; 0.0.
c) Fill in the missing probability values in column p.

11.24 Construct a histogram showing the frequency distribution of the means obtained in Exercise 11.23

11.25 Construct a probability histogram of means (substituting p for f on the ordinate).

11.26 Referring to the probability distribution constructed in Exercise 11.23, answer the following questions: Assuming you draw random samples, of $n = 2$, with replacement, calculate the probability of:
a) drawing a mean equal to or greater than 3.5.
b) drawing a mean equal to or less than 3.5.
c) drawing a mean equal to or less than 0.0.
d) drawing a mean between 1.5 and 3.5 inclusive.
e) drawing a mean as unusual as 3.0.

11.27 Calculate the variance of the frequency distribution of means you calculated in Exercise 11.23. Compare the variance of means, for $n = 2$, with the variance of the original scores (see Exercise 11.19).

11.28 A person interested in the mean amount of interest people obtained over a three-month period randomly sampled the amount of interest of 20% of the people with savings accounts in a certain bank in New York City. The mean was $25.00, with a standard deviation of 10. The most accurate statement the investigator could make would be (choose one):
a) The mean interest accumulated by people in New York City is $25.00 for a three-month period.
b) The mean interest accumulated by people in New York is $25.00 for a three-month period.
c) The mean interest accumulated by people in a certain bank in New York City is $25.00 for a three-month period.

11.29 For the above problem, indicate which of the following would be the best way for the investigator to sample 20% of the people at that bank:
a) Stand in the bank every noonhour and find the interest of each person who comes in until 20% of the people have been sampled.
b) Put the names of all the people with savings accounts in the bank in an urn and draw 20% of the names.
c) Take the first 20% of a list formed alphabetically.

11.30 Suppose a manager is recruiting people for a certain job. He prefers people who score 50 or better on a certain skills test. It has already been established that the mean for people applying for the job is 40 and the standard deviation is 10. What is the probability that a person who applies will have a score of:
a) 40 or better?
b) 50 or better?
c) between 40 and 50?
d) between 50 and 60?
e) 40 or less?
f) 75 or better?
g) 20 or less?

11.31 Referring to the above problem, assume that the manager hires only 0.25 of those persons who apply with a score of 50 or above. (After a person scores 50 or above, the score is neglected for the rest of the hiring procedure. In other words, the manager is not interested in the value of the score so long as it is 50 or above.) If a person has a score below 50, the p of being hired is 0.02. Determine the p of a person getting the job if his or her score is:

a) 60 b) 30 c) 40 d) 50

11.32 What is the p of a person obtaining a score of 60 or above and of getting hired?

11.33 For Exercise 11.31, what is the p of a person obtaining a score of 30 or below and getting hired?

11.34 Ignoring the p given in Exercise 11.31 and referring to Exercise 11.30, assume that the following p of getting hired are associated with the following score values:

Score	p
10–19	0
20–29	0
30–39	0.02
40–49	0.10
50–59	0.40
60–69	0.60
70+	0.80

a) What is the p of a person obtaining a score of 70+ and being hired?
b) What is the p of a person obtaining a score between 20 and 29 and being hired?
c) What is the p of a person obtaining a score between 50 and 59 and being hired?
d) What is the p of a person obtaining a score between 60 and 69 and being hired?

*11.35 Imagine that we have a population of the following four scores: 1, 4, 7, and 10.

a) Construct a probability distribution and histogram of all possible means when sampling with replacement, $n = 2$.
b) Construct a probability histogram of all possible means when sampling with replacement, $n = 3$. (*Hint*: The table for finding the means appears below. The values appearing in the cells represent the means of the three draws.)

1st draw		1				4				7				10			
2nd draw		1	4	7	10	1	4	7	10	1	4	7	10	1	4	7	10
3rd draw	1	1	2	3	4	2	3	4	5	3	4	5	6	4	5	6	7
	4	2	3	4	5	3	4	5	6	4	5	6	7	5	6	7	8
	7	3	4	5	6	4	5	6	7	5	6	7	8	6	7	8	9
	10	4	5	6	7	5	6	7	8	6	7	8	9	7	8	9	10

*11.36 The original population of the four scores in Exercise 11.35 was rectangular (they all had the same associated frequency of 1). Compare the probability distributions in Exercises 11.35(a) and 11.35(b) above and attempt to form a generalization about the form and the dispersion of the distribution of sample means as we increase the sample size.

*11.37 Answer the following questions based upon the probability histograms obtained in Exercise 11.35, on page 228.

a) Drawing a single sample of $n = 2$, what is the probability of obtaining a mean equal to 1? Contrast this result with the probability of randomly selecting a mean equal to 1 when $n = 3$.

b) For each distribution, determine the probability of selecting a sample with a mean as rare or as unusual as 10.

c) From each probability histogram, determine the probability of selecting a sample with a mean as low as 4.

d) From each probability histogram, determine the probability of selecting a sample mean as deviant from the population mean as a mean of 4.

*11.38 For the probability distribution of $n = 2$ [Exercise 11.35(a)] find:

a) $p(\bar{X} \leq 7)$ b) $p(\bar{X} \geq 8.5)$ c) $p(\bar{X} = 5.5)$

*11.39 For the probability distribution of $n = 3$ [Exercise 11.35(b)] find:

a) $p(4 \leq \bar{X} \leq 7)$ b) $p(5 \leq \bar{X} \leq 6)$ c) $p(\bar{X} = 3 \text{ or } \bar{X} = 9)$

*11.40 Let's now imagine a different type of sampling experiment. You have selected all possible samples of $n = 2$ from a population of scores and obtained the following means: 1, 2, 2, 3, 3, 3, 4, 4, 5. You now place paper tabs in a hat with these means written on them. You select one mean, record it, and replace it in the hat. You select a second mean, *subtract* it from the first and then replace it in the hat. The table for describing all possible *differences between means* of $n = 2$ is shown below.

Second draw of mean	First draw of mean: 1	2	2	3	3	3	4	4	5
1	0	1	1	2	2	2	3	3	4
2	−1	0	0	1	1	1	2	2	3
2	−1	0	0	1	1	1	2	2	3
3	−2	−1	−1	0	0	0	1	1	2
3	−2	−1	−1	0	0	0	1	1	2
3	−2	−1	−1	0	0	0	1	1	2
4	−3	−2	−2	−1	−1	−1	0	0	1
4	−3	−2	−2	−1	−1	−1	0	0	1
5	−4	−3	−3	−2	−2	−2	−1	−1	0

a) Construct a frequency distribution of differences between means.

b) Construct a probability distribution of differences between means.

c) Find the mean and the standard deviation of the differences between means.

*11.41 Based upon the responses to Exercise 11.40 (above), answer the following questions. Drawing two samples at random and with replacement from the population of means, and subtracting the second mean from the first, what is the probability that you will select:

a) a difference between means equal to zero?

b) a difference between means equal to or less than 1 *or* equal to or greater than −1? (*Note*: −2, −3, −4 are all less than −1.)

c) a difference between means equal to -4?
d) a difference between means as rare or as deviant as -4?
e) a difference between means equal to or greater than 3?
f) a difference between means equal to or less than -3?
g) a difference between means as rare or as unusual as -3?
h) a difference between means equal to or less than 2 or equal to or greater than -2?

Introduction to Statistical Inference

12.1 WHY SAMPLE?*

You are the leader of a religious denomination, and for the purpose of planning recruitment you want to know what proportion of the adults in the United States claim church membership. How would you go about getting this information?

You are a rat psychologist and you are interested in the relationship between strength of drive and learning. Specifically, what are the effects of duration of hunger drive on the number of trials required for a rat to learn a T-maze?

You are a sociologist and you want to study the differences in child rearing practices among parents of delinquent versus nondelinquent children.

You are a market researcher and you want to know what proportion of individuals prefer different car colors and their various combinations.

You are a park attendant and you want to determine whether the ice is sufficiently thick to permit safe skating.

You are a gambler and you want to determine whether a set of dice is "dishonest."

What do all these problems have in common? You are asking questions about the parameter of a population to which you want to generalize your answers, but you have no hope of ever studying the *entire* population. Earlier (Section 1.2) we defined a population as a *complete* set of individuals, objects, or measurements having some common observable characteristic. It is frequently impossible to study *all* the members of a given population either because the population as defined has an infinite number of members or because the population is so large that it defies exhaustive study. Consequently, when we refer to the **population** we are often dealing with a hypothetical entity.†

For example, the population of all possible outcomes resulting from tossing two dice is unlimited. Therefore the above gambler must formulate his general

* It is recommended that you reread Section 1.2 for purposes of reviewing several definitions of terms that will appear in this chapter.

† In the typical experimental situation, the actual population, or universe, does not exist. What we attempt to do is to find out something about the characteristics of that population *if it did* exist. Thus our sample groups provide us with information about the characteristics of a population if it did, in fact, exist.

conclusions based on a relatively small sample of tosses. The religious leader, as well, could not reasonably hope to obtain replies from every adult in the United States.

Since populations can rarely be studied exhaustively, we must depend on samples as a basis for arriving at a hypothesis concerning various characteristics, or parameters, of the population. Note that our interest is not in descriptive statistics, *per se*, but in making inferences from data. Thus, if we ask 100 people how they intend to vote in a forthcoming election, our primary interest is not how these 100 people will vote, but in estimating how the entire voting population will cast their ballots.

Almost all research involves the observation and the measurement of a limited number of individuals or events. These measurements are presumed to tell us something about the population. In order to understand how we are able to make inferences about a population from a sample, it is necessary to introduce the concept of sampling distributions.

12.2 THE CONCEPT OF SAMPLING DISTRIBUTIONS

In actual practice, inferences about the parameters of a population are made from statistics that are calculated from a **sample** of N observations drawn at random from this population. If we continued to draw samples of size N from this population, we should not be surprised if we found some differences among the values of the sample statistics obtained. Indeed, it is this observation that has led to the concept of **sampling distributions**.

A sampling distribution is a theoretical probability distribution of the possible values of some sample statistic which would occur if we were to draw all possible samples of a fixed size from a given population.

The sampling distribution is one of the most important concepts in inferential statistics. You are already familiar with several sampling distributions, although we have not previously named them as such. Recall the various sampling problems we have introduced throughout the earlier chapters in the text. In one example, we started with a population of seven scores and selected, with replacement, samples of $n = 2$. We obtained all possible combinations of these scores, two at a time, and then found the mean of each of these samples. We then constructed a frequency distribution and probability distribution of means drawn from that population with a fixed sample size of $n = 2$.

Recall also that in Exercise 11.35 we constructed sampling distributions based upon drawing, with replacement, all possible samples of $n = 2$ and $n = 3$ from a population of four scores (1, 4, 7, 10). Table 12.1 shows these two sampling distributions, plus the sampling distribution of the mean when $n = 4$.

Why is the concept of a sampling distribution so important? The answer is simple. Once you are able to describe the sampling distribution of *any statistic* (be it mean, standard deviation, proportion), you are in a position to entertain

Table 12.1

Sampling distributions of means drawn from a population of four scores (1, 4, 7, 10) $\mu = 5.5$, $\sigma = 3.87$ and sample sizes $n = 2$, $n = 3$ and $n = 4$

$n = 2$		$n = 3$		$n = 4$	
$\bar{X}$	$p(\bar{X})$	$\bar{X}$	$p(\bar{X})$	$\bar{X}$	$p(\bar{X})$
10.0	0.0625	10.0	0.0156	10.00	0.0039
8.5	0.1250	9.0	0.0469	9.25	0.0156
7.0	0.1875	8.0	0.0938	8.50	0.0391
5.5	0.2500	7.0	0.1562	7.75	0.0781
4.0	0.1875	6.0	0.1875	7.00	0.1211
2.5	0.1250	5.0	0.1875	6.25	0.1562
1.0	0.0625	4.0	0.1562	5.50	0.1719
		3.0	0.0938	4.75	0.1562
		2.0	0.0469	4.00	0.1211
		1.0	0.0156	3.25	0.0781
				2.50	0.0391
				1.75	0.0156
				1.00	0.0039
$\bar{X} = 5.5$		$\bar{X} = 5.5$		$\bar{X} = 5.5$	
$s_{\bar{X}} = 2.45$		$s_{\bar{X}} = 1.95$		$s_{\bar{X}} = 1.68$	

and test a wide variety of different hypotheses. For example, you draw four numbers at random from some population. You obtain a mean equal to 7.00. You ask: "Is this mean an ordinary event or is it a rare event?" In the absence of a frame of reference, this question is meaningless. However, if we know the sampling distribution for this statistic, we would have the necessary frame of reference and the answer would be straightforward. If we were to tell you that the appropriate sampling distribution is given in Table 12.1 when $n = 4$, you would have no trouble answering the question. A mean of 7.00 would be drawn about 12 percent ($p = 0.1211$) of the time; and a mean of 7 or greater would occur almost 26 percent of the time ($p = 0.1211 + 0.0781 + 0.0391 + 0.0156 + 0.0039 = 0.2578$).

Whenever we estimate a population parameter from a sample, we shall ask questions such as: "How good an estimate do I have? Can I conclude that the population parameter is identical with the sample statistic? Or is there likely to be some error? If so, how much?" To answer each of these questions, we will compare our sample results with the "expected" results. The expected results are, in turn, given by the appropriate sampling distribution. But what does the sampling distribution of a particular statistic look like? How can we ever know the form of the distribution and thus what the expected results are? Since the inferences we will be making imply knowledge of the *form* of the sampling distribution, it is necessary to set up certain idealized *models*. The normal curve and the binomial distribution are two models whose mathematical properties are known. Consequently, these two distributions are frequently employed as models to describe particular sampling distributions. Thus, for example, if we know that the sampling distribution of a particular statistic takes the form of a normal distribution, we may use the known properties of the normal distribution to make inferences and predictions about the statistic.

The following sections should serve to clarify these important points.

12.3 BINOMIAL DISTRIBUTION

Let us say that you have a favorite coin which you use constantly in everyday life as a basis of "either-or" decision making. For example, you may ask, "Should I study tonight for the statistics quiz, or should I relax at one of the local movie houses? Heads, I study, tails, I don't." Over a period of time, you have sensed that the decision has more often gone "against you" than "for you" (in other words, you have to study more often than relax!). You begin to question the accuracy and the adequacy of the coin. Does the coin come up heads more often than tails? How might you find out?

One thing is clear. The true proportion of heads and tails characteristic of this coin can never be known. You could start tossing the coin this very minute and continue for a million years (granting a long life and a remarkably durable coin) and you would not exhaust the population of possible outcomes. In this instance, the true proportion of heads and tails is unknowable because the universe, or population, is unlimited.

The fact that the *true* value is unknowable does not prevent us from trying to estimate what it is. We have already pointed out that since populations can rarely be studied exhaustively, we must depend on samples to estimate the parameters.

Returning to our problem with the coin, we clearly see that in order to determine whether or not the coin is biased, we will have to obtain a sample of the "behavior" of that coin and arrive at some generalization concerning its possible bias.

Let us define an *unbiased* coin as one in which the probability of heads is equal to the probability of tails. We may employ the symbol P to represent the probability of the occurrence of a head, and Q, the probability of the nonoccurrence of a head (i.e., the occurrence of a tail). Since we are dealing with two mutually exclusive and exhaustive outcomes, if $P = Q = \frac{1}{2}$ the coin is *unbiased*. Converse, if $P \neq Q \neq \frac{1}{2}$, the coin is *biased*.

How do we determine whether a particular coin is biased or unbiased? Suppose we conduct the following experiment. We toss the coin 10 times and obtain 9 heads and 1 tail. This may be viewed as a sample of the "behavior" of this coin. On the basis of this result, are we now justified in concluding that the coin is biased? Or is it reasonable to expect as many as 9 heads from a coin that is unbiased? Before we answer these questions, it is necessary to look at the sampling distribution of *all* possible outcomes. Let us see how we might construct this sampling distribution, employing a hypothetical coin.

12.3.1 Construction of Binomial Sampling Distributions by Enumeration

First, we must assume that this coin is unbiased, and that there are only two possible outcomes resulting from each toss of the coin: heads or tails. It will not stand on its side; it will not become lost; it cannot turn up both heads and tails at the same time.

If we toss this coin twice ($N = 2$), there are four possible ways the coin may fall: HH, HT, TH, and TT. The two middle ways may be thought of as the same outcome in that each represents one head and one tail. Thus tossing an *unbiased* coin twice results in the following theoretical frequency distribution:

Outcome	Number of ways for specified outcome to occur
$2H$	1
$1H, 1T$	2
$2T$	1
	4

Note that in N tosses of a coin, there are $N + 1$ different possible outcomes and two different ways to obtain these $N + 1$ outcomes. Thus, when $N = 2$, there are four ways of obtaining the three different outcomes.

We may calculate the probability associated with each outcome by dividing the number of ways each outcome may occur by 2^N. For example, using formula (11.1), the probability of obtaining one head and one tail in two tosses of an unbiased coin is

$$p(1H, 1T) = \tfrac{2}{4} = 0.50.$$

We have seen how we can enumerate all the possible outcomes of tossing a hypothetical coin when $N = 2$, and construct corresponding frequency and probability distributions. This probability distribution represents the sampling distribution of outcomes when $N = 2$. Similarly, we may enumerate all the possible outcomes for any number of tosses of our hypothetical coin ($N = 3$, $N = 4$, etc.) and then construct the corresponding frequency and probability distributions.

Let us illustrate the construction of a probability distribution when $N = 5$. First, all possible outcomes are enumerated, as in Table 12.2. When $N = 5$, there are 32 ways (2^5) of obtaining the six different outcomes ($5 + 1$).

By placing the six different outcomes along the baseline and representing their frequency of occurrence along the ordinate, we have constructed the theoretical frequency distribution for the various outcomes when $N = 5$ (see Fig. 12.1).

Table 12.2

All possible outcomes obtained by tossing an unbiased coin five times ($N = 5$)

Number of heads					
$0H$	$1H$	$2H$	$3H$	$4H$	$5H$
$TTTTT$	$HTTTT$	$HHTTT$	$HHHTT$	$HHHHT$	$HHHHH$
	$THTTT$	$HTHTT$	$HHTHT$	$HHHTH$	
	$TTHTT$	$HTTHT$	$HTHHT$	$HHTHH$	
	$TTTHT$	$HTTTH$	$THHHT$	$HTHHH$	
	$TTTTH$	$THHTT$	$HHTTH$	$THHHH$	
		$THTHT$	$HTHTH$		
		$THTTH$	$THHTH$		
		$TTHHT$	$HTTHH$		
		$TTHTH$	$THTHH$		
		$TTTHH$	$TTHHH$		

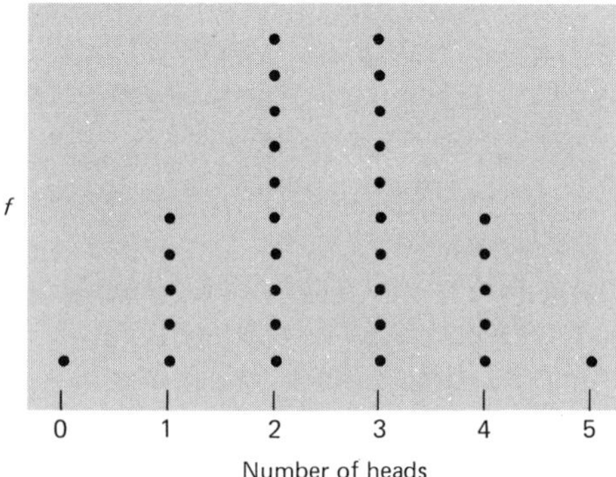

Fig. 12.1 Theoretical distribution of various numbers of heads obtained by tossing an unbiased coin five times ($N = 5$).

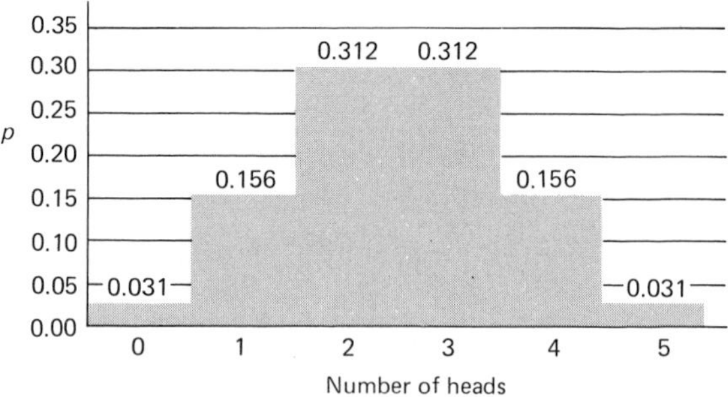

Fig. 12.2 Histogram of theoretical probability distribution of various numbers of heads obtained by tossing an unbiased coin five times ($N = 5$).

It is now possible to calculate the probability associated with each outcome. For example, using formula (11.1), the probability of obtaining 4 heads and 1 tail in 5 tosses of an unbiased coin is

$$p(4H, 1T) = \tfrac{5}{32} = 0.156.$$

A histogram of the complete probability distribution when $N = 5$ is shown in Fig. 12.2.

12.3.2 Construction of Binomial Sampling Distributions Employing the Binomial Expansion

Thus far, in order to calculate the probability associated with each outcome, we have had to enumerate all the possible ways in which various outcomes occur (see Table 12.2). As N increases, the process of enumeration becomes exceedingly laborious since the number of ways for the outcomes to occur (2^N) doubles with each additional toss of the coin.*

The simplest method for demonstrating the binomial expansion is to remember the equality $(P + Q)^N = 1.00$, in which N represents the number of trials, such as tosses of a coin:

When you toss a coin once and P represents the probability of a head, and Q represents the probability of the nonoccurrence of a head (i.e., a tail), their joint probability is:

$$(P + Q)^N = (P + Q)^1 = \tfrac{1}{2} + \tfrac{1}{2} = 1.00.$$

When you toss the coin *twice*:

$$\begin{aligned}(P + Q)^N = (P + Q)^2 &= P^2 + 2PQ + Q^2 \\ &= \tfrac{1}{4} + 2(\tfrac{1}{4}) + \tfrac{1}{4} \\ &= \tfrac{1}{4} + \tfrac{1}{2} + \tfrac{1}{4} = 1.00.\end{aligned}$$

The progression then proceeds as follows:

a) One toss: $(P + Q)^1$

b) Two tosses: $(P + Q)^2 = (P + Q)(P + Q)$; that is,

$$\begin{array}{r} P + Q \\ \times \quad P + Q \\ \hline P^2 + PQ \qquad \\ + PQ + Q^2 \\ \hline P^2 + 2PQ + Q^2 \end{array}$$

---

* In addition, when $P \neq Q \neq \frac{1}{2}$, it is not possible to obtain the probabilities associated with the various outcomes by simple enumeration. It is necessary to employ the binomial expansion to obtain these probabilities. Applications of the binomial expansion when $P \neq Q \neq \frac{1}{2}$ are presented in Chapter 18.

c) Three tosses: $(P + Q)^3 = (P + Q)^2(P + Q)^1 = [P^2 + 2PQ + Q^2][P + Q]$; that is,

$$\begin{array}{r} P^2 + 2PQ + Q^2 \\ \times \quad P + Q \\ \hline P^3 + 2P^2Q + PQ^2 \qquad\quad \\ + P^2Q + 2PQ^2 + Q^3 \\ \hline P^3 + 3P^2Q + 3PQ^2 + Q^3 \end{array}$$

d) Four tosses: $(P+Q)^4 = [P+Q]^3(P+Q)^1 = [P^3+3P^2Q+3PQ^2+Q^3][P+Q]$; that is,

$$\begin{array}{r} P^3 + 3P^2Q + 3PQ^2 + Q^3 \\ \times \quad P + Q \\ \hline P^4 + 3P^3Q + 3P^2Q^2 + PQ^3 \qquad\quad \\ + P^3Q + 3P^2Q^2 + 3PQ^3 + Q^4 \\ \hline P^4 + 4P^3Q + 6P^2Q^2 + 4PQ^3 + Q^4 \end{array}$$

If you are interested in constructing the sampling distribution of the binomial when $N = 4$, you stop at this point and substitute the appropriate value of P and Q. In the coin-tossing experiment, $P = Q = \frac{1}{2}$. Therefore,

$$\begin{aligned} (P + Q)^4 &= P^4 + 4P^3Q + 6P^2Q^2 + 4PQ^3 + Q^4 = 1.00; \\ (P + Q)^4 &= (\tfrac{1}{2})^4 + 4(\tfrac{1}{2})^3(\tfrac{1}{2}) + 6(\tfrac{1}{2})^2(\tfrac{1}{2})^2 + 4(\tfrac{1}{2})(\tfrac{1}{2})^3 + (\tfrac{1}{2})^4 \\ &= (\tfrac{1}{16}) + (\tfrac{4}{16}) + (\tfrac{6}{16}) + (\tfrac{4}{16}) + \tfrac{1}{16} \\ &= 0.0625 + 0.2500 + 0.3750 + 0.2500 + 0.0625 = 1.00. \end{aligned}$$

The last line represents the sampling distribution for the binomial when $N = 4$ and $P = Q = \frac{1}{2}$.
Formula (12.1) presents the binomial expansion in its general form:

$$(P + Q)^N = P^N + \frac{N}{1} P^{N-1}Q + \frac{N(N-1)}{(1)(2)} P^{N-2}Q^2 + \frac{N(N-1)(N-2)}{(1)(2)(3)} P^{N-3}Q^3 + \cdots + Q^N. \tag{12.1}$$

There are $N + 1$ terms to the right of the above equation, each representing a different possible outcome. The first term on the righthand side of the equation (P^N) provides the probability of all events in the P-category; the second term provides the probability of all events in the P-category, except one; and, finally, the last term (Q^N) is the probability of all events in the *Q-category*.

12.4 TESTING STATISTICAL HYPOTHESES: LEVEL OF SIGNIFICANCE

At this point, you may wonder what happened to the experiment we were about to perform to determine whether our "decision-making" coin was biased. Having learned to calculate probability values, let us now address ourselves to the experiment. We are going to toss the coin a given number of times and determine whether or not the outcome is within certain expected limits. For example, if we toss our coin 10 times and obtain 5 heads and 5 tails, would we begin to suspect our coin of being biased? Of course not, since this outcome is exactly a 50–50 split and is in agreement with the hypothesis that the coin is not biased. What if we obtained 6 heads and 4 tails? Again, this is not an unusual outcome. In fact, if we expand the binomial, we can answer the question, "Given a theoretically perfect coin, how often would we expect an outcome at least this much different from a 50–50 split?" Reference to Fig. 12.3, which represents the theoretical probability distribution of various numbers of heads when $N = 10$, reveals that departures from a 50–50 split are quite common. Indeed, whenever we obtain either 6 or more heads, or 4 or fewer heads, we are departing from a 50–50 split. Such departures will occur fully 75.4% of the time when we toss a perfect coin in a series of trials with 10 tosses per trial.

What if we obtained 9 heads and 1 tail? Clearly, we begin to suspect the honesty of the coin. Why? At what point do we change from attitudes accepting the honesty of the coin to attitudes rejecting its honesty? This question takes us to the crux of the problem of inferential statistics. We have seen that the rarer or more unusual the event, the more prone we are to look for nonchance explanations of the event.

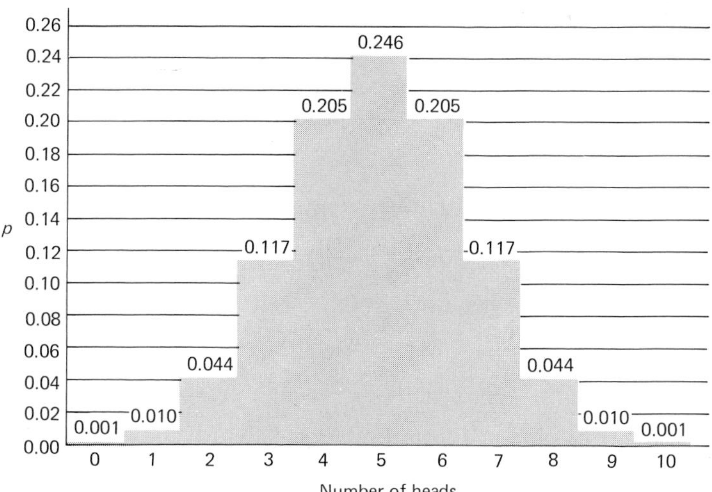

Fig. 12.3 Histogram of theoretical probability distribution of various numbers of heads obtained by tossing an unbiased coin ten times ($N = 10$).

When we obtained 6 heads in 10 tosses of our coin, we felt no necessity to find an explanation for its departure from a 50–50 split, other than to state that such a departure would occur frequently "by chance." However, when we obtained 9 heads, we had an uncomfortable feeling concerning the honesty of the coin. Nine heads out of 10 tosses is such a rare occurrence that we begin to suspect that the explanation may be found in terms of the characteristics of the coin rather than in the so-called "laws of chance." The critical question is, "Where do we draw the line which determines what inferences we make about the coin?"

The answer to this question reveals the basic nature of science: its probabilistic rather than its absolutistic orientation. In the social sciences, most researchers have adopted one of the following two cutoff points as the basis for *inferring the operation of nonchance factors.*

1. When the event or one more deviant would occur *five percent* of the time or less, *by chance*, some researchers are willing to assert that the results are due to nonchance factors. This cutoff point is known variously as the **0.05 significance level**, or the **5.00% significance level**.
2. When the event or one more deviant would occur *one percent* of the time or less, *by chance*, other researchers are willing to assert that the results are due to nonchance factors. This cutoff point is known as the **0.01 significance level**, or the **1.00% significance level**.

The level of significance set by the experimenter for inferring the operation of nonchance factors is known as the **alpha (α) level**. Thus when employing the 0.05 level of significance, $\alpha = 0.05$; when employing the 0.01 level of significance, $\alpha = 0.01$.

In order to determine whether the results were due to nonchance factors in the present coin experiment, we need to calculate the probability of obtaining an event as *rare* as 9 heads out of 10 tosses. In determining the rarity of an event, we must consider the fact that the rare event can occur in both directions and that it includes more extreme events. In other words, the probability of an event as *rare* as 9 heads out of 10 tosses is equal to

$$p(\text{9 heads}) + p(\text{10 heads}) + p(\text{1 head}) + p(\text{0 heads}).$$

Since this distribution is symmetrical,

$$p(\text{9 heads}) = p(\text{1 head}) \qquad \text{and} \qquad p(\text{10 heads}) = p(\text{0 heads}).$$

Thus

$$p(\text{9 heads}) + p(\text{10 heads}) + p(\text{1 head}) + p(\text{0 heads}) = 2[p(\text{9 heads}) + p(\text{10 heads})].$$

These p-values may be obtained from Fig. 12.3 as follows:

$$p(\text{9 heads}) = 0.010, \qquad \text{and} \qquad p(\text{10 heads}) = 0.001.$$

Therefore the two-tailed probability of an event as rare as 9 heads out of 10 tosses is

$$2(0.010 + 0.001) = 0.022 \text{ or } 2.2\%.$$

Employing the 0.05 significance level ($\alpha = 0.05$), we would conclude that the coin was biased (i.e., the results were due to nonchance factors). However, if we employed the 0.01 significance level ($\alpha = 0.01$), we would not be able to assert that these results were due to nonchance factors.

12.5 TESTING STATISTICAL HYPOTHESES: NULL HYPOTHESIS AND ALTERNATIVE HYPOTHESIS

At this point, many students become disillusioned by the arbitrary nature of decision-making in science. Let us examine the logic of statistical inference a bit further and see if we can resolve some of the doubts. Prior to the beginning of any experiment, the researcher sets up two mutually exclusive hypotheses: (1) The **null hypothesis (H_0)** which specifies hypothesized values for one or more of the *population parameters*. (2) The **alternative hypothesis (H_1)** which asserts that the *population parameter* is some value other than the one hypothesized. In the present coin experiment, these two hypotheses read as follows:

$$H_0: \text{the coin is unbiased, that is, } P = Q = \tfrac{1}{2}.$$
$$H_1: \text{the coin is biased, that is, } P \neq Q \neq \tfrac{1}{2}.$$

The alternative hypothesis may be either *directional* or *nondirectional*. When H_1 asserts *only* that the population parameter is *different from* the one hypothesized, it is referred to as a **nondirectional** or **two-tailed hypothesis** (for example, $P \neq Q \neq \frac{1}{2}$). Occasionally, H_1 is **directional** or **one-tailed**. In this instance, in addition to asserting that the population parameter is different from the one hypothesized, we assert the *direction* of that difference (for example, $P > Q$ or $P < Q$). In evaluating the outcome of an experiment, one-tailed probability values should be employed whenever our alternative hypothesis is directional.

Moreover, when the alternative hypothesis is directional, so also is the null hypothesis. For example, if the alternative hypothesis is that $P > Q$, the null hypothesis is that $P < Q$. Conversely, if H_1 is $P < Q$, H_0 reads: $P > Q$.

12.5.1 The Notion of Indirect Proof

Careful analysis of the logic of statistical inference reveals that the null hypothesis can never be proved. For example, if we had obtained exactly 5 heads on 10 tosses of a coin, would this prove that the coin was unbiased? The answer is a categorical "No!" A bias, if it existed, might be of such a small magnitude that we failed to detect it in 10 trials. But what if we tossed the coin 100 times and obtained 50 heads? Wouldn't this prove something? Again, the same considerations apply. No matter how many times we tossed the coin, we could never exhaust the population of possible outcomes. We can make the assertion, however, that *no basis exists for rejecting* the hypothesis that the coin is unbiased.

How, then, can we prove the alternative hypothesis that the coin is biased? Again, we cannot prove the alternative hypothesis directly. Think, for the moment, of the logic involved in the following problem.

Draw two lines on a paper and determine whether they are of different lengths. You compare them and say, "Well, certainly they are not equal. Therefore they must be of different lengths." By rejecting equality (in this case, the null hypothesis) you assert that there is a difference.

Statistical logic operates in exactly the same way. We cannot prove the null hypothesis, nor can we directly prove the alternative hypothesis. However, if we can *reject* the null hypothesis, we can assert its alternative, namely, that the population parameter is some value other than the one hypothesized. Applied to the coin problem, if we can reject the null hypothesis that $P = Q = \frac{1}{2}$, we can assert the alternative, namely, that $P \neq Q \neq \frac{1}{2}$. Note that the support of the alternative hypothesis is always *indirect*. We have supported it by rejecting the null hypothesis. On the other hand, since the alternative hypothesis can neither be proved nor disproved directly, we can *never prove the null hypothesis* by rejecting the alternative hypothesis. The strongest statement we are entitled to make in this respect is that we *failed to reject the null hypothesis.*

What, then, are the conditions for rejecting the null hypothesis? Simply this: when employing the 0.05 level of significance, you reject the null hypothesis when a given result occurs, by chance, 5% of the time or less. When employing the 0.01 level of significance, you reject the null hypothesis when a given result occurs, by chance, 1% of the time or less. Under these circumstances, of course, you *affirm* the alternative hypothesis.

In other words, one rejects the null hypothesis when the results occur, by chance, 5% of the time or less (or 1% of the time or less), *assuming that the null hypothesis is the true distribution.* That is, one assumes that the null hypothesis is true, calculates the probability on the basis of this assumption, and, if the probability is small, rejects the assumption.

For reasons stated above, R. A. Fisher, the eminent British statistician, has affirmed:

> In relation to any experiment we may speak of this hypothesis as the 'null hypothesis,' and it should be noted that the null hypothesis is never proved or established, but is possibly disproved, in the course of experimentation. *Every experiment may be said to exist only in order to give the facts a chance of disproving the null hypothesis.**

12.6 TESTING STATISTICAL HYPOTHESES: THE TWO TYPES OF ERROR

You may now ask, "But aren't we taking a chance that we will be wrong in rejecting the null hypothesis? Is it not possible that we have, in fact, obtained a statistically rare occurrence by chance?"

* (Italics supplied.) Fisher, R. A., *The Design of Experiments.* Edinburgh: Oliver & Boyd, 1935, p. 16.

The answer to this question must be a simple and humble "Yes." This is precisely what we mean when we say that science is probabilistic. If there is any absolute statement that scientists are entitled to make, it is that we can never assert with complete confidence that our findings or propositions are true. There are countless examples in science in which an apparently firmly established conclusion has had to be modified in the light of further evidence.

In the coin experiment, even if all the tosses had resulted in heads, it is possible that the coin was not, in fact, biased. By chance, once in every 1024 experiments, "on the average," the coin will turn up heads 10 out of 10 times. When we employ the 0.05 level of significance, approximately 5% of the time we will be wrong when we reject the null hypothesis and assert its alternative.

These are some of the basic facts of the reality of inductive reasoning to which the student must adjust. The student of behavior who insists upon absolute certainty before speaking on an issue is a student who has been mute throughout past years, and who will remain so the rest of his or her life (probably).

The above considerations have led statisticians to formulate two types of errors that may be made in statistical inference.

12.6.1 Type I Error (Type α Error)

In a **type I error**, we reject the null hypothesis when it is actually true. The probability of making a type I error is α. We have already pointed out that if we set our rejection point at the 0.05 level of significance, we will mistakenly reject H_0 approximately 5% of the time. It would seem, then, that in order to avoid this type of error, we should set the rejection level as low as possible. For example, if we were to set $\alpha =$ 0.001, we would risk a type I error only about one time in every thousand. However, the lower we set α, the greater is the likelihood that we will make a type II error.

12.6.2 Type II Error (Type β Error)

In a **type II error**, we fail to reject the null hypothesis when it is actually false. *Beta* (β) is the probability of making a type II error. This type of error is far more common than a type I error. For example, if we employ the 0.01 level of significance as the basis of rejecting the null hypothesis and then conduct an experiment in which the result we obtained would have occurred, by chance, only 2% of the time, we cannot reject the null hypothesis. Consequently, we cannot claim an experimental effect even though there may very well be one.

It is clear, then, that the lower we set the rejection level, the less is the likelihood of a type I error, and the greater is the likelihood of a type II error. Conversely, the higher we set the rejection level, the greater the likelihood of a type I error, and the smaller the likelihood of a type II error.

The fact that the rejection level is set as low as it is attests to the conservatism of scientists, i.e., the greater willingness on the part of the scientist to make an error in the direction of *failing* to claim a result than to make an error in the direction of *claiming* a result when wrong.

You may now ask, "How can we tell when we are making a type I or a type II error?" The answer is simple, "We can't." If we examine, once again, the logic of statistical inference, we shall see why. We have already stated that, with rare exceptions, we cannot or will not know the true parameters of a population. Without this knowledge, how can we know whether our sample statistics have approximated or have failed to approximate the true value? How can we know whether or not we have mistakenly rejected a null hypothesis? On the other hand, if we did know a population value, we could know whether or not we made an error. However, under these circumstances, the whole purpose of sampling statistics is vitiated. We collect samples and draw inferences from samples only because our population values are unknowable, for one reason or another. When they become known, the need for statistical inference is lost.

Is there no way, then, to know which experiments reporting significant results are accurate and which are not? The answer is a conditional "Yes." If we were to repeat the experiment and obtain similar results, we would have increased confidence that we were not making a type I error. For example, if we tossed our coin in a second series of 10 trials and obtained 9 heads, we would feel far more confident that our coin was biased. Parenthetically, repetition of experiments is one of the areas in which research in the social sciences is weakest. The general attitude is that a study is not much good unless it is "different" and is therefore making a novel contribution. Replicating experiments, when performed, frequently go unpublished. In consequence, we may feel assured that in studies employing the 0.05 significance level, approximately one out of every 20 studies which rejects the null hypothesis is making a type I error.*

CHAPTER SUMMARY

We have seen that one of the basic problems of inferential statistics involves estimating population parameters from sample statistics.

In inferential statistics, we are frequently called upon to compare our *obtained* values with *expected* values. The expected values are given by the appropriate sampling distribution, which is a theoretical probability distribution of the possible values of a sample statistic.

We have seen how to construct sampling distributions. In the present chapter, we employed the binomial distribution to construct sampling distributions for discrete, two-category variables.

We have seen that there are two mutually exclusive and exhaustive statistical hypotheses in every experiment: the null hypothesis (H_0) and the alternative hypothesis (H_1).

* The proportion is probably even higher, since our methods of accepting research reports for publication are heavily weighted in terms of the statistical significance of the results. Thus if four identical studies were conducted independently and only one obtained results which permitted rejection of the null hypothesis, *this* one would most likely be published. There is virtually no way for the general scientific public to know about the three studies which *failed* to reject the null hypothesis.

If the outcome of an experiment is rare (here "rare" is defined as some arbitrary but accepted probability value), we reject the null hypothesis and assert its alternative. If the event is not rare (i.e., the probability value is *greater* than what we have agreed upon as being significant), we fail to reject the null hypothesis. However, in no event are we permitted to claim that we have *proved* H_0.

The experimenter is faced with two types of errors in establishing a cutoff probability value which will be accepted as significant.

Type I: rejecting the null hypothesis when it is true.

Type II: accepting the null hypothesis when it is false.

The basic conservatism of the scientist causes him to establish a low level of significance, causing him to make type II errors more commonly than type I errors.

Without replication of experiments we have no basis for knowing when a type I error has been made and, even with replication, we cannot claim knowledge of absolute truth.

Finally, and perhaps most importantly, we have seen that scientific knowledge is probabilistic and not absolute.

Terms to Remember:

Population
Sample
Sampling distribution
Binomial distribution
Coefficients of the binomial
0.05 Significance level
(5% Significance level)
0.01 Significance level
(1% Significance level)
Alpha (α) level
Null hypothesis (H_0)
Alternative hypothesis (H_1)
Directional or one-tailed hypothesis
Nondirectional or two-tailed hypothesis
Type I error (type α)
Type II error (type β)

EXERCISES

12.1 Explain, in your own words, the nature of drawing inferences in behavioral science. Be sure to specify the types of risks that are taken and the ways in which the researcher attempts to keep these risks within specifiable limits.

12.2 Give examples of experimental studies in which:
a) a type I error would be considered more serious than a type II error.
b) a type II error would be considered more serious than a type I error.

12.3 After completing a study in experimental psychology, John Jones concluded, "I have proved that no difference exists between the two experimental conditions." Criticize his conclusion according to the logic of drawing inferences in science.

12.4 Explain what is meant by the following statement: "It can be said that the purpose of any experiment is to provide the occasion for rejecting the null hypothesis."

12.5 An experimental psychologist hypothesizes that drive affects running speed. Assume that she has set up a study to investigate the problem employing two different drive levels. Formulate H_0 and H_1.

12.6 In a ten-item true-false examination,
  a) what is the probability that an unprepared student will obtain all correct answers by chance?
  b) if eight correct answers constitute a passing grade, what is the probability of passing?
  c) what are the odds against passing?

12.7 Identify H_0 and H_1 in the following:
  a) The population mean in intelligence is 100.
  b) The proportion of Democrats in Watanabe County is not equal to 0.50.
  c) The population mean in intelligence is not equal to 100.
  d) The proportion of Democrats in Watanabe County is equal to 0.50.

12.8 Suppose that you are a personnel manager responsible for recommending the promotion of an employee to a high-level executive position. What type of error would you be making if:
  a) the hypothesis that she is qualified (H_0) is erroneously accepted?
  b) the hypothesis that she is qualified is erroneously rejected?
  c) the hypothesis that she is qualified is correctly accepted?
  d) the hypothesis that she is qualified is correctly rejected?

12.9 Construct a binomial sampling distribution when $N = 7$, $P = Q = \frac{1}{2}$, and answer the following questions: What is the probability that:
  a) six or more will be in the P category?
  b) four or fewer will be in the P category?
  c) one or fewer *or* 6 or more will be in the P category?
  d) two or more will be in the P category?

12.10 A stock-market analyst recommends the purchase or sale of stock by his client on the basis of hypotheses he has formulated about the future behavior of these stocks. What type of error is he making if he claims (a), (b), or (c) below?
  a) H_0: The stock will remain stable. *Fact*: It goes up precipitously.
  b) H_0: The stock will remain stable. *Fact*: It falls abruptly.
  c) H_0: The stock will remain stable. *Fact*: It shows only minor fluctuation about a central value.

12.11 An investigator sets $\alpha = 0.01$ for rejection of H_0. She conducts a study in which she obtains a p-value of 0.02 and fails to reject H_0. *Discuss*: Is it more likely that she is accepting a true or a false H_0?

12.12 *Comment*: a student of economics has collected a mass of data to test 100 different null hypotheses. On completion of the analysis he finds that 5 of the 100 comparisons yield p-values ≤ 0.05. He concludes: "Using $\alpha = 0.05$, I have found a true difference in five of the comparisons."

12.13 *Comment*: An investigator has tested 500 different individuals for evidence of extrasensory perception (ESP). Employing $\alpha = 0.01$, she concludes, "I have found 6 individuals who demonstrated ESP."

12.14 State which of the following are nondirectional hypotheses, which are directional hypotheses, and which are null hypotheses:
  a) The number of hours brand Y lightbulbs burn is greater than the number of hours brand X lightbulbs burn.
  b) There is no difference in the mileage obtained from two different brands of gasoline.

c) Students who have read this chapter have a better chance of passing a statistics exam than do students who have not read this chapter.
d) There is a difference in the shock resistance of two different brands of watches.
e) There is no difference in grades between students who have a class at 8:00 a.m. and those who have the same class at 11:00 a.m.
f) The output from a four-day work week is less than that from a five-day work week (with the same number of total hours).
g) An obnoxious advertisement will reduce the number of that product sold.

12.15 Suppose you want to test the hypothesis that there is not an equal number of male and female executives in a given large company. The appropriate null hypothesis would be:
a) There are more female than male executives.
b) The number of male and female executives are equal.
c) There are more male than female executives.

12.16 Assume that there are exactly the same number of males and females in the population of employees qualified for executive work. What is the probability of each sex becoming an executive, if the executives are selected for the jobs solely on the basis of qualification? If you sampled 30 executives, what would be the expected ratio of males to females?

12.17 To test the hypotheses in the previous two problems, you randomly sampled the sex of 12 executives and found that 10 were male and two were female. What is the probability of this distribution occurring by chance?

12.18 In Exercise 12.17, if you found six female and six male executives, could you state with absolute certainty that the number of male and female executives in that company is equal? Explain your answer.

12.19 Suppose an efficiency expert finds a significant difference between the time it takes people to read a circular dial and the time it takes to read a rectangular dial. Although $\alpha = 0.05$ or $\alpha = 0.01$ is traditionally applied as the level of significance, this choice is arbitrary. For each of the following levels of significance, state how many times in 1000 this difference would be expected to occur by chance:

a) 0.001	b) 0.01	c) 0.005	d) 0.06
e) 0.05	f) 0.095	g) 0.004	h) 0.10

12.20 With reference to Exercise 12.19, the adoption of which level of significance would be most likely to result in the following statements?
a) There is a significant difference between the reading times of the two dials.
b) It cannot be concluded that there is a significant difference between the reading times of the two dials.

12.21 Referring to Exercise 12.19, the adoption of which level of significance would be most likely to result in the following errors? Identify the type of each error.
a) There is a significant difference between the reading times of the two dials. *Fact*: There is no difference.
b) It cannot be concluded that there is a significant difference between the reading times of the two dials. *Fact*: There is a difference.

13 Statistical Inference and Continuous Variables

13.1 INTRODUCTION

In Chapter 12, we illustrated the construction of sampling distributions for a discrete two-category variable (the binomial distribution) and for all possible means when drawing samples of a fixed N from a population of four scores. Table 12.1 showed the frequency and probability distributions of means when all possible samples of a given size were selected from the population of four scores. Figure 13.1 shows probability histograms, with superimposed polygons obtained by connecting the midpoints of each bar.

Before proceeding with the discussion of sampling distributions for continuously distributed variables, examine Table 12.1 and Fig. 13.1 carefully. See if you can answer the following questions.

1. How does the mean of each sampling distribution of means compare with the mean of the population from which the samples were drawn?
2. How does the variability or dispersion of the sample means change as we increase the sample size upon which each sampling distribution is based?

Now compare your answers with ours:

1. The mean of the population of four scores is 5.5. The mean of each sampling distribution of means is 5.5. Thus the mean of a sampling distribution of means is

Fig. 13.1 Probability histograms based upon sampling distributions of means drawn with replacement from a population of four scores and sample sizes (a) $n = 2$, (b) $n = 3$, and (c) $n = 4$.

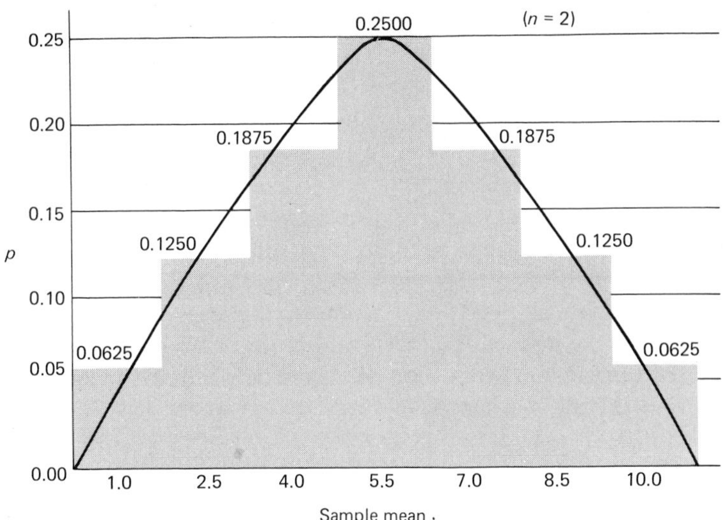
(n = 2)
0.25
0.20
0.15
0.10
0.05
0.00
p
0.2500
0.1875
0.1875
0.1250
0.1250
0.0625
0.0625
1.0
2.5
4.0
5.5
7.0
8.5
10.0
Sample mean

(a)

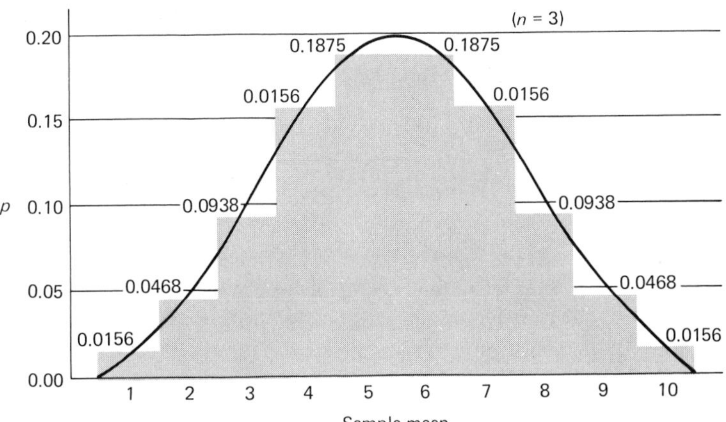
(n = 3)
0.20
0.15
0.10
0.05
0.00
p
0.1875
0.1875
0.0156
0.0156
0.0938
0.0938
0.0468
0.0468
0.0156
0.0156
1
2
3
4
5
6
7
8
9
10
Sample mean

(b)

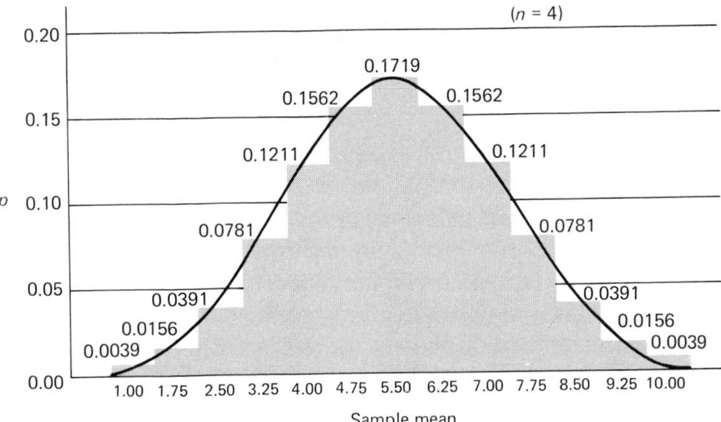
(n = 4)
0.20
0.15
0.10
0.05
0.00
p
0.1719
0.1562
0.1562
0.1211
0.1211
0.0781
0.0781
0.0391
0.0391
0.0156
0.0156
0.0039
0.0039
1.00
1.75
2.50
3.25
4.00
4.75
5.50
6.25
7.00
7.75
8.50
9.25
10.00
Sample mean

(c)

the same as the population mean from which the sample means were drawn. This statement is true for all sizes of N. In other words, the mean of the sampling distribution does not vary with the sample size.

2. As you increase the sample size, the dispersion of sample means becomes less. A greater proportion of means are close to the population mean and extreme deviations are rarer as N becomes larger. To verify these statements, note the probability of obtaining a mean as rare as 1 or 10 at different sample sizes. Note also that the proportion of means in the middle of the distribution becomes greater as sample size is increased. For example, the proportion of means between and including 4 and 7 is 0.6250 when $N = 2$, 0.6874 when $N = 3$, and 0.7265 when $N = 4$.

Finally, the standard deviation of the sample means—which we'll call the *standard error of the mean* ($s_{\bar{X}}$) from this point forward—shows that the dispersion of sample means decreases as sample size is increased.

Figure 13.2 is a line drawing showing the decreasing magnitude of $s_{\bar{X}}$ as the sample size increases. Shown are the population standard deviation of four scores and the standard error of the mean for the sampling distributions when $N = 2$, $N = 3$, and $N = 4$. Note that the decrease is not linear.

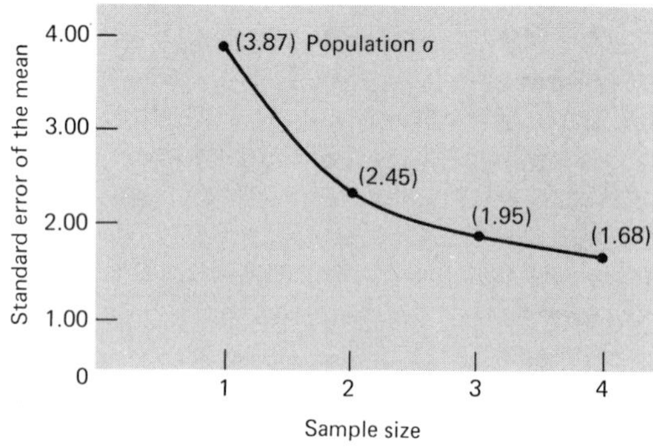

Fig. 13.2 Magnitude of the standard error of the mean as a function of sample size.

13.2 SAMPLING DISTRIBUTION OF THE MEAN

In Table O we provide several different approximately normally distributed populations. Let us conduct a hypothetical sampling experiment with one of these populations which will serve to clarify many of the concepts we shall subsequently develop.

Imagine that we randomly draw (with replacement)* a sample of two cases from the population in which $\mu = 5.00$ and $\sigma = 0.99$. For example, we might draw scores

* If the population is infinite or extremely large, the difference between sampling with or without replacement is negligible.

of 3 and 6. We calculate the sample mean and find $\overline{X} = 4.5$. Now, suppose we continue to draw samples of $N = 2$ (e.g., we might draw scores of 2, 8; 3, 7; 4, 5; 6, 6; etc.) until we obtain an indefinitely large number of samples. If we calculate the sample mean for each sample drawn, and treat each of these sample means as a raw score, we may set up a frequency distribution of these sample means.

Let us repeat the above procedures with increasingly larger sample sizes, for example, $N = 5$, $N = 15$. We now have three frequency distributions of sample means based on three different sample sizes.

Intuitively, what might we expect these distributions to look like? Since we are selecting at random from the population, we would expect the mean of the distribution of sample means to approximate the mean of the population.

How might the dispersion of these sample means compare with the variability in the original distribution of scores? In the original distribution, when $N = 1$, the probability of obtaining a score as extreme as, say, 8 is $\frac{4}{1000}$ or 0.004 (see Table O). The associated probability of obtaining a sample mean equal to 8 when $N = 2$ (i.e., drawing scores of 8, 8) is equal to $\frac{4}{1000} \times \frac{4}{1000}$ or 0.000016 (formula 11.8). Clearly, when $N = 5$, the probability of obtaining results this extreme (for example, $\overline{X} = 8$) becomes exceedingly small. In other words, the probability of drawing extreme values of the sample mean is smaller as N increases. Since the standard deviation is a direct function of the number of extreme scores (see Chapter 7), it follows that a distribution containing proportionately fewer extreme scores will have a lower standard deviation. Therefore, if we treat each of the sample means as a raw score and then calculate the standard deviation ($\sigma_{\overline{X}}$ referred to as the **standard error of the mean***), it is clear that as N increases the variability of the sample means decreases.

If these sampling experiments were actually conducted, the above frequency polygons of sample means would be obtained (Fig. 13.3).

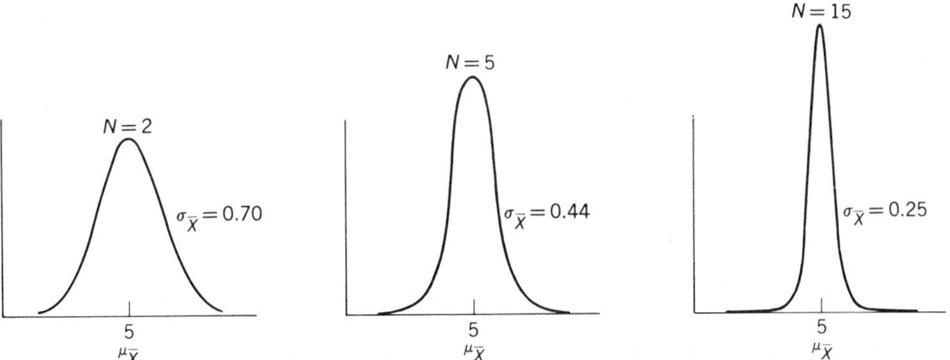

Fig. 13.3 Frequency curves of sample means drawn from a population in which $\mu = 5.00$ and $\sigma = 0.99$.

* This notation represents the standard deviation of a sampling distribution of means. This is purely a theoretical notation since, with an infinite number of sample means, it is not possible to assign a specific value to the number of sample means involved.

There are three important lessons which may be learned from a careful examination of Fig. 13.3.

1. The distribution of sample means, drawn from a normally distributed population, tends to be bell-shaped or "normal." Indeed, it can be shown that even if the underlying distribution is skewed, the distribution of sample means will tend to be normal.

2. The mean of the sample means ($\mu_{\bar{X}}$) is equal to the mean of the population (μ) from which these samples were drawn.

3. The distribution of sample means becomes more and more compact as we increase the size of the sample. This is an extremely important point in statistical inference, about which we shall have a great deal more to say shortly.

If we base our estimate of the population mean on a *single* sample drawn from the population, our approximation to the parameter is likely to be closer as we increase the size of the sample. In other words, if it is true that the dispersion of sample means decreases with increasing sample size, it also follows that the mean of any single sample is more likely to be closer to the mean of the population as the sample size increases.

These three observations illustrate a rather startling theorem which is of fundamental importance in inferential statistics, i.e., the *central-limit theorem.*

The **central-limit theorem** states: If random samples of a fixed N are drawn from *any* population (regardless of the form of the population distribution), as N becomes larger, the distribution of sample means approaches normality with the overall mean approaching μ, the variance of the sample means $\sigma_{\bar{X}}^2$ being equal to σ^2/N, and a standard error $\sigma_{\bar{X}}$ of $\sigma/\sqrt{N}$.

Stated symbolically,

$$\sigma_{\bar{X}}^2 = \frac{\sigma^2}{N}. \tag{13.1}$$

and

$$\sigma_{\bar{X}} = \frac{\sigma}{\sqrt{N}}. \tag{13.2}$$

13.3 TESTING STATISTICAL HYPOTHESES: PARAMETERS KNOWN

Let us briefly examine some of the implications of the relationships we have just discussed.

When μ and σ are *known* for a given population, it is possible to describe the form of the distribution of sample means when N is large (regardless of the form of the original distribution). It will be a normal distribution with a mean ($\mu_{\bar{X}}$) equal to μ and a standard error ($\sigma_{\bar{X}}$) equal to $\sigma/\sqrt{N}$. It now becomes possible to determine probability values in terms of areas under the normal curve. Thus we may use the known relationships of the normal probability curve to determine the probabilities associated with any sample mean (of a given N) randomly drawn from this population.

We have already seen (Section 8.3) that any normally distributed variable may be transformed into the normally distributed z-scale. We have also seen (Section 11.6) that we may establish probability values in terms of the relationships between z-scores and areas under the normal curve. That is, for any given raw score value (X) with a certain proportion of area beyond it, there is a corresponding value of z with the same proportion of area beyond it. Similarly, for any given value of a sample mean ($\bar{X}$) with a certain proportion of area beyond it, there is a corresponding value of z with the same proportion of area beyond it. Thus assuming that the form of the distribution of sample means is normal, we may establish probability values in terms of the relationships between z-scores and areas under the normal curve.

To illustrate: Given a population with $\mu = 250$ and $\sigma = 50$ from which we randomly select 100 scores ($N = 100$), what is the probability that the sample mean ($\bar{X}$) will be equal to or greater than 255? Thus H_0: $\mu = \mu_0 = 250$.

The value of z corresponding to $\bar{X} = 255$ is obtained as follows:

$$z = \frac{\bar{X} - \mu_0}{\sigma_{\bar{X}}}, \tag{13.3}$$

where μ_0 = value of the population mean under H_0,

$$\sigma_{\bar{X}} = \frac{\sigma}{\sqrt{N}} = \frac{50}{\sqrt{100}} = 5.00,$$

and

$$z = \frac{255 - 250}{5.00} = 1.00.$$

Looking up a z of 1.00 in Column C (Table A), we find that 15.87% of the sample means fall at or above $\bar{X} = 255$. Therefore there are approximately 16 chances in 100 of obtaining a sample mean equal to or greater than 255 from this population when $N = 100$.

By a simple extension of the above logic, we may entertain hypotheses concerning the values of parameters of the population from which the sample mean was drawn. For example, given a sample mean of 263 for $N = 100$, is it reasonable to assume that this sample was drawn from the above population?

Let us set up this problem in more formal statistical terms:

1. *Null hypothesis* (H_0): The mean of the population (μ) from which the sample was drawn equals 250, that is, $\mu = \mu_0 = 250$.
2. *Alternative hypothesis* (H_1): The mean of the population from which the sample was drawn does *not* equal 250; $\mu \neq \mu_0$. Note that H_1 is nondirectional; consequently, a two-tailed test of significance will be employed.
3. *Significance level*: $\alpha = 0.01$. If the difference between the sample mean and the specified population mean is so extreme that its associated probability of occurrence under H_0 is equal to or less than 0.01, we will reject H_0.

4. *Critical region for rejection of* H_0: $|z| \geq |z_{0.01}| = 2.58$*. *A* **critical region** *is that portion of area under the curve which includes those values of a statistic which lead to rejection of the null hypothesis.*

The critical region is chosen to correspond with the selected level of significance. Thus for $\alpha = 0.01$, two-tailed test, the critical region is bounded by those values of $z_{0.01}$ which mark off a total of 1% of the area. Referring to Column C (Table A), we find that the area beyond a z of 2.58 is approximately 0.005. We double 0.005 to account for both tails of the distribution. Figure 13.4 depicts the critical region for rejection of H_0 when $\alpha = 0.01$, two-tailed test.

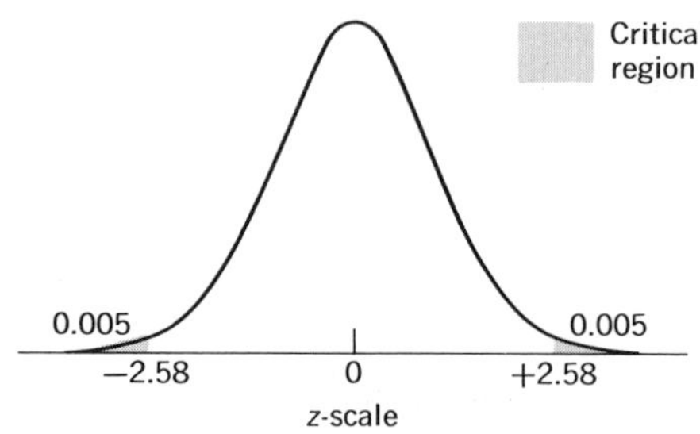

Fig. 13.4 Critical region for rejection of H_0 when $\alpha = 0.01$, two-tailed test.

It is convenient to indicate the critical region in terms of the *critical value* or values that bound it. Therefore, in order to reject H_0 at the 0.01 level of significance, the absolute value of the obtained z must be equal to or greater than $|z_{0.01}|$ or 2.58. Similarly, to reject H_0 at the 0.05 level of significance, the absolute value of the obtained z must be equal to or greater than $|z_{0.05}|$ or 1.96.

In the present example, the value of z corresponding to $\bar{X} = 263$ is

$$z = \frac{\bar{X} - \mu_0}{\sigma_{\bar{X}}} = \frac{263 - 250}{5.00} = 2.60.$$

Decision: Since the obtained z falls within the critical region (that is, $2.60 > |z_{0.01}|$), we may reject H_0 at the 0.01 level of significance.

13.4 ESTIMATION OF PARAMETERS: POINT ESTIMATION

So far, we have been concerned with testing hypotheses when the population parameters are known. However, we have taken some pains in this book to point out that population values are rarely known. The fact that we do not know the population values does not prevent us from using the above logic.

* Since $z_{0.01} = \pm 2.58$, $|z| \geq |z_{0.01}|$ is equivalent to stating $z \geq 2.58$ or $z \leq -2.58$.

Whenever we make inferences about population parameters from sample data, we compare our sample results with the expected results given by the appropriate sampling distribution. A hypothetical sampling distribution of sample means is associated with any sample mean. This distribution has a mean, $\mu_{\bar{X}}$, and a standard deviation, $\sigma_{\bar{X}}$. So far, in order to obtain the values of $\mu_{\bar{X}}$ and $\sigma_{\bar{X}}$, we have required a knowledge of μ and σ. In the absence of knowledge concerning the exact values of the parameters, we are forced to estimate μ and σ from the statistics calculated from sample data. Since, in actual practice, we rarely select more than one sample, our estimates are generally based on the statistics calculated from a single sample. All such estimates, involving the use of single sample values, are known as *point estimates.*

You will recall that the variance of a sample is defined as:

$$s^2 = \frac{\sum(X - \bar{X})^2}{N - 1}.$$

We obtained the standard deviation by finding the square root of this value. The sample variance and sample standard deviation provide the basis for estimating $\sigma_{\bar{X}}^2$ and σ_X from sample data. We shall employ the symbols $s_{\bar{X}}^2$ and $s_{\bar{X}}$ to refer to the estimated variance and the standard error of the mean, respectively.

Thus the formula for determining the variance of the mean from sample data is:

$$\text{estimated } \sigma_{\bar{X}}^2 = s_{\bar{X}}^2 = \frac{s^2}{N}. \tag{13.4}$$

We estimate the standard error of the mean by finding the square root of this value:

$$\text{estimated } \sigma_{\bar{X}} = s_{\bar{X}} = \sqrt{\frac{s^2}{N}} = \frac{s}{\sqrt{N}}. \tag{13.5}$$

Before proceeding further let us briefly review some of the symbols we have been discussing:

μ	Population mean
μ_0	Value of the population mean under H_0
$\bar{X}$	Sample mean
σ^2	Population variance
s^2	Sample variance, $\dfrac{\sum(X - \bar{X})^2}{N - 1}$
σ	Population standard deviation, $\sqrt{\sigma^2}$
$\mu_{\bar{X}}$	Mean of the sampling distribution of sample means

$\sigma_{\bar{X}}$ Standard error of the mean (theoretical), $\dfrac{\sigma}{\sqrt{N}}$

$s_{\bar{X}}$ Estimated standard error of the mean, $\dfrac{s}{\sqrt{N}}$

13.5 TESTING STATISTICAL HYPOTHESES WITH UNKNOWN PARAMETERS: STUDENT'S *t*

We previously pointed out that, when the parameters of a population are known, it is possible to describe the form of the sampling distribution of sample means. It will be a normal distribution with $\sigma_{\bar{X}}$ equal to $\sigma/\sqrt{N}$. By employing the relationship between the z-scale and the normal distribution, we were able to test hypotheses using $z = (\bar{X} - \mu_0)/\sigma_{\bar{X}}$ as a test statistic. When σ is not known, we are forced to estimate its value from sample data. Consequently, estimated $\sigma_{\bar{X}}$ (that is, $s_{\bar{X}}$) must be based on the estimated $\sigma(s)$, that is, $s_{\bar{X}} = s/\sqrt{N}$. Now, if substituting s for σ provided a reasonably good approximation to the sampling distribution of means, we could continue to use z as our test statistic and the normal curve as the model for our sampling distribution. As a matter of fact, however, this is not the case. At the turn of the century, a statistician by the name of William Gosset, who published under the pseudonym of Student, noted that the approximation of s to σ is poor, particularly for small samples. This failure of approximation is due to the fact that, with small samples, s will tend to underestimate σ more than one half of the time. Consequently, the statistic

$$\frac{\bar{X} - \mu_0}{s/\sqrt{N}}$$

will tend to be spread out more than the normal distribution.

Gosset's major contribution to statistics consisted of his description of a distribution, or rather, a family of distributions, which permits the testing of hypotheses with samples drawn from normally distributed populations, when σ is not known. These distributions are referred to variously as the ***t*-distributions** or **Student's *t***. The ratio employed in the testing of hypotheses is known as the t-ratio:

$$t = \frac{\bar{X} - \mu_0}{s_{\bar{X}}}, \tag{13.6}$$

where μ_0 is the value of the population mean under H_0.

The t-statistic is similar in many respects to the previously discussed z-statistic. Both statistics are expressed as the deviation of a sample mean from a population mean (known or hypothesized) in terms of the standard error of the mean. By reference to the appropriate sampling distribution, we may express this deviation in terms of probability. When the z-statistic is used, the standard normal curve is the appropriate sampling distribution. For the t-statistic there is a family of distributions which vary as a function of *degrees of freedom* (df).

The term **degrees of freedom** refers to the number of values which are free to vary after we have placed certain restrictions on our data. To illustrate, if we had four numbers on which we placed the restriction that the sum must equal 115, it is clear that three numbers could take on any value (i.e., are free to vary) whereas the fourth would be fixed. Thus, if three values are 30, 45, and 25, respectively, the fourth must be 15 in order to make the sum equal to 115. The number of degrees of freedom, then, is $N - 1$, or 3. Generalizing, for any given sample on which we have placed a single restriction, the number of degrees of freedom is $N - 1$.

13.5.1 Characteristics of *t*-Distributions

Let us compare the characteristics of the ***t*-distributions** with the already familiar standard normal curve. First, both distributions are symmetrical about a mean of zero. Therefore the proportion of area beyond a particular positive t-value is equal to the proportion of area below the corresponding negative t.

Secondly, the t-distributions are more spread out than the normal curve. Consequently, the proportion of area beyond a specific value of t is *greater* than the proportion of area beyond the corresponding value of z. However, the greater the df, the more the t-distributions resemble the standard normal curve. In order that you may see the contrast between the t-distributions and the normal curve we have reproduced three curves in Fig. 13.5: the sampling distributions of t when df = 3 and when df = 10, and the normal curve.

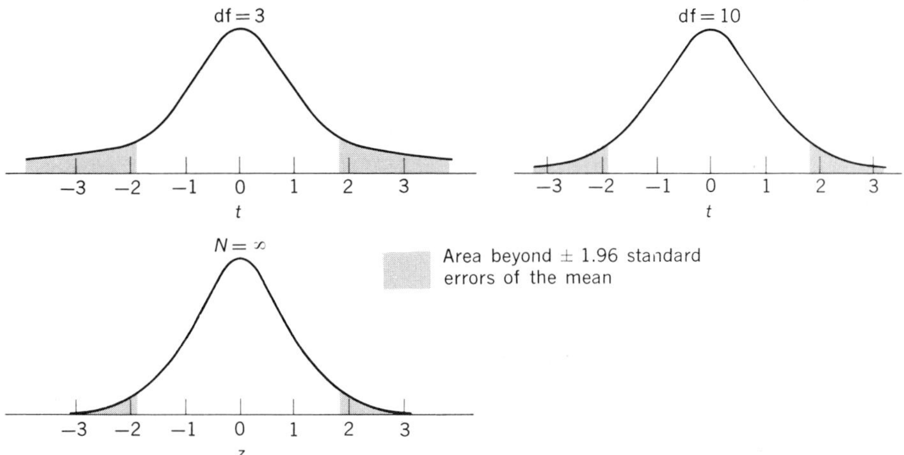

Fig. 13.5 Sampling distributions of $t = \dfrac{\bar{X} - \mu_0}{s_{\bar{X}}}$, when df = 3 and 10, compared to the standard normal curve.

Inspection of Fig. 13.5 permits several interesting observations. We have already seen that with the standard normal curve, $z \geq |1.96|$ defines the region of rejection at the 0.05 level of significance. However, when df = 3, a $t \geq +1.96$ or $t \leq -1.96$ includes approximately 15% of the total area. Consequently, if we were to employ the normal curve for testing hypotheses when N is small (therefore df is small) and σ is unknown, we would be in serious danger of making a type I error, i.e., rejecting H_0 when it is true. Obviously, a much larger value of t is required to mark off the bounds of the critical region of rejection. Indeed, when df = 3, the absolute value of the obtained t must be equal to or greater than 3.18 to reject H_0 at the 0.05 level of significance (two-tailed test). However, as df increases, the differences in the proportions of area under the normal curve and the Student t-distributions become negligible.

In contrast to our use of the normal curve, the tabled values for t (Table C) are **critical values of *t***, i.e., *those values which bound the critical rejection regions corresponding to varying levels of significance.* Thus in using the table for the distributions of t, we locate the appropriate number of degrees of freedom in the left-hand column, and then find the column corresponding to the chosen α. The tabled values represent the t-ratio required for significance. If the absolute value of our obtained t-ratio equals or exceeds this tabled value, we may reject H_0.

13.5.2 Illustrative Problem: Student's *t*

Let us now examine a hypothetical case involving a small sample.

A group of 16 ninth grade students selected on the basis of an expressed "interest" in science were given a test measuring their knowledge of basic scientific concepts. The test was constructed to yield a normal distribution with a mean for ninth graders equal to 78. The results of the study were:

$$\bar{X} = 84, \qquad s = 16, \qquad N = 16,$$

Is it reasonable to assume that these ninth graders are representative of a population in which $\mu = 78$?

Let us set up this problem in more formal statistical terms.

1. *Null hypothesis* (H_0): The mean of the population from which the sample was drawn equals 78 ($\mu = \mu_0 = 78$).
2. *Alternative hypothesis* (H_1): The mean of the population from which the sample was drawn does *not* equal 78 ($\mu \neq \mu_0$).
3. *Statistical test*: The Student t-ratio is chosen because we are dealing with a normally distributed variable in which σ is unknown.
4. *Significance level*: $\alpha = 0.05$.
5. *Sampling distribution*: The sampling distribution is the Student t-distribution with df = 15.
6. *Critical region*: $t_{0.05} = \pm 2.13$. Since H_1 is nondirectional, the critical region consists of all values of $t \geq +2.13$ or $t \leq -2.13$.

In the present example, the value of t corresponding to $\bar{X} = 84$ is

$$t = \frac{\bar{X} - \mu_0}{s_{\bar{X}}} = \frac{84 - 78}{16/\sqrt{16}} = 1.50.$$

Decision: Since the obtained t does not fall within the critical region (that is, $1.50 < |t_{0.05}|$), we accept H_0.

13.6 ESTIMATION OF PARAMETERS: INTERVAL ESTIMATION

We have repeatedly pointed out that one of the basic problems in inferential statistics is the estimation of the parameters of a population from statistics calculated from a sample. This problem, in turn, involves two subproblems: (1) **point estimation**, and (2) **interval estimation**.

When we estimate parameters employing single sample values, these estimates are known as *point estimates.* A single sample value drawn from a population provides an estimate of the population parameter. But, how good an estimate is it? If a population mean were known to be 100, would a sample mean of 60 constitute a good estimate? How about a sample mean of 130, 105, 98, or 99.4? Under what conditions do we consider an estimate good? Since we know that the population parameters are virtually never known and that we generally employ samples to estimate these parameters, is there any way to determine the amount of error we are likely to make? The answer to this question is a negative one. However, it is possible, not only to estimate the population parameter (*point estimation*), but also to state a range of values within which we are confident that the parameter falls (*interval estimation*). Moreover, we may express our confidence in terms of probability theory.

Let us say that we are to estimate the weight of a man based on physical inspection. Let us assume that we are unable to place him on a scale, and that we cannot ask him his weight. This problem is similar to many we have faced throughout this text. We cannot know the population value (the man's true weight) and hence we are forced to estimate it. Let us say that we have the impression that he weighs about 200 pounds. If we are asked, "How confident are you that he weighs *exactly* 200 pounds?," we would probably reply, "I doubt that he weighs exactly 200 pounds. If he does, you can credit me with a fantastically lucky guess. However, I feel reasonably confident that he weighs between 190 and 210 pounds." In doing this, we have stated the interval within which we feel confident that the true weight falls. After a moment's reflection, we might hedge slightly, "Well, he is almost certainly between 180 and 220 pounds. In any event, I feel perfectly confident that his true weight falls somewhere between 170 and 230 pounds." Note that the greater the size of the interval, the greater is our feeling of certitude that the true value is encompassed between these limits. Note also that in stating these confidence limits, we are, in effect, making two statements: (1) We are stating the limits between which we feel our subject's true weight falls, and (2) we are rejecting the possibility that his true weight falls outside of these limits. Thus if someone asks, "Is it conceivable that our subject weighs as much as 240 pounds or as little as 160 pounds?," our reply would be a negative one.

13.7 CONFIDENCE INTERVALS AND CONFIDENCE LIMITS

In our preceding example, we were, in a sense, concerning ourselves with the problem of estimating *confidence limits.* In effect, we were attempting to determine the interval within which any hypotheses concerning the weight of the man might be considered tenable and outside which any hypotheses would be considered untenable. The interval within which we consider hypotheses tenable is known as the **confidence interval**, and the limits defining the interval are referred to as **confidence limits**.

Let us look at a sample problem and apply our statistical concepts to the estimation of confidence intervals.

A school district is trying to decide on the feasibility of setting up a vocational training program in its public high school curriculum. In part, the decision will depend on estimates of the average I.Q. of high school students within the district. With only one school psychologist in the district, it is impossible to administer an individual test to each student. Consequently, we must content ourselves with testing a random sample of students and base our estimates on this sample. We administer this test to a random sample of 25 students and obtain the following results:

$$\bar{X} = 108,$$

$$s = 15,$$

$$N = 25.$$

Our best estimate of the population mean (i.e., the mean I.Q. of children within the school district) is 108. However, even though our sample statistics provide our best estimates of population values, we recognize that such estimates are subject to error. As with the weight problem, we would be fantastically lucky if the mean I.Q. of the high school population were actually 108. On the other hand, if we have employed truly random selection procedures, we have a right to believe that our sample value is fairly close to the population mean. The critical question becomes: Between what limits will we entertain, as tenable, hypotheses concerning the value of the population mean (μ) in intelligence?

We have seen that the mean of the sampling distribution of sample means ($\mu_{\bar{X}}$) is equal to the mean of the population. We have also seen that, since, for any given N, we may determine how far sample means are likely to deviate from any given or hypothesized value of μ, we may determine the likelihood that a particular $\bar{X}$ could have been drawn from a population with a mean of μ_0, where μ_0 represents the value of the population mean under H_0. Now, since we do not know the value of the population mean, we are free to hypothesize *any* value we desire.

It should be clear that we could entertain an unlimited number of hypotheses concerning the population mean and subsequently reject them, or fail to reject them, on the basis of the size of the t-ratios. For example, in the present problem, let us select a number of hypothetical population means. We may employ the 0.05 level of significance (two-tailed test), and test the hypothesis that $\mu_0 = 98$. The value of t corresponding to $\bar{X} = 108$ is

$$t = \frac{108 - 98}{15/\sqrt{25}} = 3.333.$$

In Table C, we find that $t_{0.05}$ for 24 df is 2.06. Since our obtained t is greater than this critical value, we reject the hypothesis that $\mu_0 = 98$. In other words, it is unlikely that $\bar{X} = 108$ was drawn from a population with a mean of 98. Our null hypothesis is that $\mu_0 = 100$, which gives a t of:

$$t = \frac{108 - 100}{3} = 2.667.$$

Since $2.667 > t_{0.05}$ (or 2.06), we may reject the hypothesis that the population mean is 100.

If we hypothesize $\mu_0 = 102$, the resulting t-ratio of 2.000 is less than $t_{.005}$. Consequently, we may consider the hypothesis that $\mu_0 = 102$ tenable. Similarly, if we obtained the appropriate t-ratios, we would find that the hypothesis $\mu_0 = 114$ is tenable, whereas hypotheses of values greater than 114 are untenable. Thus $\bar{X} = 108$ was probably drawn from a population whose mean falls in the interval 102–114 (note that these limits, 102 and 114, represent approximate limits, i.e., the closest integers). The hypothesis that $\bar{X} = 108$ was drawn from a population with $\mu < 102$ or $\mu > 114$ may be rejected at the 0.05 level of significance. The interval within which the population mean probably lies is called the *confidence interval.* We refer to the limits of this interval as the *confidence limits.* Since we have been employing $\alpha = 0.05$, we call it the **95% confidence interval**. Similarly, if we employed $\alpha = 0.01$, we could obtain the **99% confidence interval**.

It is not necessary to perform all the above calculations to establish the confidence limits. We may calculate the exact limits of the 95% confidence interval directly.

To determine the upper limit for the 95% confidence interval:

$$\text{upper limit } \mu_0 = \bar{X} + t_{0.05}(s_{\bar{X}}). \tag{13.7}$$

Similarly, for the lower limit,

$$\text{lower limit } \mu_0 = \bar{X} - t_{0.05}(s_{\bar{X}}). \tag{13.8}$$

For the 99% confidence interval, merely substitute $t_{0.01}$ in the above formulas.

You will note that these formulas are derived algebraically from formula (13.9):

$$t_{0.05} = \frac{\bar{X} - \mu_0}{s_{\bar{X}}}, \qquad \text{therefore } \mu_0 = \bar{X} + t_{0.05}(s_X).$$

Employing formula (13.7), we find that the upper 95% confidence limit in the above problem is

$$\text{upper limit } \mu_0 = 108 + (2.06)(3.0) = 108 + 6.18 = 114.18.$$

Similarly, employing formula (13.8), we find that the lower confidence limit is

$$\text{lower limit } \mu_0 = 108 - (2.06)(3.0) = 108 - 6.18 = 101.82.$$

Having established the lower and the upper limits as 101.82 and 114.18, respectively, we may now conclude: On the basis of our obtained mean and standard deviation, which were computed from scores drawn from a population in which the the true mean is unknown, we assert that the population mean probably falls within the interval which we have established. Since the probability that the population mean lies outside these limits is 5% ($\alpha = 0.05$), the probability that this interval contains the population mean is 95%. In other words, in the long run, we will be correct 95% of the time when we state that the true mean lies within the 95% confidence interval.

Some words of caution in interpreting the confidence interval. In establishing the interval, within which we believe the population mean falls, we have *not* established any probability that our obtained mean is correct. In other words, we cannot claim that the chances are 95 in 100 that the population mean is 108. Our statements are valid only with respect to the interval and not with respect to any particular value of the sample mean. In addition, since the population mean is a fixed value and does not have a distribution, our probability statements never refer to μ. The probability we assert is about the interval, i.e., the probability that the interval contains μ.

Finally, when we have established the confidence interval of the mean, we are not stating that the probability is 0.95 that the particular interval we have calculated contains the population mean. It should be clear that, if we were to select repeated samples from a population, both the sample means and the standard deviations would differ from sample to sample. Consequently, our estimates of the confidence interval would also vary from sample to sample. When we have established the 95% confidence interval of the mean, then we are stating that if repeated samples of a given size are drawn from the population, 95% of the interval estimates will include the population mean.

13.8 TEST OF SIGNIFICANCE FOR PEARSON *r*, ONE-SAMPLE CASE

In Chapter 9 we discussed the calculation of two statistics—the Pearson r and r_{rho}—commonly employed to describe the extent of the relationship between two variables. It will be recalled that the coefficient of correlation varies between ± 1.00, with $r = 0.00$ indicating the absence of a relationship. It is easy to overlook the fact that correlation coefficients based on sample data are only estimates of the corresponding population parameter and, as such, will distribute themselves about the population value. Thus it is quite possible that a sample drawn from a population in which the true correlation is zero may yield a high positive or negative correlation *by* chance. The null hypothesis most often investigated in the one-sample case is that the *population correlation coefficient* (ρ) is zero.

It is clear that a test of significance is called for. However, the test is complicated by the fact that the sampling distribution of ρ is usually nonnormal, particularly as as ρ approaches the limiting values of ± 1.00. Consider the case in which ρ equals $+0.80$. It is clear that sample correlation coefficients drawn from this population will distribute themselves around $+0.80$ and can take on any value from -1.00 to $+1.00$. It is equally clear, however, that there is a definite restriction in the range of values that sample statistics greater than $+0.80$ can assume whereas there is no similar restriction for values less than $+0.80$. The result is a negatively skewed sampling distribution. The departure from normality will, in general, be less as the number of paired scores in the sample increases. When the population correlation from which the sample is drawn is equal to zero, the sampling distribution is more likely to be normal. These relationships are demonstrated in Fig. 13.6 which illustrates the sampling distribution of the correlation coefficient when $\rho = -0.80$, 0, and $+0.80$.

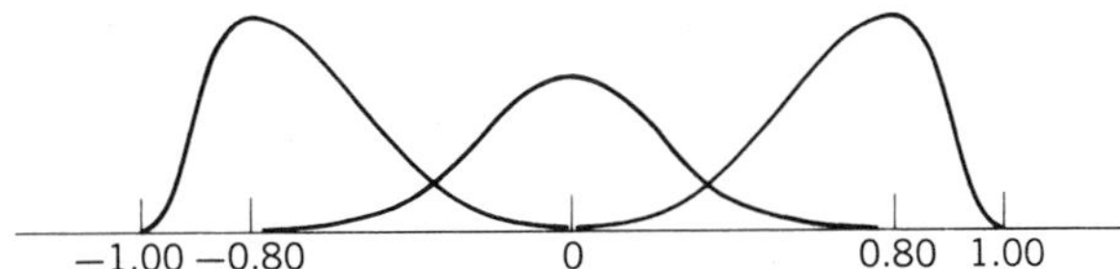

Fig. 13.6 Illustrative sampling distributions of correlation coefficients when $\rho = -0.80$ and $+0.80$.

Fisher has described a procedure for transforming sample r's to a statistic z_r, which yields a sampling distribution more closely approximating the normal curve, even for samples employing small n's. The transformation to z_r's involves the use of the following formula:

$$z_r = \tfrac{1}{2}\log_e(1 + r) - \tfrac{1}{2}\log_e(1 - r).$$

The test statistic is the normal deviate, z, in which

$$z = \frac{z_r - Z_r}{\sqrt{\dfrac{1}{n - 3}}}, \qquad (13.9)$$

where

z_r = the transformed value of the sample r, and

Z_r = the transformed value of the population correlation coefficient specified under H_0.

Obtaining z_r is greatly simplified by Table F which shows the value of z_r corresponding to each value of r, in steps of 0.01, between 0.00 and 0.99. Thus, referring to Table F, we see that an r of 0.60, for example, has a corresponding z_r of 0.693.

Let us look at an illustrative example: A sociologist has developed a scale which purports to measure the degree of submission to authority. He correlates the scores made on the scale by 15 subjects with their scores on an inventory which reveals the degree of prejudice felt toward minority groups. He obtains a Pearson r of 0.60. May he conclude that the obtained correlation is not likely to have been drawn from a population in which the true correlation is zero?

Let us set up this problem in formal statistical terms.

1. *Null hypothesis* (H_0): The population correlation coefficient from which this sample was drawn equals 0.00 ($\rho = 0.00$).
2. *Alternative hypotheses* (H_1): The population correlation coefficient from which the sample was drawn does *not* equal 0.00 ($\rho \neq 0.00$).
3. *Statistical test*: The z-test based on Fisher's transformation of r and ρ to z_r and Z_r respectively.
4. *Significance level*: $\alpha = 0.05$, two-tailed test.
5. *Sampling distribution*: The normal probability curve.
6. *Critical region*: $z_{0.05} = \pm 1.96$. Since H_1 is nondirectional, the critical region consists of all values of $z \geq 1.96$ and $z \leq -1.96$.

In the present example, the value of z_r corresponding to $r = 0.60$ is 0.693 and $Z_r = 0.00$. Thus

$$z = \frac{0.693 - 0.00}{\sqrt{\dfrac{1}{15 - 3}}} = \frac{0.693}{0.289} = 2.40.$$

Decision: Since the obtained $z > |z_{0.05}|$, it falls within the critical region for rejecting H_0. Thus it may be concluded that the sample was drawn from a population in which $\rho > 0.00$.

13.8.1 Test of Significance of r_{rho}, One-Sample Case

Table G presents the critical values of r_{rho}, one- and two-tailed tests, for selected values of n from 5 to 30.

In Section 9.5 we demonstrated the calculation of r_{rho} from data consisting of 15 pairs of ranked scores. A correlation of 0.64 was found.

Since $n = 15$ is not listed in Table G, it is necessary to interpolate, employing the critical values for $n = 14$ and $n = 16$. The critical value at the 0.05 level, two-tailed test, for $n = 14$ is 0.544; at $n = 16$, it is 0.506. By employing linear interpolation, we may roughly approximate the critical value corresponding to $n = 15$. Linear interpolation involves subtracting the latter from the former, dividing by 2 (since

$n = 15$ is half-way between $n = 14$ and $n = 16$) and adding to the latter. Thus

$$r_{\text{rho}(0.05)} = 0.506 + \frac{(0.544 - 0.506)}{2} = 0.525.$$

Since our obtained r_{rho} of 0.65 exceeds the critical value at the 0.05 level, we may conclude that the population value of the Spearman correlation coefficient from which the sample was drawn is greater than 0.00.

CHAPTER SUMMARY

We have seen that if we take a number of samples from a given population, then

1. the distribution of sample means tends to be normal,
2. the mean of these sample means ($\mu_{\bar{X}}$) is equal to the mean of the population (μ), and
3. the standard error of the mean ($\sigma_{\bar{X}}$) is equal to $\sigma/\sqrt{N}$. As N increases, the variability decreases.

We used these relationships in the testing of hypotheses (for example, $\mu = \mu_0$), when the standard deviation of a population was known, employing the familiar z-statistic and the standard normal curve.

When σ is not known, we demonstrated the use of sample statistics to estimate these parameters. We used these estimates of the parameters to test hypotheses, employing the Student t-ratio and the corresponding sampling distributions. We compared these t-distributions, which vary as a function of degrees of freedom (df), with the standard normal curve.

We employed the t-ratio as a basis for establishing confidence intervals.

Finally, we demonstrated the test of significance for the Pearson r and the Spearman r_{rho}, one-sample case.

Terms to Remember:

Standard error of the mean
Central-limit theorem
Law of large numbers
Critical region
Unbiased estimate of a parameter
$\hat{s}^2$ ***and*** $\hat{s}$
t-distributions
Student's t-ratio
degrees of freedom (df_4)
Critical values of t
Point estimation
Interval estimation
Confidence limits
95% *Confidence interval*
Confidence limits
99% *Confidence interval*
z_r
Z_r
ρ

EXERCISES

13.1 Describe what happens to the distribution of sample means when you
a) increase the size of each sample.
b) increase the number of samples.

13.2 Explain why the standard deviation of a sample will usually underestimate the standard deviation of a population. Give an example.

13.3 Given that $\bar{X} = 24$ and $s = 4$ for $N = 14$ use the t-distribution to find
a) the 95% confidence limits for μ,
b) the 99% confidence limits for μ.

13.4 Given that $\bar{X} = 24$ and $s = 4$ for $N = 121$, use the t-distribution to find
a) the 95% confidence limits for μ,
b) the 99% confidence limits for μ.
Compare the results with those of Exercise 13.2.

13.5 An instructor gives his class an examination which, as he knows from years of experience, yields $\mu = 78$ and $\sigma = 7$. His present class of 22 obtains a mean of 82. Is he correct in assuming that this is a superior class? Employ $\alpha = 0.01$, two-tailed test.

13.6 An instructor gives her class an examination which, as she knows from years of experience, yields $\mu = 78$. Her present class of 22 obtains $\bar{X} = 82$ and $s = 7$. Is she correct in assuming that this is a superior class? Employ $\alpha = 0.01$, two-tailed test.

13.7 Explain the difference between Exercises 13.5 and 13.6. What test statistic is employed in each case and why? Why is the decision different? Generalize: What is the effect of knowing σ upon the likelihood of a type II error?

13.8 Superintendent X of Zuba school district claims that the children in his district are brighter, on the average, than the general population of students. In order to determine the I.Q. of school children in the district, a study was conducted. The results were as follows.

Test scores	Test scores
105	115
109	103
115	110
112	125
124	99

The mean of the general population of school children is 106. Set this up in formal statistical terms (that is, H_0, H_1, etc.), and draw the appropriate conclusions. Employ a one-tailed test, $\alpha = 0.05$.

13.9 For a particular population with $\mu = 28.5$ and $\sigma = 5.5$, what is the probability that, in a sample of 100, the $\bar{X}$ will be:
a) equal to or less than 30.0
b) equal to or less than 28.0?
c) equal to or more than 29.5?
d) between 28.0 and 29.0?

13.10 Given that $\bar{X} = 40$ for $N = 24$ from a population in which $\sigma = 8$, find:

a) the 95% confidence limits for μ,

b) the 99% confidence limits for μ.

13.11 It is axiomatic that when pairs of individuals are selected at random and the intelligence test scores of the first members of the pairs are correlated with the second members, $\rho = 0.00$.

a) Thirty-nine pairs of siblings are randomly selected and an $r = +0.27$ is obtained between members of the pairs for intelligence. Are siblings more alike in intelligence than unrelated individuals?

b) A study of 28 pairs of monozygotic twins yields $r = +0.91$ on intelligence test scores. What do you conclude?

13.12 *ABC* University claims that because of its superior facilities and close faculty supervision, its students complete the Ph.D program earlier than is usual. They base this assertion on the fact that the national mean age for completion is 32.11, whereas the mean age of their 25 Ph.D.'s is 29.61 with $s = 6.00$. Test the validity of their assumption.

13.13 Employing the data in Exercise 13.12 find the interval within which you are 95% confident that the true population mean (average age for Ph.D.'s at *ABC* University) probably falls.

13.14 A sociologist asserts that the average length of courtship is longer before a second marriage than before a first. She bases this assertion on the fact that the average for first marriages is 265 days, whereas the average for second marriages (e.g., for her 625 subjects) is 265.5 days, with $s = 50$. Test the validity of her assumption.

13.15 Employing the data in Exercise 13.14, find the interval within which you are 99% confident that the true population mean (average courtship days for a second marriage) probably falls.

13.16 Random samples of size 2 are selected from the following finite population of scores: 1, 3, 5, 7, 9, and 11.

a) Construct a histogram showing the sampling distribution of means when $n = 2$. Employ sampling *without* replacement.

b) Construct a histogram showing the means of all possible samples that can be drawn employing sampling *with* replacement.

13.17 Employing Exercise 13.16(a), answer the following: Selecting a sample with $n = 2$, what is the probability that:

a) a mean as high as 10 will be obtained?

b) a mean as low as 2 will be obtained?

c) a mean as deviant as 8 will be obtained?

d) a mean as low as 1 will be obtained?

13.18 Employing Exercise 13.16(b), answer the following: Selecting a sample with $n = 2$, what is the probability that:

a) a mean as high as 10 will be obtained?

b) a mean as low as 2 will be obtained?

c) a mean as deviant as 8 will be obtained?

d) a mean as low as 1 will be obtained?

13.19 A stock analyst claims that she has an unusually accurate method for forecasting price gains of listed common stock. During a given period, stocks she advocated showed the following price gains: \$1.25, \$2.50, \$1.75, \$2.25, \$3.25, \$3.00, \$2.00, \$2.00. During the same period, the market as a whole showed a mean price gain of \$1.83. Set up and test the H_0 that the stocks she selected have been randomly selected from the population of stock gains during the specified period.

13.20 In a test of a gasoline additive, a group of carefully engineered automobiles were run at a testing site under rigorously supervised conditions. The number of miles obtained from a single gallon of gasoline were: 15, 12, 13, 16, 17, 11, 14, 15, 13, 14. Thousands of prior trials with the same gasoline minus the additive had yielded the expectation of 12.62 miles per gallon. Using $\alpha = 0.05$, can it be concluded that the additive improved gasoline mileage?

13.21 A restaurant owner ranked his 17 waiters in terms of their speed and efficiency on the job. He correlated these ranks with the total amount of tips each of these waiters received for a one-week period. The obtained $r_{\text{rho}} = 0.438$. What do you conclude?

13.22 The owner of a car-leasing company ranked 25 of her customers on their neatness and general care of their rented cars during a three-month period. She correlated these ranks with the number of miles each customer drove during this same period. The obtained $r_{\text{rho}} = -0.397$. Employing $\alpha = 0.05$, two-tailed test, what do you conclude?

13.23 As a requirement for admission to *ABC* University, a candidate must take a standardized entrance examination. The correlation between performance on this examination and college grades is 0.43.

a) The director of admissions claims that a better way to predict college success is by using high school averages. To test his claim, he randomly selects 52 students and correlates their college grades with their high school averages. He obtains $r = 0.54$. What do you conclude?

b) The director's assistant constructs a test which he claims is better for predicting college success than the one currently used. He randomly selects 67 students and correlates their grade point averages with performance on his test. The obtained $r = 0.61$. What do you conclude?

13.24 Referring to Exercise 9.21, suppose a person states that she believes that the correlation between home runs and runs batted in is really zero. Employing a significance level of 0.05, could you reject her statement? What is z?

13.25 Assume that the mean number of minutes it takes to read a circular dial is 1.2 sec, with a σ of 0.2. If it takes a mean of 1.5 sec for 16 people to read a rectangular dial, can it be concluded that a circular dial is read more quickly? What is z?

13.26 A company measured the mean number of identical units produced by workers per day and found a μ of 60 and a σ of 10. Another company states that their method of producing the same units is superior, because the mean number of units produced is 62 for 49 days. How would you settle the debate between the two companies? What is the value of z?

13.27 It was stated in the chapter that the distribution of sample means approaches normality even though the population distribution may be skewed. As an informal demonstration of this, assume that the following numbers comprise the entire population, and that the scale of measurement is continuous:

2	2.5	3.5	4	4	4.5	4.5	5	5	5.5	5.5	5.5
5.5	5.5	6	6	6	6	6	6	6	6	6	6
7	7	7	7	7	7	7	7	7	7	7	7
7.5	7.5	7.5	7.5	7.5	7.5	7.5	7.5	8			

Represent each measurement of the population on a piece of paper.

a) Randomly draw 10 samples of $N = 2$ and construct a polygon of the distribution of sample means.
b) Draw 10 samples of $N = 5$ and construct a polygon of the distribution of sample means.
c) Draw 10 samples of $N = 10$ and construct a polygon of the distribution of sample means.

13.28 Assume that the mean number of words typed per minute by secretaries is 50. A group of people who attended a certain secretarial school show the following typing speeds (words per minute):

55	50	45	75	80	75	80	80
85	60	50	65	60	50	80	50

Calculate the value of t. Can you conclude that the school produced secretaries with superior typing abilities?

13.29 A sports fan found a r_{rho} of 0.15 between batting averages of 20 baseball players and the amount of vitamins in their breakfast cereals. Can he conclude that this correlation is significant?

13.30 Ms. Smith stated that her training program for selling life insurance enables a company to sell more insurance than the "average" company. The mean amount of life insurance sold by all salespersons per month is $100,000. A sample of ten people who have been through the training program show monthly selling rates (in thousands) of:

$100	$120	$130	$120	$125
$ 90	$130	$135	$140	$110

If you were an insurance sales supervisor, would you adopt Ms. Smith's training program? Calculate the value of t.

14 Statistical Inference with Two Independent Samples

BOX 14.1

Throughout the text we have looked at mileage data on 1976 model automobiles in order to illustrate important aspects of descriptive statistics.

Using the Table of Random Digits (Table R), we selected ten cars under 3000 pounds and ten over 4000 pounds. We obtained both the city and the highway miles-per-gallon for each car. The accompanying table shows the raw scores obtained by cars in both weight classes.

	Under 3000 pounds			Over 4000 pounds	
Car	City mpg	Highway mpg	Car	City mpg	Highway mpg
Gremlin	16	26	Buick Estate Wagon	11	15
Vega Kamback	19	30	Cadillac Seville	13	19
Astre	19	30	Chevrolet Wagon	10	15
Porsche 911S	16	23	Monaco	11	17
Audi 100	18	26	Cougar	12	17
Starfire	17	26	Delta 88	12	16
Dasher Wagon	24	36	Lincoln Continental	11	15
Jeep	13	18	Gran Fury	11	17
Subaru	24	34	Mercury	12	17
Datsun 710 Wagon	21	30	Rolls Royce Silver Shadow	10	13
$\bar{X}$	18.7	27.9	$\bar{X}$	11.3	16.1
s	3.53	5.26	s	0.95	1.66

The figure which follows shows the mean miles-per-gallon, in both highway and city tests, for automobiles under 3000 pounds.

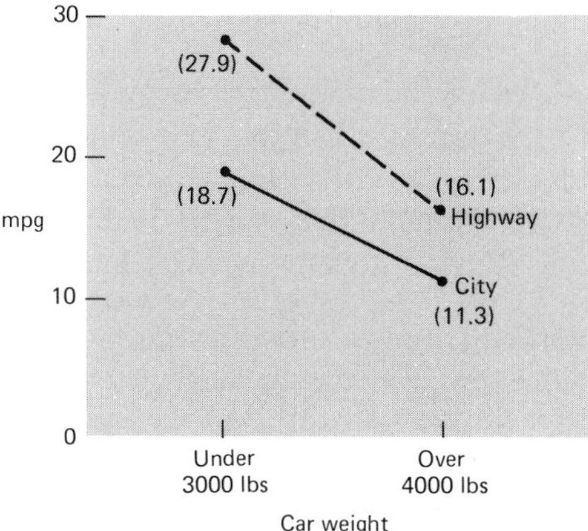

Some interesting questions involving inferential statistics may be raised:

1. Is there a statistically significant difference in the city and highway mpg between cars of the two weight classes?
2. Is there a statistically significant difference in city and highway mpg for automobiles in each weight class?

14.1 SAMPLING DISTRIBUTION OF THE DIFFERENCE BETWEEN MEANS

What is the effect of drug versus no drug on maze-learning?

Does the recidivism rate of juvenile offenders who are provided with "father figures" differ from those without "father figures?"

Do students in the ungraded classroom differ in performance on standardized achievement tests from students in the straight-age classroom setting?

Each of the above problems involves the comparison of at least two samples. Thus far, for heuristic purposes, we have restricted our examination of hypothesis testing to one sample case. However, most research involves the comparison of two or more samples to determine whether or not these samples might have reasonably been drawn from the same population. If the means of two samples differ, must we conclude that these samples were drawn from two different populations?

Recall our previous discussion on the sampling distribution of sample means (Section 13.2). We saw that some variability in the sample statistics is to be expected, even when these samples are drawn from the same population. We were able to describe this variability in terms of the sampling distribution of sample means. To conceptualize this distribution, we imagined drawing an extremely large number of samples, of a fixed N, from a population to obtain the distribution of sample means. In the two sample case, we should imagine drawing pairs of samples, finding the difference between the means of each pair, and obtaining a distribution of these differences. The resulting distribution would be the **sampling distribution of the difference between means**.

To illustrate, imagine that we randomly draw (with replacement) two samples at a time from the population described in Table O in which $\mu = 5.00$ and $\sigma = 0.99$. For illustrative purposes, let us draw two cases for the first sample (that is, $n_1 = 2$) and three cases for the second sample (that is, $n_2 = 3$). For example, we might draw scores of 5, 6 for our first sample and scores of 4, 4, 7 for our second sample. Thus, since $\bar{X}_1 = 5.5$ and $\bar{X}_2 = 5.0$, $\bar{X}_1 - \bar{X}_2 = 0.5$. Now suppose we continue to draw samples of $n_1 = 2$ and $n_2 = 3$ until we obtain an indefinitely large number of pairs of samples. If we calculate the differences between these pairs of sample means and treat each of these differences as a raw score, we may set up a frequency distribution of these differences.

Intuitively, what might we expect this distribution to look like? Since we are selecting pairs of samples at random from the *same* population, we would expect a normal distribution with a mean of zero.

Going one step further, we may describe the distribution of the difference between pairs of sample means, even when these samples are *not* drawn from the same population. It will be a normal distribution with a mean ($\mu_{\bar{X}_1 - \bar{X}_2}$) equal to $\mu_1 - \mu_2$ and a standard deviation ($\sigma_{\bar{X}_1 - \bar{X}_2}$, which we refer to as the **standard error of the difference between means**) equal to $\sqrt{\sigma^2_{\bar{X}_1} + \sigma^2_{\bar{X}_2}}$.

Thus the sampling distribution of the statistic

$$z = \frac{(\bar{X}_1 - \bar{X}_2) - (\mu_1 - \mu_2)}{\sigma_{\bar{X}_1 - \bar{X}_2}}$$

is normal and therefore permits us to employ the standard normal curve in the testing of hypotheses.

14.2 ESTIMATION OF $\sigma_{\bar{X}_1 - \bar{X}_2}$ FROM SAMPLE DATA

The statistic z is employed only when the population standard deviations are known. Since it is rare that these parameters are known, we are once again forced to estimate the standard error in which we are interested, that is, $\sigma_{\bar{X}_1 - \bar{X}_2}$.

If we draw a random sample of n_1 observations from a population with unknown variance and a second random sample of n_2 observations from a population with unknown variance, the standard error of the difference between means may be estimated by

$$s_{\bar{X}_1 - \bar{X}_2} = \sqrt{s^2_{\bar{X}_1} + s^2_{\bar{X}_2}}, \qquad (14.1)$$

or

$$s_{\bar{X}_1 - \bar{X}_2} = \sqrt{\frac{s_1^2}{n_1} + \frac{s_2^2}{n_2}}. \qquad (14.2)$$

However, if $n_1 = n_2 = n$, formula (14.2) simplifies to

$$s_{\bar{X}_1-\bar{X}_2} = \sqrt{\frac{s_1^2 + s_2^2}{n}}, \tag{14.3}$$

where n = the number in either sample.

When n's are not equal, formulas (14.1) and (14.2) represent biased estimates of the standard error of the difference between means. However, as n becomes larger, they approach the unbiased estimate. The unbiased estimate, which assumes that the two samples are drawn from a population with the same variance, pools the sums of squares and degrees of freedom of the two samples to obtain a pooled estimate of the standard error of the difference. Hence

$$s_{\bar{X}_1-\bar{X}_2} = \sqrt{\left(\frac{\sum x_1^2 + \sum x_2^2}{n_1 + n_2 - 2}\right)\left(\frac{1}{n_1} + \frac{1}{n_2}\right)}. \tag{14.4}^*$$

However, if $n_1 = n_2 = n$, formula (14.4) simplifies to

$$s_{\bar{X}_1-\bar{X}_2} = \sqrt{\frac{\sum x_1^2 + \sum x_2^2}{n(n-1)}}. \tag{14.5}^*$$

In order to obtain the sum of squares for our two groups, we apply formula (7.7) to each sample and obtain

$$\sum x_1^2 = \sum X_1^2 - \frac{(\sum X_1)^2}{n_1} \quad \text{and} \quad \sum x_2^2 = \sum X_2^2 - \frac{(\sum X_2)^2}{n_2}.$$

When n's are equal, formulas (14.1), (14.2), and (14.3) are algebraically identical to formulas (14.4) and (14.5) [pooled estimates of $s_{\bar{X}_1-\bar{X}_2}$]. Thus, for equal n's, any of the five formulas may be employed, depending upon computational ease.

14.3 TESTING STATISTICAL HYPOTHESES: STUDENT'S *t*

The statistic employed in the testing of hypotheses, when population standard deviations are not known, is the familiar *t*-ratio

$$t = \frac{(\bar{X}_1 - \bar{X}_2) - (\mu_1 - \mu_2)}{s_{\bar{X}_1-\bar{X}_2}}, \tag{14.6}$$

* Appendix III presents this formula in terms of raw scores.

in which $(\mu_1 - \mu_2)$ is the expected value as stated in the null hypothesis.* The Student t-ratio requires an unbiased estimate of $\sigma_{\bar{X}_1 - \bar{X}_2}$. Therefore formula (14.3), (14.4), or (14.5) is used to obtain the value of $s_{\bar{X}_1 - \bar{X}_2}$.

You will recall that Table C provides the critical values of t required for significance at various levels of α. Since the degrees of freedom for each sample is $n_1 - 1$ and $n_2 - 1$, the total df in the two sample case is $n_1 + n_2 - 2$.

14.3.1 Illustrative Problem: Student's *t*

A researcher wants to determine whether or not a given drug has any effect on the scores of human subjects performing a task of psychomotor coordination. Nine subjects in group 1 (experimental group) receive an oral administration of the drug prior to being tested. Ten subjects in group 2 (control group) receive a placebo at the same time.

Let us set up this problem in formal statistical terms. The results of the experiment are shown in Table 14.1.

Table 14.1

Scores of two groups of subjects on a test of psychomotor coordination

Group 1 Experimental		Group 2 Control	
X_1	X_1^2	X_2	X_2^2
12	144	21	441
14	196	18	324
10	100	14	196
8	64	20	400
16	256	11	121
5	25	19	361
3	9	8	64
9	81	12	144
11	121	13	169
		15	225
$\sum$ 88	996	$\sum$ 151	2445

$n_1 = 9$ $\qquad$ $n_2 = 10$

$\bar{X}_1 = 9.778$ $\qquad$ $\bar{X}_2 = 15.100$

* The most common null hypothesis tested is that both samples come from the same population, that is, $(\mu_1 - \mu_2) = 0$.

1. *Null hypothesis* (H_0): There is no difference between the population means of the drug group and the no-drug group on the test of psychomotor coordination, that is, $\mu_1 = \mu_2$, or $\mu_1 - \mu_2 = 0$.
2. *Alternative hypothesis* (H_1): There is a difference between the population means of the two groups on the test of psychomotor coordination. Note that our alternative hypothesis is nondirectional. Consequently, a two-tailed test of significance will be employed, that is, $\mu_1 \neq \mu_2$.
3. *Statistical test*: Since we are comparing two sample means presumed to be drawn from normally distributed populations with equal variances, the **Student *t*-ratio two-sample case** is appropriate.
4. *Significance level*: $\alpha = 0.05$.
5. *Sampling distribution*: The sampling distribution is the Student t-distribution with df $= n_1 + n_2 - 2$, or $9 + 10 - 2 = 17$.
6. *Critical region*: $t_{0.05} = \pm 2.110$. Since H_1 is nondirectional, the critical region consists of all the values of $t \geq 2.110$ and $t \leq -2.110$.

Since $n_1 \neq n_2$ and the population variances are assumed to be equal, we shall employ formula (14.4) to estimate the standard error of the difference between means. The sum of squares for group 1 is

$$\sum x_1^2 = 996 - \frac{(88)^2}{9} = 135.56.$$

Similarly, the sum of squares for group 2 is

$$\sum x_2^2 = 2445 - \frac{(151)^2}{10} = 164.90.$$

The value of t in the present problem is

$$t = \frac{(\bar{X}_1 - \bar{X}_2) - (\mu_1 - \mu_2)}{\sqrt{\left(\dfrac{\sum x_1^2 + \sum x_2^2}{n_1 + n_2 - 2}\right)\left(\dfrac{1}{n_1} + \dfrac{1}{n_2}\right)}}$$

$$= \frac{(9.778 - 15.100) - 0}{\sqrt{\left(\dfrac{135.56 + 164.90}{17}\right)\left(\dfrac{1}{9} + \dfrac{1}{10}\right)}} = \frac{-5.322}{1.93} = -2.758.$$

Decision: Since the obtained t falls within the critical region (that is, $|-2.758| > 2.110$, or $-2.758 < -2.110$), we reject H_0. The negative t-ratio simply means that the mean for group 2 is greater than the mean for group 1. When referring to Table C, we ignore the sign of the obtained t-ratio.

14.4 THE t-RATIO AND HOMOGENEITY OF VARIANCE

The assumptions underlying the use of the t-distributions are as follows:

1) The sampling distribution of the difference between means is normally distributed.
2) Estimated $\sigma_{\bar{X}_1 - \bar{X}_2}$ (that is, $s_{\bar{X}_1 - \bar{X}_2}$) is based on the unbiased estimate of the population variance.
3) Both samples are drawn from populations whose variances are equal. This assumption is referred to as *homogeneity of variance.*

Occasionally, for reasons which may not be very clear, the scores of one group may be far more widely distributed than the scores of another group. This may indicate that we are sampling two different distributions, but the critical question becomes: Two different distributions of what? . . . means or variances?

To determine whether or not two variances differ significantly from one another, we must make reference to yet another distribution: the F-distribution. Named after R. A. Fisher, the statistician who first described it, the F-distribution is unlike any other we have encountered in the text; it is tridimensional in nature. To employ the F Table (Table D), we must begin at the entry stating the number of degrees of freedom of the group with the larger variance, and move down the column until we find the entry for the number of degrees of freedom of the group with the smaller variance. At that point, we will find the critical value of F required for rejecting the null hypothesis of no difference in variances. F, itself, is defined as follows:

$$F = \frac{s^2 \text{ (larger variance)}}{s^2 \text{ (smaller variance)}}. \tag{14.7}$$

In the preceding sample problem, the variance for group 1 is 135.56/8 or 16.94, and for group 2, 164.9/9 or 18.32. The F-ratio becomes

$$F = 18.32/16.94 = 1.08, \qquad \text{df} = 9/8.$$

Referring to Table D, under 9 and 8 df, we find that an F-ratio of 3.39 or larger is required for significance at the 0.05 point or the 0.10 level. If we desire to employ $\alpha = 0.05$, we refer to Table D_1 which provides critical values for F at the 0.025 point or the 0.05 level. Referring to this table, under 9 and 8 df, we find that an $F \geq 4.36$ is significant at the 0.05 level. We may therefore conclude that it is reasonable to assume that both samples were drawn from a population with the same variances.

What if we found a significant difference in variances? Would it have increased our likelihood of rejecting the null hypothesis of no difference between means? Probably not. If anything, a significant difference in variances would have lowered the likelihood of rejecting the null hypothesis. Why, then, do we concern ourselves with an analysis of the variances? Frequently, a significant difference in variances (particularly, when the variance of the experimental group is significantly greater than that of the control group) is indicative of a *dual* effect of the experimental conditions. A larger variance indicates more extreme scores *at both ends* of a distribution. The alert researcher will seize upon these facts as a basis for probing into the possibility of dual effects. For example, years ago the experimental question, "Does anxiety improve or hinder performance on complex psychological tasks?," was thoroughly studied with rather ambiguous results.

More recently, we have come to recognize that anxiety-induced conditions have dual effects, depending on a host of factors, e.g., personality variables. An increase in anxiety causes some individuals to become better oriented to the task at hand, while increases in anxiety cause others to "blow up," so to speak. A study of the variances of our experimental groups may facilitate the uncovering of such interesting and theoretically important dual effects.

CHAPTER SUMMARY

We have seen that if we take a number of pairs of samples either from the same population or from two different populations, then

1. the distribution of differences between pairs of sample means tends to be normal;
2. the mean of these differences between means ($\mu_{\bar{X}_1 - \bar{X}_2}$) is equal to the difference between the population means, that is, $\mu_1 - \mu_2$;
3. the standard error of the difference between means ($\sigma_{\bar{X}_1 - \bar{X}_2}$) is equal to $\sqrt{\sigma^2_{\bar{X}_1} + \sigma^2_{\bar{X}_2}}$.

We presented several different formulas for estimating $\sigma_{\bar{X}_1 - \bar{X}_2}$ from sample data. Employing estimated $\sigma_{\bar{X}_1 - \bar{X}_2}$ (that is, $s_{\bar{X}_1 - \bar{X}_2}$), we demonstrated the use of the Student's t to test hypotheses in the two-sample case.

An important assumption underlying the use of the t-distributions is that both samples are drawn from populations with equal variances. Although failure to find homogeneity of variance will probably not seriously affect our interpretations, the fact of heterogeneity of variance may have important theoretical implications.

Terms to Remember:

Sampling distribution of the difference between means
Standard error of the difference between means
Student's t-ratio, two-sample case
F-ratio
Homogeneity of variance
Heterogeneity of variance

EXERCISES

14.1 Two statistics classes of 25 students each obtained the following results on the final examination: $\bar{X}_1 = 82$, $\sum x_1^2 = 384.16$; $\bar{X}_2 = 77$, $\sum x_2^2 = 1536.64$. Test the hypothesis that the two classes are equal in ability, employing $\alpha = 0.01$.

14.2 In an experiment on the effects of a particular drug on the number of errors in maze-learning behavior of rats, the following results were obtained:

Drug group	Placebo group
$\sum X_1 = 324$	$\sum X_2 = 256$
$\sum X_1^2 = 6516$	$\sum X_2^2 = 4352$
$n_1 = 18$	$n_2 = 16$

Set this experiment up in formal statistical terms, employing $\alpha = 0.05$, and draw the appropriate conclusions concerning the effect of the drug on errors.

14.3 On a psychomotor task involving two target sizes, the following results were obtained:

Group 1	Group 2
9	6
6	7
8	7
8	9
9	8

Set this experiment up in formal statistical terms, employing $\alpha = 0.05$, and draw the appropriate conclusions.

14.4 A study was undertaken to determine whether or not the acquisition of a response is influenced by a drug. The criterion variable was the number of trials required to master the task (X_1 is the experimental group and X_2 is the control group).

X_1	6	8	14	9	10	4	7	
X_2	4	5	3	7	4	2	1	3

a) Set this study up in formal statistical terms, and state the appropriate conclusions, employing $\alpha = 0.01$.
b) Is there evidence of heterogeneity of variance?

14.5 Given two normal populations,

$$\mu_1 = 80, \quad \sigma_1 = 6; \quad \mu_2 = 77, \quad \sigma_2 = 6.$$

If a sample of 36 cases is drawn from population 1 and a sample of 36 cases from population 2, what is the probability that

a) $\bar{X}_1 - \bar{X}_2 \geq 5$?
b) $\bar{X}_1 - \bar{X}_2 \geq 0$?
c) $\bar{X}_1 - \bar{X}_2 \leq 0$?
d) $\bar{X}_1 - \bar{X}_2 \leq -5$?

14.6 Assuming the same two populations as in Problem 5, calculate the probability that $\bar{X}_1 - \bar{X}_2 \geq 0$, when

a) $n_1 = n_2 = 4$,
b) $n_1 = n_2 = 9$,
c) $n_1 = n_2 = 16$,
d) $n_1 = n_2 = 25$.

14.7 Graph the above probabilities as a function of n. Can you formulate any generalization about the probability of finding a difference in the correct direction between sample means (that is, $\bar{X}_1 - \bar{X}_2 \geq 0$, when $\mu_1 > \mu_2$) as a function of n?

14.8 A gasoline manufacturer runs tests to determine the relative performance of automobiles employing two different additives. The results are as follows (expressed in terms of miles per gallon of gasoline).

Additive 1: 12, 17, 15, 13, 11, 10, 14, 12
Additive 2: 16, 14, 18, 19, 17, 13, 11, 18

Set up and test the appropriate null hypothesis.

14.9 Each of two market analysts claims that his ability to forecast price gains in common stock is better than that of his rival. Over a specified period, each selected ten common stocks that he predicted would show a gain. The results were as follows:

Analyst 1: \$1.25, \$2.50, \$1.75, \$2.25, \$2.00, \$1.75, \$2.25, \$1.00, \$1.75, \$2.00
Analyst 2: \$1.25, \$0.75, \$1.00, \$1.50, \$2.00, \$1.75, \$0.50, \$1.50, \$0.25, \$1.25

Set up and test the appropriate null hypothesis.

14.10 A toothpaste manufacturer claims that children brushing their teeth daily with her product (brand A) will have fewer cavities than children employing brand X. In a carefully supervised study, the number of cavities in a sample of children using her toothpaste was compared with the number of cavities among children using brand X. The results were as follows:

Brand A: 1, 2, 0, 3, 0, 2, 1, 4, 2, 3, 1, 2, 1, 1
Brand X: 3, 1, 2, 4, 1, 5, 2, 0, 5, 6, 3, 2, 4, 3

Test the manufacturer's claim.

14.11 The random samples of fifteen manufacturing concerns with total assets under \$5,000,000 showed an average after-tax profit of 2.2% of sales and a standard deviation of 0.5%. A random sample of 12 manufacturing concerns with assets between \$5,000,000 and \$10,000,000 yielded an average after-tax profit of 2.5% of sales and a standard deviation of 0.6%. Is the difference attributable to chance variations which will occur whenever we base our conclusion on samples? Use the 0.05 significance level.

14.12 A college maintains that it has made vast strides in raising the standards of admission for its entering freshmen. It cites the fact that the mean high school average of 80 entering freshmen was 82.53 for last year with a standard deviation of 2.53. For the present year these statistics are 83.04 and 2.58, respectively, for 84 entering freshmen. Set up and test the appropriate null hypothesis, employing $\alpha = 0.01$.

14.13 A training director in a large industrial firm claims that employees taking his training course perform better on the job than those not receiving training. Of 30 recently hired employees, 15 are randomly selected to receive training. The remaining 15 are employed as controls. Six months later, on-the-job test evaluations yield the following statistics for the training group: $\bar{X} = 24.63$, $s = 3.53$. The controls obtain $\bar{X} = 21.45$ and $s = 4.02$.

Set up and test the appropriate null hypothesis, employing $\alpha = 4.02$.

14.14 A publisher claims that students who receive instruction in mathematics based on her newly constructed text will score at least five points higher on end-of-term grades than those instructed using the previous text. Thirty-six students are randomly assigned to two classes: the experimental group employs the new text for instruction, and the control group uses the previous text. Students in the experimental group achieve $\bar{X} = 83.05$ and $s = 6.04$ as final grades, whereas the controls obtain $\bar{X} = 76.85$ and $s = 5.95$.

Set up and test the appropriate null hypothesis employing $\alpha = 0.01$, one-tailed test. [*Note*: Remember that the numerator in the test statistic is: $(\bar{X}_1 - \bar{X}_2) - (\mu_1 - \mu_2)$.]

14.15 A manufacturer of carpets is considering the replacement of his looms with new machines which are expected to produce carpets at a higher rate per hour. His cost analyst informs him that the exchange is inadvisable unless the new looms produce at a rate that is 10% greater than the old looms. In a test involving 12 old and 10 new looms, the mean number of carpets produced in one hour was found to be 12.30 and 14.45, respectively, with standard deviations of 3.45 and 2.96. Employing $\alpha = 0.01$, one-tailed test, decide whether or not the manufacturer should make the change-over to the new looms.

14.16 If we found a significant difference between means at the 5% level of significance, it would also be true that (true or false):

a) This difference is significant at the 1% level of significance.
b) This difference is significant at the 10% level of significance.
c) The difference observed between means is the true difference.

For Exercises 14.17 through 14.21, the following two finite populations of scores are given:

Population 1: 2, 4, 6, 8
Population 2: 1, 3, 5, 7

14.17 Random samples of size 2 are selected (*without replacement*) from each population. Construct a histogram showing the sampling distribution of differences between the sample means.

14.18 What is the probability that:

a) $\bar{X}_1 - \bar{X}_2 \geq 0$? b) $\bar{X}_1 - \bar{X}_2 \geq 1$? c) $\bar{X}_1 - \bar{X}_2 \leq 0$?
d) $\bar{X}_1 - \bar{X}_2 \leq -1$? e) $\bar{X}_1 - \bar{X}_2 \geq 4$? f) $\bar{X}_1 - \bar{X}_2 \leq -4$?
g) $\bar{X}_1 - \bar{X}_2 \geq 2$ or $\bar{X}_1 - \bar{X}_2 \leq -2$?

14.19 A manufacturer sampled the number of dresses produced per day for 10 days by a group of 26 workers (Group A) who were operating on a fixed-wage plan. She then introduced a wage incentive plan to 26 other employees (Group B) and recorded their output for 10 days. The number of dresses produced per day were as follows:

Group A:	75	72	73	76	78	72	80	74	76	75
Group B:	80	83	84	78	79	81	84	85	78	86

The wages paid to each group are equal. Can the manufacturer conclude that the wage incentive plan is more efficient? Calculate the value of t.

14.20 Company A finds that the mean number of burning hours of its lightbulbs is 1200, with $\sigma_{\bar{X}_1}^2 = 100$. Company B shows a mean of 1250 burning hours, with $\sigma_{\bar{X}_2}^2 = 125$. Is it probable that the lightbulbs at the two companies are from the same population of lightbulbs? What is the value of z?

14.21 From records of past employees, two large companies sampled the number of years that secretaries stayed with the company. Using a sample of 25 employees each, company A found that $\sum x_1^2 = 42$ and $\bar{X}_1 = 40$, and company B found that $\sum x_2^2 = 58$ and $\bar{X}_2 = 50$. Is this difference significant at the 0.05 level? What is the value of t?

14.22 A manager finds that the number of errors that employees make increases as the day progresses, reaching a peak between 3:00 and 5:00. He divides a sample of 20 employees into two groups. One group proceeds on the same work schedule as before, but the other group gets a 15 min coffee break from 2:45 to 3:00. The subsequent number of errors made between 3:00 and 5:00 are listed at the top of page 281.

No break group:	5	6	7	4	8	9	6	5	7	6
Break group:	2	3	4	3	4	4	3	1	5	4

Does the break significantly reduce the number of errors? Calculate the value of t.

14.23 Determine if the variances are homogeneous in the above problem. Calculate the value of F.

14.24 Banks A and B use two different forms for recording checks written. The banks found that the following number of checks had bounced for 15 customers during the last 10 years:

Bank A:	4	8	3	0	3	5	3	4	0	5	2	4	6	2	0
Bank B:	2	0	1	2	1	1	3	3	4	3	1	4	0	5	0

Determine whether there is a significant difference between the number of checks bounced at each bank. Employ $\alpha = 0.05$.

14.25 State A finds that the mean number of cigarettes smoked by persons of that state is 120, with a $\sigma^2_{\bar{X}_1}$ of 10. State B's cigarette tax is \$0.05 greater per pack. The mean number smoked per day is 110, with a $\sigma^2_{\bar{X}_2}$ of 9. Is there a significant difference in the number of cigarettes smoked between the two states? What is the value of z?

14.26 Two grocery store managers find that they have an overstock of spaghetti sauce. The price and usual amount sold in the two stores are identical. Manager A keeps the sauce in its regular place, while manager B piles the cans close to the check-out counters for a month. The managers record the number of cans sold during 10 days, with the following results:

A:	19	20	20	21	18	20	19	21	23	17
B:	26	24	25	23	25	24	22	26	27	25

Is there a significant difference in the number of cans sold? What is the value of t?

14.27 Determine if the variances are homogeneous in Exercise 14.29. Calculate F.

14.28 The eyes have been referred to as the "windows of the soul." Research has demonstrated that changes in the size of the pupils may be taken as an indicator of "what turns people on." One study compared the pupil responses of heterosexual males and homosexual males when viewing pictures of men and women.

The table below shows the change in pupil size of five heterosexual and five homosexual males when viewing pictures of a male.

Subject	Heterosexuals	Subject	Homosexuals
1	−00.4	6	+18.8
2	−54.5	7	−04.6
3	+12.5	8	+18.9
4	+06.3	9	+18.2
5	−01.5	10	+15.8

Source: Hess, E. H., A. L. Seltzer, and J. M. Shlien, "Pupil responses of hetero- and homosexual males to pictures of men and women: A pilot study," *J. Abn. Psychol.*, **70**, 1965, 165–168.

Formulate H_0 and H_1, two-tailed test. Using the Student t-ratio for independent samples, determine whether H_0 may be rejected at $\alpha = 0.05$. [*Hint*: To facilitate calculations when negative numbers are involved, algebraically add 55 to each score. This procedure eliminates all the negative values and makes use of the generalization shown in Section 2.3.]

To calculate the mean for each group:

$$\frac{\sum_{i=1}^{5} X_i}{n} = \frac{\sum_{i=1}^{5} X_i - 5(55)}{n}.$$

Note: Adding 55 to all scores will not change the difference between means and the standard error of the difference, since the relative differences among scores are maintained. However, the mean for each group will increase by 55.

14.29 The table shows the change in pupil size of five heterosexual and five homosexual males when viewing pictures of a female.

Subject	Heterosexuals	Subject	Homosexuals
1	+05.9	6	+11.2
2	−22.4	7	−38.0
3	+19.2	8	+18.1
4	+39.0	9	−05.6
5	+23.1	10	+21.5

Formulate H_0 and H_1, two-tailed test. Using the Student t-ratio for independent samples, determine whether H_0 may be rejected at $\alpha = 0.05$. [*Hint*: To facilitate calculations when negative numbers are involved, algebraically add 39 to each score.]

14.30 Employing the data presented in Box 14.1, answer the following questions, using $\alpha = 0.05$ [two-tailed test].

a) For cars under 3000 pounds, is there a statistically significant difference between city and highway mpg?

b) For cars over 4000 pounds, is there a statistically significant difference between city and highway mpg?

c) For city driving, is there a statistically significant difference between the two weight classes?

d) For highway driving, is there a statistically significant difference between the two weight classes?

Statistical Inference with Correlated Samples

BOX 15.1

In Chapter 14 we saw that $s_{\bar{X}_1 - \bar{X}_2}$ is a sort of benchmark against which we evaluate the difference between means. If $s_{\bar{X}_1 - \bar{X}_2}$ is large relative to the mean difference, we are less likely to reject H_0 than if $s_{\bar{X}_1 - \bar{X}_2}$ is small relative to the same difference between means.

If there were some way to legitimately reduce the size of the error term ($s_{\bar{X}_1 - \bar{X}_2}$), our benchmark would be more sensitive. Just such an opportunity is provided by correlational analysis. Recall from Chapter 10 that the higher the correlation between two variables, the greater is the *explained variation* and the less the *unexplained* or random variation. Since the unexplained variation represents error, correlational analysis provides a legitimate way to lessen unexplained variation and thereby reduce error.

The table shows the highway miles-per-gallon scores achieved by twenty cars with manual transmission and twenty cars with automatic transmission. Each pair represents the same model car. In other words, both members of each pair are identical except for the type of transmission. When correlated samples are employed, the *difference* between the paired scores is used as the basis for calculating the error term. If the correlation is low, these differences are large. So also is the error term based on these differences. When r is high, the differences are small. The higher the correlation, the smaller the differences between the paired scores. The error term is similarly reduced.

The accompanying scatter diagram suggests that the correlation between the paired observations is very high. As a matter of actual fact, $r = 0.89$. The resulting error term, referred to as the standard error of the mean difference (s_{MD}), is much lower than the error term ($s_{\bar{X}_1 - \bar{X}_2}$) we would have used in the absence of correlated samples.

Car Model	Type of transmission: Manual	Automatic	Difference
Gremlin	26	19	7
Hornet	16	20	−4
Audi Fox	36	32	4
Skyhawk	26	24	2
Vega	30	27	3
Datsun 210	41	35	6
Datsun 280	25	22	3
Fiat 131	30	25	5
Mustang II	24	20	4
Grinada	20	21	−1
Capri II	27	23	4
Monarch	20	21	−1
Starfire	26	24	2
Valiant/Dust	23	20	3
Sunbird	26	24	2
Lemans	21	18	3
Corolla	35	26	9
Corona MK. II	22	18	4
Volvo	27	23	4
LUV pickup	32	28	4

Highway mileage of twenty pairs of 1976 automobiles. Each pair consists of one car with manual transmission and the same model car with automatic transmission.

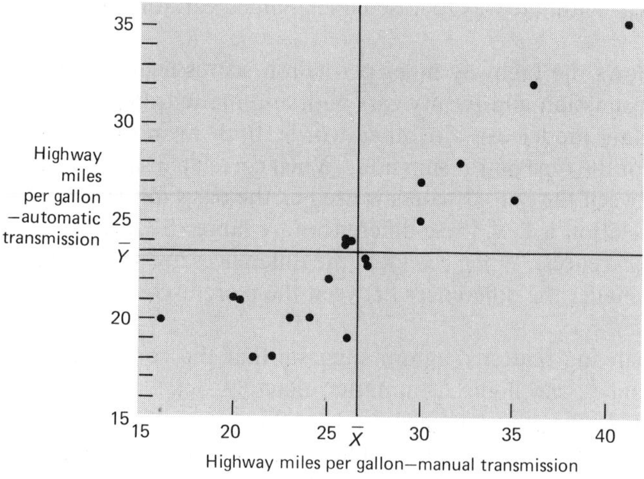

15.1 INTRODUCTION

One of the fundamental problems which frequently confronts the researcher is the extreme variability of data. Indeed, it is *because* of this variability that he or she is so concerned with the field of inferential statistics.

When an experiment is conducted, data comparing two or more groups are obtained, a difference in some measure of central tendency is found, and then we raise the question: Is the difference of such magnitude that it is unlikely to be due to chance factors? As we have seen, a visual inspection of the data is not usually sufficient to answer this question because there is so much overlapping of the experimental groups. The overlapping, in turn, is due to the fact that the experimental subjects themselves manifest widely varying aptitudes and proficiencies relative to the criterion measure. In an experiment, the score of any subject on the criterion variable may be thought to reflect at least three factors: (1) the subject's ability and/or proficiency on the criterion task; (2) the effects of the experimental variable; and (3) random error due to a wide variety of different causes, e.g., minor variations from time to time in experimental procedures or conditions, momentary fluctuations in such things as attention span, motivation of the experimental subjects, etc. There is little we can do about *random error* except to maintain as close control over experimental conditions as possible. The *effects of the experimental variable* are, of course, what we are interested in assessing. In most studies the *individual differences among subjects* is, by and large, the most significant factor contributing to the scores and the variability of scores on the criterion variable. Anything we can do to take this factor into account or "statistically remove" its effects will improve our ability to estimate the effects of the experimental variable on the criterion scores. This chapter is concerned with a technique that is commonly employed to accomplish this very objective: the employment of correlated samples.

15.2 STANDARD ERROR OF THE DIFFERENCE BETWEEN MEANS FOR CORRELATED GROUPS

In our earlier discussion of Student's t-ratio, we presented the formula for the unpooled estimate of the **standard error of the mean difference** (14.1) as

$$s_{\bar{X}_1-\bar{X}_2} = \sqrt{s^2_{\bar{X}_1} + s^2_{\bar{X}_2}}.$$

Actually, this is not the most general formula for the standard error of the difference. The most general formula is

$$s_{\bar{X}_1-\bar{X}_2} = \sqrt{s^2_{\bar{X}_1} + s^2_{\bar{X}_2} - 2rs_{\bar{X}_1}s_{\bar{X}_2}}. \tag{15.1}$$

We drop the last term whenever our sample subjects are assigned to experimental conditions at random for the simple reason that when scores are paired at random, the correlation between the two samples will average zero. Any observed correlation will be spurious since it will represent a chance association. Consequently, when subjects are assigned to experimental conditions at random, the last term reduces to zero (since $r = 0$).

However, there are many experimental situations in which we do not assign our experimental subjects at random. Most of these situations can be placed in one of two classes.

1. **Before-after design.** A reading on the *same* subjects is taken both before and after the introduction of the experimental variable. It is presumed that each individual will remain relatively consistent with himself. Thus there will be a correlation between the before sample and the after sample.

2. **Matched-group design.** Individuals in both experimental and control groups are matched on some variable known to be correlated to the criterion or dependent variable. Thus, if we wanted to determine the effect of some drug on learning the solution to a mathematical problem, we might match individuals on the basis of I.Q. estimates, amount of mathematical training, grades in statistics, or performance on other mathematics problems. Such a design has two advantages:
   a) It ensures that the experimental groups are "equivalent" in initial ability.
   b) It permits us to take advantage of the correlation based on initial ability and allows us, in effect, to remove one source of error from our measurements.

To understand the advantage of employing correlated samples, let us look at a sample problem and calculate the standard error of the difference between means using formula (14.1) based upon unmatched groups and formula (15.1) which takes the correlation into account. Table 15.1 presents data for two groups of subjects matched on a variable known to be correlated with the criterion variable. The members comprising each pair are assigned at random to the experimental conditions.

Table 15.1

Scores of two groups of subjects in an experiment employing matched group design (hypothetical data)

Matched pairs	X_1	X_1^2	X_2	X_2^2	X_1X_2
A	2	4	4	16	8
B	3	9	3	9	9
C	4	16	5	25	20
D	5	25	6	36	30
	$\sum 14$	54	18	86	67

The following steps are employed in the calculation of the standard error of the difference between means for *unmatched* groups.

Step 1. The sum of squares for group 1 is

$$\sum x_1^2 = 54 - (14)^2/4 = 5.$$

Step 2. The standard deviation for group 1 is

$$s_1 = \sqrt{5/3} = 1.2923.$$

Step 3. The standard error of the mean for group 1 is

$$s_{\bar{X}_1} = s_1/\sqrt{n_1} = 1.2923/\sqrt{4} = 0.6462.^*$$

Step 4. Similarly, the standard error of the mean for group 2 is

$$s_{\bar{X}_2} = 0.6462.$$

Step 5. The standard error of the difference between means for independent groups is

$$s_{\bar{X}_1 - \bar{X}_2} = \sqrt{s_{\bar{X}_1}^2 + s_{\bar{X}_2}^2} = 0.9138.$$

To calculate the standard error of the difference between means for *matched* groups, the following steps are employed.

Step 1. Employing formula (9.2),[†] we find that the correlation between the two groups is

$$r = \frac{\sum x_1 x_2}{\sqrt{(\sum x_1^2)(\sum x_2^2)}} = \frac{4}{\sqrt{(5)(5)}} = 0.80.$$

Step 2. The standard error of the difference between means for matched groups is

$$\begin{aligned} s_{\bar{X}_1 - \bar{X}_2} &= \sqrt{s_{\bar{X}_1}^2 + s_{\bar{X}_2}^2 - 2rs_{\bar{X}_1}s_{\bar{X}_2}} \\ &= \sqrt{0.4167 + 0.4167 - 2(0.80)(0.6455)(0.6455)} = 0.41. \end{aligned}$$

You will note that formula (15.1), which takes correlation into account, provides a markedly reduced error term for assessing the significance of the difference between means. In other words, it provides a more sensitive test of this difference and is more likely to lead to the rejection of the null hypothesis when it is false. In the language of inferential statistics, it is a more *powerful* test. Of course, the greater power, or sensitivity, of formula (15.1) is directly related to our success in matching subjects on a variable that is correlated with the criterion variable. When r is large, $s_{\bar{X}_1 - \bar{X}_2}$ will

* The standard error of the mean may be obtained directly by employing formula (13.8), that is,

$$s_{\bar{X}_1} = \sqrt{\frac{\sum x_1^2}{n_1(n_1 - 1)}} = \sqrt{\frac{5}{4(3)}} = 0.6455.$$

[†] Note that x_2 replaces y in formula (9.2).

be correspondingly small. As r approaches zero, the advantage of employing correlated samples becomes progressively smaller.

Balanced against the increased sensitivity of the standard error of the difference between means when r is large, is the loss of degrees of freedom. Whereas the number of degrees of freedom for unmatched samples is $n_1 + n_2 - 2$, the number of degrees of freedom when correlated samples are employed is the number of pairs minus one $(n - 1)$. This difference can be critical when the number of degrees of freedom is small, since, as we saw in Section 13.5.1, larger t-ratios are required for significance when the degrees of freedom are small.

15.3 THE DIRECT-DIFFERENCE METHOD: STUDENT'S *t*-RATIO

Fortunately, it is not necessary to actually determine the correlation between samples in order to find $s_{\bar{X}_1 - \bar{X}_2}$. Another method is available which permits the direct calculation of the standard error of the difference. We shall refer to this method as the **direct-difference method** and represent it symbolically as s_D.

In brief, the direct-difference method consists of finding the differences between the criterion scores obtained by each pair of matched subjects, and treating these differences as if they were raw scores. The null hypothesis is that the obtained mean of the difference scores ($\sum D/N$, symbolized as $\bar{D}$) comes from a population in which the mean difference (μ_D) is some specified value. The t-ratio employed to test H_0: $\mu_D = 0$ is

$$t = \frac{\bar{D} - \mu_D}{s_D} = \frac{\bar{D}}{s_{\bar{D}}}. \qquad (15.2)$$

The raw score formula for calculating the sum of squares of the difference scores is

$$\sum d^2 = \sum D^2 - (\sum D)^2/n, \qquad (15.3)$$

where D is the difference between paired scores and d is the deviation of a difference score (D) from $\bar{D}$. It follows, then, that the standard deviation of the difference scores is

$$s_D = \sqrt{\frac{\sum d^2}{(n - 1)}}. \qquad (15.4)$$

Furthermore the standard error of the mean difference may be obtained by dividing formula (15.4) by $\sqrt{N}$. Thus

$$s_{\bar{D}} = \sqrt{\frac{\sum d^2}{n(n - 1)}} \qquad (15.5)$$

or

$$s_{\bar{D}} = s_D/\sqrt{n}. \qquad (15.6)$$

15.3.1 Testing the Hypothesis

In Box 15.1 we showed the miles-per-gallon scores of twenty pairs of 1976 automobiles. Recall that each car in the pair is identical except for type of transmission. It has often been claimed that cars with manual transmissions obtain better mileage performance than those with automatic transmissions. Let us test this claim.

1. *Null hypothesis* (H_0): The difference in miles-per-gallon scores between manual and automatic transmissions is equal to or less than zero, that is $\mu_D \leq 0$.
2. *Alternative hypothesis* (H_1): Automobiles with manual transmissions achieve better gas mileage than those with automatic transmissions, that is, $\mu_D > 0$. Note that our alternative hypothesis is directional; consequently, a one-tailed test of significance is employed.
3. *Statistical test*: Since we are employing correlated samples, the Student t-ratio for correlated samples is appropriate.
4. *Significance level*: $\alpha = 0.01$
5. *Sampling distribution*: The sampling distribution is the Student's t-distribution with df $= n - 1$, or $20 - 1 = 19$.
6. *Critical region*: $t_{0.01} = 2.539$. Since H_1 predicts that the scores in the *manual* condition will be higher than those in the *automatic* condition, we expect the difference scores to be positive. Therefore the critical region consists of all values of $t \geq 2.539$. The following steps are employed in the direct-difference method.

Step 1. The sum of squares of the difference scores is

$$\sum d^2 = 353 - (63)^2/20 = 154.55.$$

Step 2. The standard error of the mean difference is

$$s_{\bar{D}} = \sqrt{154.55/20(19)} = 0.64.$$

Step 3. The value of $\bar{D}$ is $\bar{D} = 63/20 = 3.15$. (To check the accuracy of $\sum D$ we subtract $\sum X_2$ from $\sum X_1$, that is, $\sum X_1 - \sum X_2 = \sum D$, $553 - 470 = 63$.)

Step 4. The value of t in the present problem is

$$t = \bar{D}/s_{\bar{D}} = 3.15/0.64 = 4.922.$$

Decision: Since the obtained t does fall within the critical region (that is, $4.922 > t_{0.01}$), we reject H_0.

15.4 SANDLER'S *A*-STATISTIC

In recent years, a psychologist, Joseph Sandler, has demonstrated an extremely simple procedure for arriving at probability values in all situations for which the Student t-ratio for correlated samples is appropriate. Indeed, since Sandler's statistic, A, is rigorously derived from Student's t-ratio, the probability values are identical with Student's p-values.

The **Sandler A-statistic** is defined as follows:

$$A = \frac{\text{the sum of the squares of the differences}}{\text{the square of the sum of the differences}} = \frac{\sum D^2}{(\sum D)^2}. \quad (15.7)$$

By making reference to the table of A (Table E) under $n - 1$ degrees of freedom, we can determine whether our obtained A is *equal to* or *less than* the tabled values at various levels of significance.

Let us illustrate the calculation of A from our previous example. It will be recalled that $\sum D^2 = 353$ and $\sum D = 63$. The value of A becomes

$$A = 353/(63)^2 = 0.089.$$

Referring to Table E under 9 degrees of freedom, we find that an A *equal to or less than* 0.197 is required for significance at the 0.01 level (one-tailed test). Since 0.089 is less than the tabled value, we reject the null hypothesis. It will be noted that our conclusion is precisely the same as the one we arrived at by employing the Student t-ratio. This is correct, of course, since the two are mathematically equivalent. Since the calculation of A requires far less time and labor than the determination of t, Sandler's A-distribution can replace Student's t whenever correlated samples are employed.

CHAPTER SUMMARY

The most general formula for the standard error of the difference is

$$s_{\bar{X}_1 - \bar{X}_2} = \sqrt{s_{\bar{X}_1}^2 + s_{\bar{X}_2}^2 - 2rs_{\bar{X}_1}s_{\bar{X}_2}}.$$

It is obvious that by matching samples on a variable correlated with the criterion variable, we may reduce the magnitude of the standard error of the difference and thereby provide a more sensitive test of the difference between means. The higher the correlation, of course, the greater the reduction in the standard error of the difference.

We demonstrated the use of the direct-difference method for determining the significance of the difference between the means of correlated samples.

Finally, we demonstrated the use of a mathematically equivalent test, the Sandler A-statistic. Due to its computational ease, the Sandler test will unquestionably replace the Student t-ratio when correlated samples are employed.

Terms to Remember:

Before-after design
Matched group design
Direct-difference method
Standard error of the mean difference
Sandler A-statistic

EXERCISES

15.1 An investigator employs an experimental procedure wherein each subject (S) performs a task which requires the cooperation of a partner. By prearrangement, the partner plays the role of a complaining rejecting teammate. When the task is completed, S is asked to recall any remarks made by the partner. Prior to the experiment, S's have been tested and classified as either secure or insecure in interpersonal relations and then matched on the basis of intelligence.

The following statistics were derived from data indicating the number of remarks recalled.

Secure S's	Insecure S's
$\bar{X}_1 = 14.2$	$\bar{X}_2 = 12.9$
$s_1 = 2.0$	$s_2 = 1.5$
$n_1 = 16$	$n_2 = 16$

$r = 0.55$ between number of remarks recalled and intelligence.

Set this task up in formal statistical terms, employing $\alpha = 0.01$ (two-tailed test) and state the appropriate conclusions.

15.2 Had we not employed a matched-group design (that is, $r = 0.00$), would our conclusion have been any different? Explain. [*Hint*: In calculating $s_{\bar{X}_1 - \bar{X}_2}$, employ formula (14.1), (14.2), or (14.3).]

15.3 Arthur Diamond, manager of a Little League team in the American League, has said that the American League is more powerful than the National League. Determine the validity of this statement from the home-run (HR) figures given below. Employ Student's t-ratio for correlated samples *and* Sandler's A, $\alpha = 0.05$.

Final standing in respective league	American League	Number of HR	National League	Number of HR
1	Minnesota	16	Los Angeles	18
2	Chicago	17	San Francisco	11
3	Baltimore	15	Pittsburgh	14
4	Detroit	12	Cincinnati	10
5	Cleveland	11	Milwaukee	12
6	New York	9	Philadelphia	13
7	California	13	St. Louis	8
8	Washington	16	Chicago	10
9	Boston	18	Houston	9
10	Kansas City	14	New York	15

15.4 In Exercise 15.3, the teams were paired on the basis of final standings in their respective leagues. The matching technique assumes a correlation between final standing and number of home runs. Is this assumption valid?

15.5 Perform a Student's t-ratio for independent samples on the data in Exercise 15.3. Why is the obtained t-ratio closer to the rejection region than the answer to Exercise 15.3?

15.6 Experimenter Ned designs a study in which he matches subjects on a variable which he believes to be correlated with the criterion variable. Assuming H_0 to be false, what is the likelihood of a type II error (compared to the use of Student's t-ratio for uncorrelated samples) in the following situations:
a) the matching variable is uncorrelated with the criterion variable;
b) the matching variable is highly correlated with the criterion variable.

15.7 It has often been stated that women have a higher life expectancy than men. Employ $\alpha = 0.05$ to determine the validity of this statement for
a) white Americans,
b) nonwhite Americans,
c) white males compared to nonwhite females.

Expectation of life in the United States*

	White		Nonwhite (Chiefly negro)			White		Nonwhite (Chiefly negro)	
Age	Male	Female	Male	Female	Age	Male	Female	Male	Female
0	67.5	74.4	60.9	66.5	11	58.6	65.2	53.5	58.7
1	68.2	74.8	62.9	68.0	12	57.6	64.2	52.6	57.7
2	67.3	73.9	62.1	67.2	13	56.7	63.2	51.6	56.7
3	66.4	73.0	61.2	66.3	14	55.7	62.2	50.7	55.7
4	65.4	72.0	60.3	65.4	15	54.7	61.3	49.7	54.8
5	64.4	71.0	59.3	64.5	16	53.8	60.3	48.8	53.8
6	63.5	70.1	58.4	63.5	17	52.8	59.3	47.8	52.8
7	62.5	69.1	57.4	62.5	18	51.9	58.3	46.9	51.9
8	61.6	68.1	56.5	61.6	19	51.0	57.4	46.0	50.9
9	60.6	67.1	55.5	60.6	20	50.1	56.4	45.1	50.0
10	59.6	66.2	54.5	59.6					

* *Source.* Reader's Digest 1966 Almanac, p. 492. New York: Reader's Digest Association, with permission.

15.8 Numerous consumer organizations have criticized the automobile industry for employing odometers which show large variations from one instrument to another and from one manufacturer to another. To test whether odometers from two competing manufacturers may be considered to have been drawn from a common population, eleven different cars were equipped with two odometers each, one from each manufacturer. All automobiles were driven over a measured course of 100 miles and their odometer readings were tabulated. Apply the appropriate test for the significance of the difference between the odometer readings of each manufacturer, employing $\alpha = 0.01$.

Automobile	Manufacturer A	Manufacturer B	Automobile	Manufacturer A	Manufacturer B
1	104	102	7	97	99
2	112	106	8	107	102
3	103	107	9	100	98
4	115	110	10	104	101
5	99	93	11	108	102
6	104	101			

15.9 Another complaint by consumer organizations is that the odometers are purposely constructed to overestimate the distance traveled in order to inflate the motorist's estimates of gasoline mileage. As a review of Chapter 13, conduct a one-sample test of $H_0: \mu_0 = 100$ miles for the product of each manufacturer.

15.10 Runyon (1968) showed that a simple extension of the Sandler A-test may be employed as an algebraically equivalent substitute for the Student t-ratio, one-sample case. The technique consists of subtracting the value of the mean hypothesized under H_0 from each score, summing the differences and squaring $[(\sum D)^2]$, squaring the differences and summing $[\sum D^2]$, and substituting these values in the formula for the A-statistic, that is, $A = \sum D^2/(\sum D)^2$. Degrees of freedom equal $n - 1$. Conduct a one-sample test of $H_0: \mu_0 = 100$ miles for the odometers of each manufacturer. Compare the results of this analysis with the results of the foregoing analysis.

15.11 In a study aimed at determining the effectiveness of a new diet, an insurance company selects a sample of 12 overweight men between age 40 and 50 and obtains their weight measurements both before initiating the diet and sixty days later. Set up and test the appropriate null hypothesis, employing $\alpha = 0.05$.

Subject	Weight Before	Weight After	Subject	Weight Before	Weight After
1	202	180	7	209	205
2	237	221	8	191	196
3	173	175	9	200	185
4	161	158	10	189	187
5	185	180	11	177	172
6	210	197	12	184	186

15.12 A difficulty with interpreting the results of Exercise 15.11 is that a control group is lacking. It is possible that a random selection of overweight men who are *not* on a diet will reveal weight losses over a two-month period. A control group, matched in weight with the experimental subjects in Exercise 15.11, demonstrated the following before-after changes.

Subject	Weight Before	Weight After	Subject	Weight Before	Weight After
1	203	207	7	209	215
2	235	231	8	192	215
3	175	172	9	201	184
4	159	164	10	187	196
5	183	187	11	178	184
6	210	204	12	185	173

Set up and test the appropriate null hypothesis for the control subjects. Employ $\alpha = 0.05$.

15.13 Employing the "after" weight only for the matched subjects in Exercises 15.11 and 15.12, test for the significance of the difference between the two conditions, employing $\alpha = 0.05$.

15.14 Obtain the before-after difference score for each subject in Exercises 15.11 and 15.12. Conduct a "matched-pairs" analysis of the difference scores, employing $\alpha = 0.05$. Compare the results of this analysis with the foregoing analysis.

15.15 A large discount house advertises that its prices are lower than those of its largest competitor. To test the validity of this claim, the prices of 15 randomly selected items are compared. The results are as follows:

Discount house	Competitor	Discount house	Competitor
$3.77	$3.95	$2.99	$2.95
7.50	7.75	1.98	2.49
4.95	4.99	0.49	0.52
3.18	3.25	5.50	5.62
5.77	5.98	0.99	0.98
2.49	2.39	6.49	6.66
8.77	9.49	5.49	5.55
6.99	6.49		

What do you conclude?

15.16 A company has just switched to a four-day work week. It measured the number of units produced per week for 10 employees before and after the change:

	A	B	C	D	E	F	G	H	I	J
Before:	25	26	27	22	29	25	29	30	25	28
After:	23	24	26	23	30	24	26	32	25	29

Using the Sandler A test, test the null hypothesis at $\alpha = 0.05$ level of significance.

15.17 Referring to Exercise 14.25, assume that the two groups had been matched on their abilities before the coffee break was instituted. Assume that the pairs are in identical order for the two groups. Determine the standard error of the difference between means and the t-scores. Compare the obtained value with that of Exercise 14.22.

15.18 A store owner wants to increase the number of people walking into his store. For a week, he records the number coming in per day. He then hires a designer to set up the window display and records the number the following week. The records show the following:

	Mon	Tue	Wed	Th	Fri	Sat
Before:	150	175	140	180	170	160
After:	200	180	180	175	190	175

Did the new display help the owner? What is the value of A?

15.19 In an attempt to increase record sales, a manager advertises that five records are on sale. For a sample of 10 albums not on sale, she finds the following amounts were sold a day before the sale and on a day during the sale:

	A	B	C	D	E	F	G	H	I	J
Before:	25	15	10	25	30	5	0	40	50	35
During:	30	17	13	30	25	5	1	45	45	40

Did the sale significantly increase the number of other records sold? What is the value of A?

15.20 Referring to Exercise 3.22, determine if there is a significant difference between the output produced in rooms A and B. Calculate the value of A.

15.21 Referring to Exercise 3.23, assume that the two records of room B are made with each employee occupying the same position on each record. Is there a significant difference at the 0.05 level between the output before and after the introduction of music, carpeting, and better illumination? Calculate the value of A.

15.22 A merchant recorded the number of packs of cigarettes sold for five brands on three different occasions: one day just one month before a $0.04 per pack tax increase, the day after the increase, and one day a month after the increase.

	Before tax	One day after	One month after
Brand A	50	45	49
Brand B	40	35	42
Brand C	60	57	59
Brand D	65	67	67
Brand E	50	40	47

Is there a significant difference between the number of packs sold before the tax and one day after? What is the value of A?

15.23 Referring to the above problem, is there a significant difference between the number sold one month before and one month after the tax was increased? What is the value of A?

15.24 Again referring to Exercise 15.22, is there a significant difference between the number sold one day and one month after the tax was increased? Calculate the value of A.

15.25 We have already seen that cars with manual transmissions achieve significantly better gas mileage on the highway than matched cars with automatic transmissions. City driving is different from highway driving in that there is more stop and go. Using the data presented below see whether a significant difference ($\alpha = 0.01$) exists between cars with manual and automatic transmissions in city driving.

	Type of transmission	
Car model	Manual	Automatic
Gremlin	16	15
Hornet	13	12
Audi Fox	24	23
Skyhawk	17	18
Vega	19	19
Datsun 210	29	27
Datsun 280	17	17
Fiat 131	17	19
Mustang II	15	15
Granada	15	15
Capri II	16	17
Monarch	15	15
Starfire	17	18
Valiant/Duster	16	15
Sunbird	17	18
LeMans	14	13
Corolla	20	21
Corona MK II	16	16
Volvo	18	18
LUV Pickup	21	19

An Introduction to the Analysis of Variance 16

16.1 MULTIGROUP COMPARISONS

We have reviewed the classic design of experiments on several different occasions in this text. The classic study consists of two groups, an experimental and a control. The purpose of statistical inference is to test a specific hypothesis, e.g., whether or not both groups could have reasonably been drawn from the same population (see Chapter 14).

Although this classical research design is still employed in many studies, its limitations should be apparent to you. To restrict our observations to two groups, on all occasions, is to overlook the wonderful complexity of the phenomena which the scientist investigates. Rarely do events in nature conveniently order themselves into two groups, an experimental and a control. More commonly, the questions we pose to nature are: Which of several alternative schedules of reinforcement leads to the greatest resistance to experimental extinction? Which of five different methods of teaching the concept of fractions to the primary grades leads to the greatest learning gains? Which form of psychotherapy leads to the greatest incidence of patient recovery?

Obviously, the research design necessary to provide experimental answers to the above questions would require comparison of more than two groups. You may wonder: But why should multigroup comparisons provide any obstacles? Can we not simply compare the mean of each group with the mean of every other group and obtain a Student t-ratio for each comparison? For example, if we had four experimental groups, A, B, C, D, could we not calculate Student t-ratios comparing A with B, C and D; B with C and D; and C with D?

If you will think for a moment of the errors in inference, which we have so frequently discussed, you will recall that our greatest concern has been to avoid type I errors. When we establish the region of rejection at the 0.05 level, we are, in effect, acknowledging our willingness to take the risk of being wrong as often as 5% of the time in our rejection of the null hypothesis. Now, what happens when we have numerous comparisons to make? For an extreme example, let us imagine that we have conducted a study involving the calculation of 1000 separate Student

t-ratios. Would we be terribly impressed if, say, 50 of the t's proved to be significant at the 0.05 level? Of course not. Indeed, we would probably murmur something to the effect that, "With 1000 comparisons, we would be surprised if we didn't obtain approximately 50 comparisons that are significant *by chance* (i.e., due to predictable sampling error)."

The **analysis of variance** is a technique of statistical analysis which permits us to overcome the ambiguity involved in assessing significant differences when more than one comparison is made. It allows us to answer the question: Is there an overall indication that the experimental treatments are producing differences among the means of the various groups? Although the analysis of variance may be used in the two-sample case (in which event it yields precisely the same probability values as the Student t-ratio), it is most commonly employed when three or more groups are involved. Indeed, it has its greatest usefulness when two or more independent variables are studied. However, in this text we shall restrict our introductory treatment of analysis of variance to several levels of a single independent variable. For purposes of exposing the fundamental characteristics of the analysis of variance, our initial illustrative material will involve the two-sample case.

16.2 THE CONCEPT OF SUMS OF SQUARES

You will recall that we previously defined the **variance** as

$$s^2 = \frac{\sum x^2}{N - 1}.$$

It will be recalled that, when the deviation of scores from the mean $(X - \bar{X})$ or x, is large, the variance, and therefore the variability of scores, is also large. When the deviations are small, the variance is correspondingly small.

Now, if we think back to the Student t-ratio for a moment, we shall note that both the numerator and denominator give us some estimate of variability:

$$t = \frac{\bar{X}_1 - \bar{X}_2}{s_{\bar{X}_1 - \bar{X}_2}}.$$

It will be recalled that the denominator, which we referred to as the standard error of the difference between means, is based on the pooled estimate of the variability within each experimental group, that is,

$$s_{\bar{X}_1 - \bar{X}_2} = \sqrt{\frac{\sum x_1^2 + \sum x_2^2}{n_1 + n_2 - 2}\left(\frac{1}{n_1} + \frac{1}{n_2}\right)}.$$

However, the numerator is also a measure of variability, i.e., the variability between means. When the difference between means is large relative to $s_{\bar{X}_1 - \bar{X}_2}$, the Student

t-ratio is large. When the difference between means is small relative to $s_{\bar{X}_1 - \bar{X}_2}$, the Student t-ratio is also small.

The analysis of variance consists of obtaining two independent estimates of variance, one based upon variability between groups (**between-group variance**) and the other based upon the variability within groups (**within-group variance**). The significance of the difference between these two variance estimates is provided by Fisher's F-distributions. (We are already familiar with F-distributions from our prior discussion of homogeneity of variance, Section 14.4.) If the *between-group variance* is large (i.e., the difference between means is large) relative to the *within-group variance*, the F-ratio is large. Conversely, if the *between group variance* is small relative to the *within-group variance*, the F-ratio will be small.

A basic concept in the analysis of variance is the sum of squares. We have already encountered the **sum of squares** in calculating the standard deviation, the variance, and the standard error of the difference between means. It is simply the numerator in the formula for variance, that is, $\sum x^2$. As you will recall, the raw score formula for calculating the sum of squares is

$$\sum x^2 = \sum X^2 - (\sum X)^2/N.$$

The advantage of the analysis of variance technique is that we can partial the total sum squares ($\sum x_{\text{tot}}^2$) into two components, the *within-group sum squares* ($\sum x_W^2$) and the *between-group sum squares* ($\sum x_B^2$). Before proceeding any further, let us clarify each of these concepts with a simple example.

Imagine that we have completed a study comparing two experimental treatments and obtained the scores listed in Table 16.1.

Table 16.1

Scores of two groups of subjects in a hypothetical experiment

Group 1		Group 2	
X_1	X_1^2	X_2	X_2^2
1	1	6	36
2	4	7	49
5	25	9	81
8	64	10	100
$\sum 16$	94	32	266

$n_1 = 4,\quad \bar{X}_1 = 4\quad n_2 = 4,\quad \bar{X}_2 = 8$

$\sum X_{\text{tot}} = 48,\quad N = 8,\quad \bar{X}_{\text{tot}} = 6.$

The mean for group 1 is 4; the mean for group 2 is 8. The overall mean, $\bar{X}_{\text{tot}}$, is $\frac{48}{8}$ or 6. Now, if we were to subtract the overall mean from each score and square, we would obtain the **total sum of squares**:

$$\sum x_{\text{tot}}^2 = \sum(X - \bar{X}_{\text{tot}})^2. \tag{16.1}$$

The alternative raw score formula is

$$\sum x_{\text{tot}}^2 = \sum X_{\text{tot}}^2 - (\sum X_{\text{tot}})^2/N. \tag{16.2}$$

For the data in Table 16.1, the total sum squares is

$$\begin{aligned}\sum x_{\text{tot}}^2 &= 360 - (48)^2/8 \\ &= 360 - 288 = 72.\end{aligned}$$

The **within-group sum of squares** is merely the sum of the sum squares obtained within each group, that is,

$$\sum x_W^2 = \sum x_1^2 + \sum x_2^2 \tag{16.3}$$

$$\begin{aligned}\sum x_1^2 &= \sum X_1^2 - (\sum X_1)^2/n_1 \\ &= 94 - (16)^2/4 = 94 - 64 = 30, \\ \sum x_2^2 &= \sum X_2^2 - (\sum X_2)^2/n_2 \\ &= 266 - (32)^2/4 = 266 - 256 = 10, \\ \sum x_W^2 &= 30 + 10 = 40.\end{aligned}$$

Finally, the **between-group sum of squares** ($\sum x_B^2$) may be obtained by subtracting the overall mean from each group mean, squaring the result, multiplying by the n in each group, and summing across all the groups. Thus

$$\sum x_B^2 = \sum n_i(\bar{X}_i - \bar{X}_{\text{tot}})^2, \tag{16.4}$$

where n_i is the number in the ith group and $\bar{X}_i$ is the mean of the ith group,

$$\sum x_B^2 = 4(4 - 6)^2 + 4(8 - 6)^2 = 32.$$

The raw-score formula for calculating the between-group sum squares is

$$\sum x_B^2 = \sum \frac{(\sum X_i)^2}{n_i} - \frac{(\sum X_{\text{tot}})^2}{N}, \tag{16.5}$$

and

$$\begin{aligned}\sum x_B^2 &= (16)^2/4 + (32)^2/4 - (48)^2/8 \\ &= 64 + 256 - 288 \\ &= 320 - 288 = 32.\end{aligned}$$

It will be noted that the total sum squares is equal to the sum of the between-group sum squares and the within-group sum squares. In other words,

$$\sum x_{\text{tot}}^2 = \sum x_W^2 + \sum x_B^2. \tag{16.6}$$

In the above example, $\sum x_{\text{tot}}^2 = 72$, $\sum x_W^2 = 40$, and $\sum x_B^2 = 32$. Thus $72 = 40 + 32$.

16.3 OBTAINING VARIANCE ESTIMATES

Now, to arrive at variance estimates from between- and within-group sum squares, all we need to do is to divide each by the appropriate number of degrees of freedom. The degrees of freedom of the between-group is simply the number of groups (k) minus 1.

$$\text{df}_B = k - 1. \tag{16.7}$$

With two groups ($k = 2$), df $= 2 - 1 = 1$. Thus our between-group variance estimate for the problem at hand is

$$s_B^2 = \sum x_B^2/\text{df}_B = \tfrac{32}{1} = 32, \qquad \text{df} = 1. \tag{16.8}$$

The number of degrees of freedom of the within-groups is the total N minus the number of groups. Thus

$$\text{df}_W = N - k. \tag{16.9}$$

In the present problem, $\text{df}_W = 8 - 2 = 6$ and our within-group variance estimate becomes

$$s_W^2 = \sum x_W^2/\text{df}_W = \tfrac{40}{6} = 6.67, \qquad \text{df} = 6. \tag{16.10}$$

Now, all that is left is to calculate the F-ratio and determine whether or not our two variance estimates could have reasonably been drawn from the same population. If not, we shall conclude that the significantly larger between-group variance is due to the operation of the experimental conditions. In other words, we shall conclude that the experimental treatments produced a significant difference in means. The F-ratio, in analysis of variance, is the between-group variance estimate divided by the within-group variance estimate. Symbolically,

$$F = s_B^2/s_W^2. \tag{16.11}$$

For the above problem our F-ratio is

$$F = 32/6.67 = 4.80, \qquad \text{df} = 1/6.$$

Looking up the F-ratio under 1 and 6 degrees of freedom, Table D, we find that an F-ratio of 5.99 or larger is required for significance at the 0.05 level. For the present problem, then, we cannot reject the null hypothesis.

16.4 FUNDAMENTAL CONCEPTS OF ANALYSIS OF VARIANCE

In these few pages, we have examine all the basic concepts necessary to understand simple analysis of variance. Before proceeding with an example involving three groups, let us briefly review these fundamental concepts.

1. We have seen that, in an experiment involving two or more groups, it is possible to identify two different bases for estimating the population variance: the between-group and the within-group.
   a) The **between-group variance estimate** reflects the magnitude of the difference between and/or among the group means. The larger the difference between means, the larger the between-group variance.
   b) The **within-group variance estimate** reflects the dispersion of scores within each treatment group. The within-group variance is analogous to $s_{\bar{X}_1 - \bar{X}_2}$ in the Student t-ratio. It is often referred to as the error term.

2. The null hypothesis is that the two independent variance estimates may be regarded as estimates of the same population value. In other words, H_0 is that the samples were drawn from the same population, or that $\mu_1 = \mu_2 = \cdots = \mu_k$.

3. The ***F*-ratio** consists of the between-group variance estimate divided by the within-group variance estimate. By consulting Table D of the distribution of F we can determine whether or not the null hypothesis of equal population variance can reasonably be entertained. In the event of a significant F-ratio, we may conclude that the groups were not drawn from populations with the same means.

4. In the two-sample case, the F-ratio yields probability values identical to those of the Student t-ratio. Indeed, in the one-degree-of-freedom situation (that is, $k = 2$), $t = \sqrt{F}$ or $t^2 = F$. You may check this statement by calculating the Student t-ratio for the sample problem we have just completed.

16.5 AN EXAMPLE INVOLVING THREE GROUPS

Let us imagine that you have just completed a study concerned with the determination of the effectiveness of three different methods of instruction for the basic principles of arithmetic. Twenty-seven primary grade children were randomly assigned to three equal groups employing one of the methods for achieving randomness described in Section 11.2. Following the completion of their instruction, all children were tested on an "Inventory of Basic Arithmetic." The results of this hypothetical study are presented in Table 16.2.

Table 16.2

Scores of three groups of subjects in a hypothetical experiment

Group 1		Group 2		Group 3	
X_1	X_1^2	X_2	X_2^2	X_3	X_3^2
4	16	12	144	1	1
5	25	8	64	3	9
4	16	10	100	4	16
3	9	5	25	6	36
6	36	7	49	8	64
10	100	9	81	5	25
1	1	14	196	3	9
8	64	9	81	2	4
5	25	4	16	2	4
$\sum$46	292	78	756	34	168

$n_1 = 9, \quad \bar{X}_1 = 5.11 \quad n_2 = 9, \quad \bar{X}_2 = 8.67 \quad n_3 = 9, \quad \bar{X}_3 = 3.78$

$$\sum X_{\text{tot}} = 46 + 78 + 34 = 158,$$
$$\sum X_{\text{tot}}^2 = 292 + 756 + 168 = 1216,$$
$$N = 27.$$

The following steps are employed in a three-group analysis of variance:

Step 1. Employing formula (16.2), the total sum squares is

$$\sum x_{\text{tot}}^2 = 1216 - (158)^2/27 = 291.41.$$

Step 2. Employing formula (16.5) for three groups, the between-group sum squares is

$$\sum x_B^2 = (46)^2/9 + (78)^2/9 + (34)^2/9 - (158)^2/27 = 114.96.$$

Step 3. The within-group sum squares may be obtained by employing formula (16.3) for three groups:

$$\sum x_W^2 = (292 - (46)^2/9) + (756 - (78)^2/9) + (168 - (34)^2/9) = 176.45.$$

You may obtain the within-group sum squares by subtraction, that is,

$$\sum x_W^2 = \sum x_{\text{tot}}^2 - \sum x_B^2 = 291.41 - 114.96 = 176.45.$$

Step 4. The between-group variance estimate is

$$\text{df}_B = k - 1 = 2, \qquad s_B^2 = 114.96/2 = 57.48.$$

Step 5. The within-group variance estimate is

$$df_W = N - k = 24, \qquad s_W^2 = 176.45/24 = 7.35.$$

Step 6. Employing formula (16.11) we find that the value of F is

$$F = 57.48/7.35 = 7.82, \qquad df = 2/24.$$

To summarize these steps, we employ the format shown in Table 16.3.

Table 16.3

Summary table for representing the relevant statistics in analysis-of-variance problems

Source of variation	Sum squares	Degrees of freedom	Variance estimate*	F
Between-groups	114.96	2	57.48	7.82
Within-groups	176.45	24	7.35	
Total	291.41	26		

* In many texts, the term "mean square" appears in this box. However, the authors prefer the term "variance estimate" since this term accurately describes the nature of the entries in the column.

By employing the format recommended in Table 16.3, you have a final check upon your calculation of sum squares and your assignment of degrees of freedom. Thus, $\sum x_B^2 + \sum x_W^2$ must equal $\sum x_{\text{tot}}^2$. The degrees of freedom of the total are found by

$$df_{\text{tot}} = N - 1. \tag{16.12}$$

In the present example, the number of degrees of freedom for the total is

$$df_{\text{tot}} = 27 - 1 = 26.$$

16.6 THE INTERPRETATION OF *F*

When we look up the F required for significance with 2 and 24 degrees of freedom, we find that an F of 3.40 or larger is significant at the 0.05 level.

Since our F of 7.86 exceeds this value, we may conclude that there is a significant difference among the means of our three experimental groups. Now, do we stop at this point? After all, are we not interested in determining whether or not one of the three methods of instructing the fundamentals of arithmetic is superior to the other two? The answer to the first question is negative, and the answer to the second is affirmative.

The truth of the matter is that our finding an overall significant F-ratio now permits us to investigate specific hypotheses. In the absence of a significant F-ratio, any significant differences between specific comparisons would have to be regarded as suspicious—very possibly representing a chance difference.

Over the past ten years, a large number of tests have been developed which permit the researcher to investigate specific hypotheses concerning population parameters. Two broad classes of such tests exist:

1. **A priori** or **planned comparisons:** When comparisons are planned in advance of the investigation, an *a priori* test is appropriate. For *a priori* tests, it is not necessary that the overall F-ratio be significant.

2. **A posteriori comparisons:** When the comparisons are not planned in advance, an *a posteriori* test is appropriate.

In the present example, we shall illustrate the use of an *a posteriori* test for making pairwise comparisons among means.

Tukey (1953) has developed such a test which he named the HSD (honestly significant difference) test. To employ this test, the overall F-ratio must be significant.

A difference between two means is significant, at a given α-level, it it equals or exceeds HSD, which is:

$$\text{HSD} = q_\alpha \sqrt{\frac{s_W^2}{n}}, \qquad (16.13)$$

in which

s_W^2 = the within-group variance estimate,

n = number of subjects in each condition,

q_α = tabled value for a given α-level found in Table P for df_W

and

k = number of means.

16.6.1 A Worked Example

Let us employ the data from Section 16.5 to illustrate the application of the HSD test. We shall employ $\alpha = 0.05$ for testing the significance of the difference between each pair of means.

Table 16.4
Differences among means

	$\bar{X}_1$	$\bar{X}_2$	$\bar{X}_3$
$\bar{X}_1 = 5.11$	...	3.56	1.33
$\bar{X}_2 = 8.67$	...	...	4.89
$\bar{X}_3 = 3.78$	...	...	...

Step 1. Prepare a matrix showing the mean of each condition and the differences between pairs of means. This is shown in Table 16.4.

Step 2. Referring to Table P under error df = 24, $k = 3$ at $\alpha = 0.05$, we find $q_{0.05} = 3.53$.

Step 3. Find HSD by multiplying $q_{0.05}$ by $\sqrt{\frac{s_w^2}{n}}$. The quantity s_w^2 is found in Table 16.3 under within-group variance estimate. The n per condition is 9. Thus:

$$\text{HSD} = 3.53\sqrt{\frac{7.35}{9}}$$

$$= 3.53(0.90)$$

$$= 3.18$$

Step 4. Referring back to Table 16.4, we find that the differences between $\bar{X}_1$ vs. $\bar{X}_2$ and $\bar{X}_2$ vs. $\bar{X}_3$ both exceed HSD = 3.18. We may therefore conclude that these differences are statistically significant at $\alpha = 0.05$.

16.7 WITHIN-GROUP VARIANCE AND HOMOGENEITY

When discussing the Student t-ratio in Chapter 14, we noted that a fundamental assumption underlying the use of the Student t-ratio is that the variances for each group must be homogeneous, i.e., drawn from the same population of variances. The same assumption holds true for the analysis of variances. In other words, a basic assumption underlying the analysis of variance is that the treatment variances (which, when summed together, makes up s_W^2) are homogeneous. As with the Student t-ratio, there is a test for determining whether or not the hypothesis of identical variances is tenable. However, it is beyond the scope of this introductory text to delve into Bartlett's test of **homogeneity of variances**. Application of this test is described in Edwards (1968).

CHAPTER SUMMARY

We began this chapter with the observation that the scientist is frequently interested in conducting studies which are more extensive than the classical two-group design. However, when more than two groups are involved in a study, we increase the risk of making a type I error if we accept, as significant, *any* comparison which falls within the rejection region. In multigroup studies, it is desirable to know whether or not there is an indication of an overall effect of the experimental treatments before we investigate specific hypotheses. The analysis of variance technique provides such a test.

In this chapter we presented a mere introduction to the complexities of analysis of variance. We showed that total sum squares can be partitioned into two component sum squares: the within-group and the between-group. These two component sum squares provide us, in turn, with independent estimates of the population variance. A between-group variance estimate which is large, relative to the within-group variance, suggests that the experimental treatments are responsible for the large differences among the group means. The significance of the difference in variance estimates is obtained by reference to the F-table (Table D).

When the overall F-ratio is found to be statistically significant, we are free to investigate specific hypotheses, employing a multiple-comparison test.

Finally, we pointed out a basic assumption of the analysis of variance technique, i.e., homogeneity of variances among treatment groups.

Terms to Remember:

Analysis of variance

Sum of squares

Variance

Between-group variance

Within-group variance

Total sum of squares

Within-group sum of squares

Between-group sum of squares

Between-group variance estimate

Within-group variance estimate

F-ratio

Homogeneity of variances

A priori or planned comparisons

A posteriori comparisons

EXERCISES

16.1 Using the following data, derived from the 10-year period 1955–1964, determine whether there is a significant difference, at the 0.01 level, in death rate among the various seasons. (*Note*: Assume death rates for any given year to be independent.)

Winter	Spring	Summer	Fall
9.8	9.0	8.8	9.4
9.9	9.3	8.7	9.4
9.8	9.3	8.8	10.3
10.6	9.2	8.6	9.8
9.9	9.4	8.7	9.4
10.7	9.1	8.3	9.6
9.7	9.2	8.8	9.5
10.2	8.9	8.8	9.6
10.9	9.3	8.7	9.5
10.0	9.3	8.9	9.4

16.2 Conduct an HSD test, comparing the death rates of each season with every other season. Employ the 0.01 level, two-tailed test, for each comparison.

16.3 Conduct separate *t*-analyses comparing the death rates of each season with every other season. Employ the 0.01 level, two-tailed test, for each comparison.

16.4 Conduct an analysis of variance on the data in Exercise 14.3. Verify that, in the two-group condition, $F = t^2$.

16.5 Manufacturer Baum negotiates contracts with 10 different independent research organizations to compare the effectiveness of his product with that of his leading competitor. A significant difference (0.05 level) in favor of Mr. Baum's product is found in one of the 10 studies. He subsequently advertised that independent research has demonstrated the superiority of his product over the leading competitor. Criticize this conclusion.

16.6 Various drug companies make the claim that they manufacture an analgesic which releases its active ingredient "faster." A random selection of the products of each manufacturer revealed the following times, in seconds, required for the release of 50% of the analgesic agent. Test the null hypothesis that all the analgesics are drawn from a common population of means.

Brand *A*	Brand *B*	Brand *C*	Brand *D*
28	34	29	22
19	23	24	31
30	20	33	18
25	16	21	24

16.7 A consumer organization randomly selects several gas clothes dryers of three leading manufacturers for study. The time required for each machine to dry a standard load of clothes was tabulated. Set up and test the appropriate null hypothesis.

Brand A	Brand B	Brand C
42	52	38
36	48	44
47	43	33
43	49	35
38	51	32

16.8 Automobile tires, selected at random from six different brands, required the following braking distances, in feet, when moving at 25 mi/hr. Set up and test the appropriate null hypothesis.

Brand A	Brand B	Brand C	Brand D	Brand E	Brand F
22	25	17	21	27	20
20	23	19	24	29	14
24	26	15	25	24	17
18	22	18	23	25	15

16.9 Auto gasoline taxes (cents per gallon), 1970

New England		Mideast		Far West	
Maine	8	New York	7	Wash.	9
N. H.	7	New Jersey	7	Oregon	7
Vermont	8	Penn.	8	Nevada	6
Mass.	6.5	Delaware	7	Calif.	7
R. I.	8	Maryland	7	Alaska	8
Conn.	8	D. C.	7	Hawaii	5

Source: Federal Highway Administration. *The 1972 World Almanac and Book of Facts.*, L. H. Long (ed.). New York: Newspaper Enterprise Association, Inc., 1971, p. 122.

Determine if there is a significant difference in state gasoline taxes among the three geographical areas given.

16.10 Perform an analysis of variance, using the data given in Exercise 14.27.

16.11 Suppose typing speeds (words per minute) are compared for a sample of 24 people who attended four different secretarial schools. The following data are obtained:

School A	School B	School C	School D
50	55	50	70
50	60	65	80
55	65	75	65
60	55	55	70
45	70	60	75
55	65	65	60

Is there a significant difference in the typing speeds among the four groups?

16.12 A nutritional expert divides a sample of bicyclers into three groups. Group B is given a vitamin supplement and Group C is given a diet of health foods. Group A is instructed to eat as they normally do. The expert subsequently records the number of minutes it takes each person to ride six miles:

A	B	C
15	14	13
16	13	12
14	15	11
17	16	14
15	14	11

Set up the appropriate hypothesis and conduct an analysis of variance.

16.13 For Exercise 16.12, determine which diet or diets are superior.

16.14 In 1971, three baseball clubs showed the following number of home runs made by their players:

Chicago Cubs:	2	16	28	2	21	2	2	19	0	4	8	6
Houston Astros:	9	2	1	10	13	1	12	0	1	2	7	7
Cincinnati Reds:	13	39	25	9	3	0	13	27	5	1	2	

Test the hypothesis that there is no difference in the number of home runs hit by the three ball clubs. (See *The 1972 World Almanac*, p. 910–911.)

16.15 Assume that there are only four manufacturers of house paint, and manufacturer A states that her paint is superior to all the other paints. To test this claim, an investigator samples the number of years between the time a house is painted and the time it needs repainting, with the following results:

A	B	C	D
4.5	4.0	2.5	3.5
5.5	4.5	3.0	3.0
5.0	4.0	3.0	4.0
5.5	3.5	3.5	3.0
6.0	3.0	4.0	4.5
5.0	4.5	2.5	3.0

Is manufacturer A's claim supported by the data? Compare the records of A and B.

Section II

Inferential Statistics

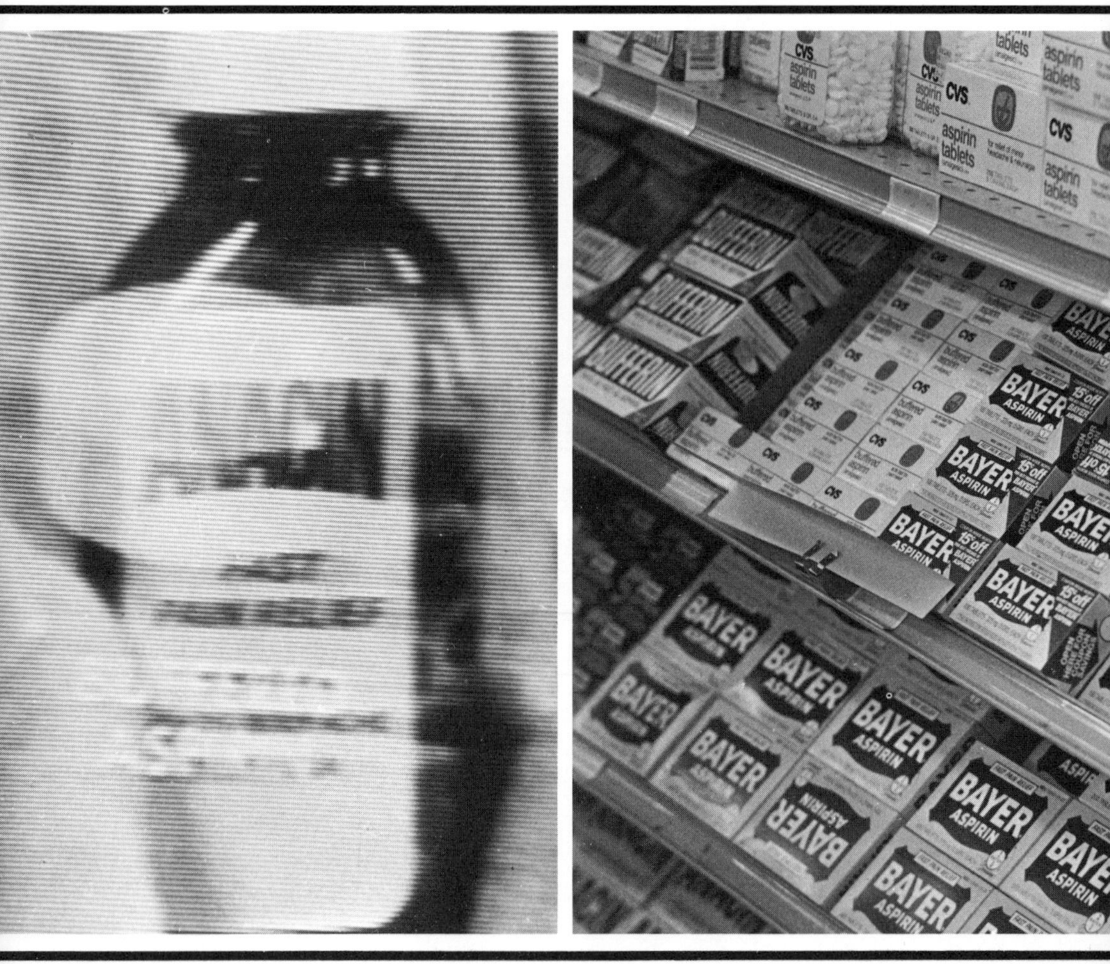

Part B

NONPARAMETRIC TESTS OF SIGNIFICANCE

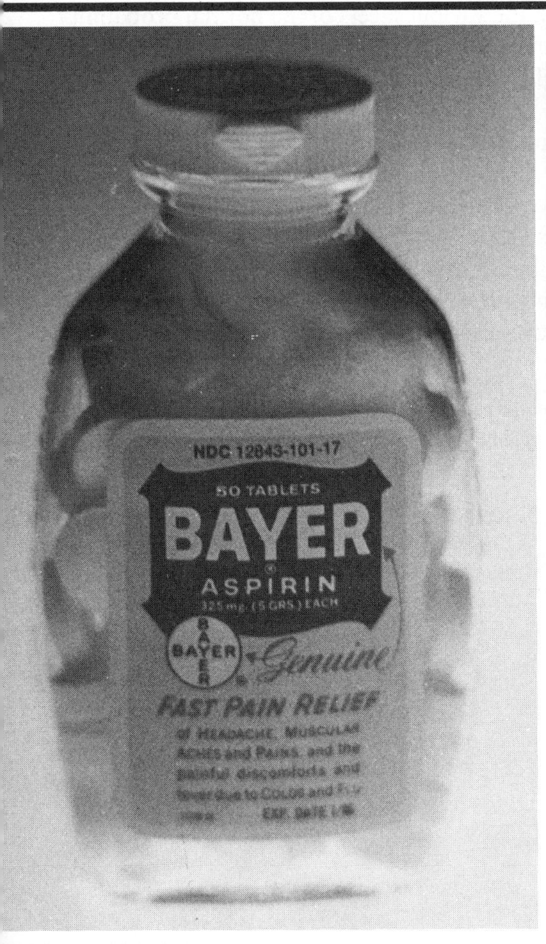

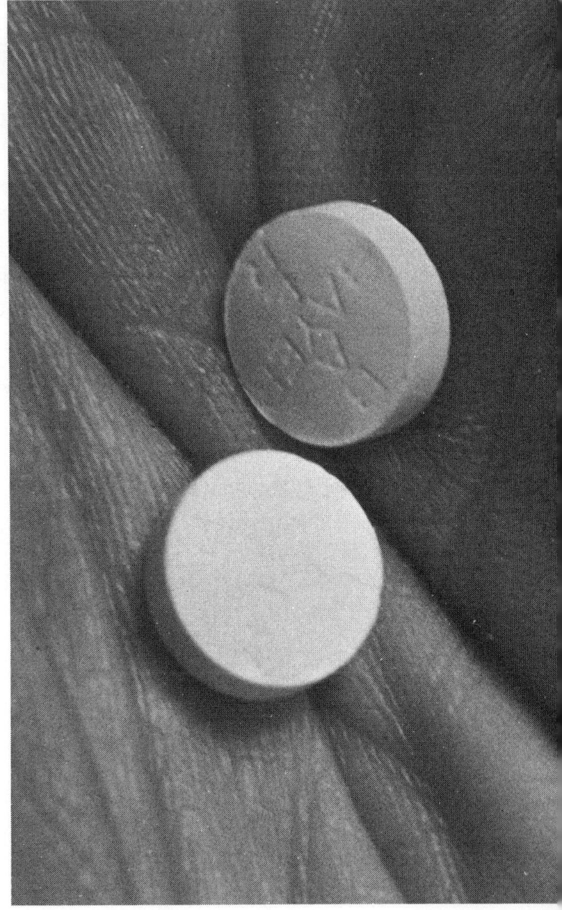

So far we have concerned ourselves with the various facets of the normal probability curve as applied to descriptive and inferential statistics. Statistical tests of inference which make use of the normal probability model are referred to as *parametric tests of significance.*

Many data are collected which either do not lend themselves to analysis in terms of the normal probability curve, or fail to meet the basic assumptions for its application. Consider a study in which the data collected consist of ranks (e.g., ranking students in terms of cooperativeness). The resulting ordinal values are non-quantitative and are, of necessity, distributed in a rectangular fashion. Clearly, the normal probability statistics do not apply. Recent years have seen the development of a remarkable variety of nonparametric tests which may be employed with such data.

> *A nonparametric test of significance is defined as one which makes no assumptions concerning the shape of the parent distribution or population, and accordingly is commonly referred to as a distribution-free test of significance.*

In this final section of the text we shall describe several of the more important non-parametric statistical tests.

Statistical Inference with Categorical Variables 17

17.1 INTRODUCTION

In recent years there has been a broadening in both the scope and the penetration of research in the behavioral and social sciences. Much provocative and stimulating research has been initiated in such diverse areas as personality, psychotherapy, group processes, economic forecasts, etc. New variables have been added to the arsenal of the researcher, many of which do not lend themselves to traditional parametric statistical treatment, either because of the scales of measurement employed or because of flagrant violations of the assumptions of these parametric tests. For these reasons, many new statistical techniques have been developed.

Parametric techniques are usually preferable because of their greater sensitivity. This generalization is not true, however, when the underlying assumptions are seriously violated. Indeed, under certain circumstances (e.g., badly skewed distributions, particularly with small n's), a nonparametric test may well be as powerful as its parametric counterpart.* Consequently, the researcher is frequently faced with the difficult choice of the statistical test appropriate to the data.

At this point, let us interject a word of caution with respect to the choice of statistical tests of inference. For heuristic purposes, we will take a few sample problems and subject them to statistical analyses employing several different tests of significance. This procedure will serve to clarify points of differences among the various tests. However, you may inadvertently draw an erroneous conclusion, namely, that the researcher first collects data, and then "shops around" for the statistical test most sensitive to any existing differences. Actually, nothing could be further from the truth. The null hypothesis, alternative hypothesis, statistical test, sampling distribution, and level of significance should all be specified in *advance* of the collection of data. If one "shops around," so to speak, after the collection of the

* Numerous investigators have demonstrated the robustness of the t and F tests; i.e., even substantial departures from the assumptions underlying parametric tests do not seriously affect the validity of statistical inferences. For articles dealing with this topic, see Haber, Runyon, and Badia, *Readings in Statistics*, Reading, Mass.: Addison-Wesley Publishing Co., Inc., 1970.

data, he tends to maximize the effects of any chance differences which favor one test over another. As a result, the possibility of a type I error (rejecting the null hypothesis when it is true) is substantially increased.

Usually, we do not have the problem of choosing statistical tests in relation to categorical variables because nonparametric tests alone are suitable for enumerative data.

The problem of choosing a statistical treatment usually arises when we employ small samples* and/or when there is doubt concerning the normality of the underlying population distribution. In Chapter 18 we shall demonstrate several nonparametric statistical tests employed when such doubts arise.

17.2 THE BINOMIAL TEST

In Section 2.5, when discussing various scales of measurement, we pointed out that the observation of unordered variables constitutes the lowest level of measurement. In turn, the simplest form of nominal scale is one which contains only two classes or categories, and is referred to variously as a **two-category** or **dichotomous population**. Examples of two-category populations are numerous, e.g., male and female, right and wrong on a test item, married and single, juvenile delinquent and nondelinquent, literate and illiterate. Some of these populations may be thought of as inherently dichotomous (e.g., male vs. female) and therefore not subject to measurement on a higher-level numerical scale, whereas others (e.g., literate and illiterate) may be thought of as continuous, varying from the absence of the quality under observation to different degrees of its manifestation. Obviously, whenever possible, the data we collect should be at the highest level of measurement that we can achieve. However, for a variety of reasons, we cannot always scale a variable at the ordinal level or higher. Nevertheless, we are called upon to collect nominally scaled data and to draw inferences from these data.

In a two-category population, we define P as the proportion of cases in one class and $Q = 1 - P$ as the proportion in the other class.

The probabilities associated with specific outcomes may be obtained by employing formula (17.1):

$$p(x) = \frac{N!}{x!(N-x)!} P^x Q^{N-x}, \tag{17.1}$$

where

$x =$ the number of objects in one category or the number of successes,

$N - x =$ the number of objects in the remaining category or the number of failures,

* When large samples are employed, the parametric tests are almost always appropriate because of the *central-limit theorem* (Section 13.2).

N = the total number of objects or total number of trials,
$p(x)$ = the probability of x objects in one category,
! = the factorial sign directs us to multiply the indicated value by all integers less than it but greater than zero, e.g., if $N = 5$, $N! = 5 \cdot 4 \cdot 3 \cdot 2 \cdot 1$.*

A sample problem should serve to illustrate the use of formula (17.1) in the testing of hypotheses.

17.2.1 Sample Problem

The dean of students in a large university claims that ever since the sale of cigarettes was prohibited on campus, the proportion of students who smoke has dropped to 0.30. However, previous observations at other institutions, where the sale of cigarettes has been banned, have found little effect on smoking behavior; i.e., far greater than 0.30 of the students continue to smoke.

Nine students, selected at random, are asked to indicate whether or not they smoke. Six of these students respond in the affirmative.

To test the validity of the dean's claim, we will let P represent the proportion of students in the population who smoke, and Q, the proportion of students who do not smoke.

1. *Null hypothesis* (H_0): $P = 0.30$, $Q = 0.70$.
2. *Alternative hypothesis* (H_1): $P > 0.30$, $Q < 0.70$. Note that H_1 is directional.
3. *Statistical test*: Since we are dealing with a two-category population, the binomial test is appropriate.
4. *Significance level*: $\alpha = 0.05$.
5. *Sampling distribution*: The sampling distribution is given by the binomial expansion, formula (12.1).
6. *Critical region*: The critical region consists of all values of x which are so large that the probability of their occurrence under H_0 is less than or equal to 0.05. Since H_1 is directional, the critical region is one-tailed.

In order to determine whether the obtained x lies in the critical region, we must obtain the sum of the probabilities associated with $x = 9$, $x = 8$, $x = 7$, and $x = 6$.

To illustrate, employing formula (17.1), we find that the probability of $x = 9$, that is, if $P = 0.3$ (under H_0) the probability that all of the 9 students smoke, is

$$\begin{aligned} p(9) &= \frac{9!}{9!(9-9)!}(0.3)^9(0.7)^{9-9} \\ &= (0.3)^9 = 0.000019683. \end{aligned}$$

* In obtaining the binomial probabilities, it is important to remember that $0! = 1$ and any value other than zero raised to the zero power equals 1, i.e., $X^0 = 1$.

The probability of $x = 8$ is

$$p(8) = 9(0.3)^8(0.7)^1 = 0.000413343.$$

The probability of $x = 7$ is

$$p(7) = 36(0.3)^7(0.7)^2 = 0.003857868.$$

Finally, the probability of $x = 6$ is

$$p(6) = 84(0.3)^6(0.7)^3 = 0.021003948.$$

Thus the probability that at least 6 out of 9 students smoke when $P = 0.30$ is the sum of the above probabilities, that is, $p(x \geq 6) = 0.025$.

Decision: Since the obtained probability is less than 0.05, we may reject H_0.

The above example was offered to illustrate the application of formula (17.1) when $P \neq Q \neq \frac{1}{2}$. However, when $N \leq 20$, and $P \neq Q \neq \frac{1}{2}$, Table N may be employed to obtain the critical values of x directly for selected values of P and Q. Referring to Table N, for $N = 9$ and $P = 0.30$, we find that x must be equal to or greater than 6 to reject H_0 at $\alpha = 0.05$.

Incidentally, when $P = Q = \frac{1}{2}$ and $N \leq 25$, Table M provides one-tailed probability values of x when x is defined as the smaller of the observed frequencies. For example, if we toss a coin ten times ($N = 10$), we find that the probability of obtaining two or fewer heads ($x \leq 2$) is 0.055.

17.3 NORMAL CURVE APPROXIMATION TO BINOMIAL VALUES

As N increases, the binomial distribution approaches the normal distribution. The approximation is more rapid as P and Q approach $\frac{1}{2}$. On the other hand, as P or Q approach zero, the approximation to the normal curve becomes poorer for any given N. A good rule of thumb to follow, when considering the **normal curve approximation to the binomial**, is that the product NPQ should equal at least 9 when P approaches 0 or 1. Accepting these restrictions, we see that the sampling distribution of x, defined as the number of objects in one category, is normal with a mean equal to NP and a standard deviation equal to $\sqrt{NPQ}$.

To test the null hypothesis, we put x into standardized form:

$$z = \frac{x - NP}{\sqrt{NPQ}}. \tag{17.2}$$

The distribution of the z-scores is approximately normal with a mean equal to zero and a standard deviation equal to one. Thus the probability of any given x equals the probability of its corresponding z-score. However, the normal distribution is based on a continuous variable whereas the binomial distribution is discontinuous. The approximation to the normal distribution becomes better if a **correction**

for continuity is made. This correction consists of subtracting the constant 0.5 from the absolute value of $x - NP$. Thus z becomes

$$z = \frac{|x - NP| - 0.5}{\sqrt{NPQ}}. \tag{17.3}$$

Let us look at an example in which $x = 5$, $N = 20$, and $P = Q = \frac{1}{2}$. To obtain the probability of $x \leq 5$, we put x into standardized form, and employing the correction for continuity, we have

$$z = \frac{|5 - 10| - 0.5}{\sqrt{(20)(0.5)(0.5)}} = \frac{4.50}{\sqrt{5.00}} = \frac{4.50}{2.24} = 2.01.$$

Locating this value in Column C (Table A), we find a probability value of 0.0222. Note that if we had looked up the result in Table M under $x = 5$ and $N = 20$, we would obtain a one-tailed p-value of 0.021. In other words, the approximation is extremely good. To obtain the two-tailed p-value we would, of course, double the obtained p. Thus in the current example, $p = 0.04$.

17.4 THE χ^2 ONE-VARIABLE CASE

Let us suppose that you are a market researcher hired by a soap manufacturer to conduct research on the packaging of his product. Realizing that color may be an important determinant of consumer selection, you conduct the following study. Six hundred housewives, selected by an accepted sampling technique, are each given three differently packaged cakes of the same brand of soap. They are told that the three soaps are made according to different formulas and that the distinctive coloring of the packages is merely to aid their identification of each soap. One month later, you inform each housewife that she is to receive a free case of soap of her own choosing. Their selections are listed below.

	Color of wrapper		
	Red	White	Brown
Number of housewives selecting	200	300	100

This is the type of problem for which the χ^2* **one-variable test** is ideally suited. In single-variable applications, the χ^2 test has been described as a "**goodness-of-fit**"

* The symbol χ^2 will be used to denote the test of significance as well as the quantity obtained from applying the test to observed frequencies whereas the word, "chi-square," will refer to the theoretical chi-square distribution.

technique: it permits us to determine whether or not a significant difference exists between the *observed* number of cases falling into each category, and the *expected* number of cases, based on the null hypothesis. In other words, it permits us to answer the question, "How well does our observed distribution fit the theoretical distribution?"

What we require, then, is a null hypothesis which allows us to specify the frequencies that would be expected in each category and, secondly, a test of this null hypothesis. The null hypothesis may be tested by

$$\chi^2 = \sum_{i=1}^{k} \frac{(f_o - f_e)^2}{f_e}, \tag{17.4}$$

where

f_o = the observed number in a given category,

f_e = the expected number in that category,

$\sum_{i=1}^{k}$ = directs us to sum this ratio over all k categories.

As is readily apparent, if there is close agreement between the observed frequencies and the expected frequencies, the resulting χ^2 will be small, leading to a failure to reject the null hypothesis. As the discrepancy $(f_o - f_e)$ increases, the value of χ^2 increases. The larger the χ^2, the more likely we are to reject the null hypothesis.

In the above example, the null hypothesis would be that there is an equal preference for each color, i.e., 200 is the expected frequency in each category. Thus

$$\begin{aligned}\chi^2 &= (200 - 200)^2/200 + (300 - 200)^2/200 + (100 - 200)^2/200 \\ &= 0 + 50.00 + 50.00 \\ &= 100.00.\end{aligned}$$

In studying the Student t-ratio (Section 13.5) we saw that the sampling distributions of t varied as a function of degrees of freedom. The same is true for χ^2. However, assignment of degrees of freedom with the Student t-ratio are based on N, whereas, for χ^2, the degrees of freedom are a function of the number of categories (k). In the one-variable case, df $= k - 1$.* Table B lists the critical values of χ^2 for various α-levels. If the obtained χ^2 value exceeds the critical value at a given probability level, the null hypothesis may be rejected at that level of significance.

In the above example, $k = 3$. Therefore, df $= 2$. Employing $\alpha = 0.01$ (two-tailed test), Table B indicates that a χ^2 value of 9.21 is required for significance. Since our obtained value of 100.00 is greater than 9.21, we may reject the null hypothesis and assert, instead, that color is a significant determinant for soap preference among women.

* Since the marginal total is fixed, only $k - 1$ categories are free to vary.

17.5 THE χ^2 TEST OF THE INDEPENDENCE OF CATEGORICAL VARIABLES

So far in this chapter, we have been concerned with the one-variable case. In practice, employing categorical variables, we do not encounter the one-variable case too frequently. More often, we ask questions concerning the interrelationships between and among variables. For example, we may ask:

Is there a difference in the crime rate of children coming from different socio-economic backgrounds?

Is there a difference in the recovery rates of patients undergoing various forms of psychotherapeutic treatment?

If we are conducting an opinion poll, can we determine whether there is a difference between males and females in their opinions about a given issue?

These are but a few examples of problems for which the χ^2 **test of independence** technique has an application. You could undoubtedly extend this list to include many campus activities, such as attitudes of fraternity brothers vs. nonfraternity students toward certain basic issues (e.g., cheating on exams), differences in grading practices among professors in various departments of study. All the above problems have some things in common: (1) They deal with two or more nominal categories in which (2) the data consist of a frequency count which is tabulated and placed in the appropriate cells.

These examples also share an additional and more important characteristic: (3) There is no immediately obvious way to assign expected frequency values to each category. However, as we shall point out, shortly, what we must do is base our expected frequencies on the obtained frequencies themselves.

Let us take a look at a hypothetical example. It is frequently claimed that men select cars primarily by performance characteristics, whereas women select cars primarily by appearance characteristics. Two hundred men and 250 women were asked the question, "What is the one characteristic of your present automobile which is most satisfying to you?" The responses to the question were placed in one of two subgroups, depending on whether performance characteristics or appearance was mentioned. The results are found in Table 17.1. Inspection of the table reveals that the results confirm the claim.

Table 17.1

2×2 contingency table showing the number of men and women indicating characteristics of automobiles they like best

	Response to question		
Sex of respondents	Appearance	Performance	Row marginal
Male	(a) 75	(b) 125	200
Female	(c) 150	(d) 100	250
Column marginal	225	225	450

We must now apply a test of significance. In formal statistical terms,

1. *Null hypothesis* (H_0): There is no difference between men and women in their preferences concerning "most liked" automobile characteristics.
2. *Alternative hypothesis* (H_1): There is a difference between men and women in their preferences concerning "most liked" characteristic.
3. *Statistical test*: Since the two groups (male and female) are independent, and the data are in terms of frequencies in discrete categories, the χ^2 test of independence is the appropriate statistical test.
4. *Significance level*: $\alpha = 0.05$.
5. *Sampling distribution*: The sampling distribution is the chi-square distribution with df $= (r - 1)(c - 1)$.

   Since marginal totals are fixed, the frequency of only one cell is free to vary. Therefore we have a one-degree-of-freedom situation. The general rule for finding df in the two-variable case is $(r - 1)\ (c - 1)$, in which $r =$ number of rows and $c =$ number of columns. Thus in the present example, df $= (2 - 1)(2 - 1) = 1$.
6. *Critical region*: Table B (in Appendix V) shows that for df $= 1$, $\alpha = 0.05$, the critical region consists of all values of $\chi^2 \geq 3.84$. χ^2 is calculated from the formula

$$\chi^2 = \sum_{r=1}^{r} \sum_{c=1}^{c} \frac{(f_o - f_e)^2}{f_e}, \tag{17.5}$$

where $\sum_{r=1}^{r} \sum_{c=1}^{c}$ directs us to sum this ratio over both rows and columns.

The *main problem* now is to decide on a basis for determining the expected cell frequencies. Let us concentrate for a moment on cell (a). The two marginal totals common to cell (a) are row 1 marginal and column 1 marginal. If the null hypothesis is correct, we would expect the same proportion of men and of women to cite appearance as a deciding factor in automobile preference. Since 200 of the total sample of 450 are men, we would expect that $\frac{200}{450} \times 225$ men would be found in cell (a). This figure comes to 100. Since we have a one-degree-of-freedom situation, the expected frequencies in all the remaining cells are determined as soon as we have calculated one expected cell frequency. Consequently, we may obtain all the remaining expected cell frequencies by subtracting from the appropriate marginal totals. Thus cell (b) is $200 - 100$ or 100; cell (d) is $225 - 100$ or 125; and cell (c) is $225 - 125$ or 100. To make certain that no error was made in our original calculation of the expected frequency for cell (a), it is wise to independently calculate the expected frequency for *any* of the remaining cells. If this figure agrees with the result we obtained by subtraction, we may feel confident that we made no error. Thus the expected frequency for cell (d), obtained through direct calculation, is

$$\frac{250}{450} \times 225 = 125.$$

Since this figure does agree with the result obtained by subtraction, we may proceed with our calculation of the χ^2 value.

Incidentally, you may have noted that there is a simple rule which may be followed in determining expected frequency of a given cell; you multiply the marginal frequencies common to that cell and divide by N.

Table 17.2 presents the obtained data, with the expected cell frequencies in the lower right-hand corner of each cell.

Table 17.2

Number of men and women indicating characteristics of automobiles they prefer most (expected frequencies within parentheses)

	Response to question		
Sex of respondents	Appearance	Performance	Row marginal
Male	75 (100)	125 (100)	200
Female	150 (125)	100 (125)	250

Now, all that remains to calculate is the χ^2 value. However, as in the normal approximation to the binomial, the empirical distributions of categorical variables are discrete, whereas the theoretical distributions of chi-square are continuous. Consequently, in the one-degree-of-freedom situation, a correction for continuity is required to obtain a closer approximation of the obtained χ^2 values to the theoretical distribution. This correction consists of subtracting 0.5 from the absolute difference $|f_o - f_e|$. Thus, in the one-degree-of-freedom situation, the formula for calculating χ^2 becomes

$$\chi^2 = \sum_{r=1}^{2} \sum_{c=1}^{2} \frac{(|f_o - f_e| - 0.5)^2}{f_e}, \tag{17.6}$$

where

$$\chi^2 = \frac{(|75 - 100| - 0.5)^2}{100} + \frac{(|125 - 100| - 0.5)^2}{100} + \frac{(|150 - 125| - 0.5)^2}{125} + \frac{(|100 - 125| - 0.5)^2}{125}$$

$$= \frac{600.25}{100} + \frac{600.25}{100} + \frac{600.25}{125} + \frac{600.25}{125} = 21.60.$$

Since the χ^2 value 21.60 is greater than 3.84 required for significance at the 0.05 level (one-tailed test), we may reject H_0. In other words, we may conclude that men and women show a differential basis for car preference and, more specifically, women state their preference on the basis of appearance characteristics more often than men.

In research, we often find that we have more than two subgroups within a nominal class. For example, we might have three categories in one scale and four in another, resulting in a 3×4 contingency table. The procedure for obtaining the expected frequencies is the same as the one for the 2×2 contingency table. Of course, the degrees of freedom will be greater than 1 (e.g., 3×4 contingency table, df = 6). Thus no correction for continuity is necessary.

17.6 LIMITATIONS IN THE USE OF χ^2

A fundamental assumption in the use of χ^2 is that each observation or frequency is independent of all other observations. Consequently, one may not make several observations on the same individual and treat each as though it were independent of all the other observations. Such an error produces what is referred to as an **inflated** *N*, that is, you are treating the data as though you had a greater number of independent observations than you actually have. This error is extremely serious and may easily lead to the rejection of the null hypothesis when it is, in fact, true.

Consider the following hypothetical example. Imagine that you are a student in a sociology course and, as a class project, you decide to poll the student body to determine whether male and female students differ in their opinions on some issue of contemporary significance. Each of 15 members of the class is asked to obtain replies from 10 respondents, 5 male and 5 female. The results are listed in Table 17.3.

Employing $\alpha = 0.05$, we find that the critical region consists of all the values of $\chi^2 \geq 3.84$. Since the obtained χ^2 of $9.67 > 3.84$, you reject the null hypothesis of no difference in the opinions of male and female students on the issue in question. You conclude, instead, that approval of the issue is dependent on the sex of the respondent.

Table 17.3

	Response to question		
Sex	Approve	Disapprove	
Male	30 (40)	45 (35)	75
Female	50 (40)	25 (35)	75
	80	70	150
	$\chi^2 = 9.67$		

Subsequent to the study, you discover that a number of students were inadvertently polled as many as two or three times by different members of the class. Consequently, the frequencies within the cells are not independent since some individuals had contributed as many as two or three responses. In a reanalysis of the data, in which only one frequency per respondent was permitted, we obtained the results shown in Table 17.4.

Table 17.4

	Response to question		
Sex	Approve	Disapprove	
Male	28 (32.5)	37 (32.5)	65
Female	32 (27.5)	23 (27.5)	55
	32	60	120
	$\chi^2 = 2.15$		

Note that now the obtained χ^2 of $2.15 < 3.84$; thus you must accept H_0. The failure to achieve independence of responses resulted in a serious error in the original conclusion. Incidentally, you should note that the requirement of independence within a cell or condition is basic to *all* statistical tests. We have mentioned this specifically in connection with the χ^2 test because violations may be very subtle and not easily recognized.

When discussing the normal approximation to the binomial, we pointed out that the extent of the approximation becomes less as P and Q diverge from $\frac{1}{2}$. We stated, as a rule of thumb, that NPQ should equal at least 9 to justify the normal probability model. A similar stricture applies to the χ^2 test. With small N's or when the expected proportion in any cell is small, the approximation of the sample statistics to the chi-square distribution may not be very close. A rule which has been generally adopted, in the one-degree-of-freedom situation, is that the expected frequency in all cells should be equal to or greater than 5. When df > 1, the expected frequency should be equal to or greater than 5 in at least 80% of the cells. When these requirements are not met, other statistical tests are available. (See Siegel, 1956.)

CHAPTER SUMMARY

In this chapter we have discussed four tests of significance employed with categorical variables, i.e., the binomial test, the normal approximation to the binomial, the χ^2 one-variable test and the χ^2 two-variable test.

1. We saw that the binomial may be employed to test null hypotheses when frequency counts are distributed between two categories or cells. When $P = Q = \frac{1}{2}$ and when $N \geq 25$, Table M provides the one-tailed probabilities directly. When $P \neq Q$, probabilities may be obtained by employing formula (17.1), which is based on the binomial expansion. When $N \geq 10$, Table N provides the critical values for the selected values of P and Q. A rule of thumb is that NPQ must equal or exceed 9 as P approaches 0 or 1 to permit the use of the normal approximation to the binomial.

2. The χ^2 two-variable case may be employed to determine whether two variables are related or independent. If the χ^2 value is significant, we may conclude that the variables are interdependent, or related. In this chapter, we restricted our discussion to the 2×2 contingency table. However, the procedures are easily extended to include more than two categories within each variable.

3. Finally, we discussed two limitations on the use of the χ^2 test. In the one degree-of-freedom situation, the expected frequency should equal or exceed 5 to permit the use of the χ^2 test. When df > 1, the expected frequency in 80% of the cells should equal or exceed 5. A second, and most important, restriction is that the frequency counts must be independent of one another. Failure to meet this requirement results in an error known as the *inflated* N and may well lead to the rejection of the null hypothesis when it is true.

Terms to Remember:

Two category (dichotomous) populations
Binomial test
Binomial expansion
Normal curve approximation to binomial
Correction for continuity
χ^2 one-variable case
"Goodness-of-fit" test
χ^2 test of independence
Inflated N

EXERCISES

17.1 In 9 tosses of a single coin,
a) what is the probability of obtaining exactly 7 heads?
b) what is the probability of obtaining as many as 8 heads?
c) what is the probability of obtaining a result as rare as 8 heads?

17.2 A relevation to many students is the surprisingly low probability of obtaining a passing grade on multiple choice examinations when the selection of alternatives is made purely on a chance basis (i.e., without any knowledge of the material covered on the exam). If 60% is considered a passing grade on a 40-item multiple choice test, determine the probability of passing when there are
a) four alternatives, b) three alternatives, c) two alternatives.

17.3 In reference to Exercise 16.4, Chapter 16, show how the binomial expansion might be employed to determine the probability of one success (i.e., rejection of H_0 at $\alpha = 0.05$) out of ten attempts.

17.4 A study was conducted in which three groups of rats (5 per group) were reinforced under three different schedules of reinforcement (100% reinforcement, 50% reinforcement, 25% reinforcement). The number of bar-pressing responses obtained during extinction are shown below.

100%	50%	25%
615	843	545

Criticize the use of chi-square as the appropriate statistical technique.

17.5 The World Series may last from four to seven games. During the period 1922–1965, the distribution of the number of games played per series was as follows:

Number of games	4	5	6	7
Frequency of occurrence	9	8	9	18

For these data, test the hypothesis that each number of games is equally likely to occur.

17.6 In a large eastern university, a study of the composition of the student council reveals that 6 of its 8 members are political science majors. In the entire student body of 1200 students, 400 are political science majors. Set this study up in formal statistical terms and draw the appropriate conclusions.

17.7 A study was conducted to determine if there is a relationship between socioeconomic class and attitudes toward a new urban-renewal program. The results are listed below.

		Disapprove	Approve
Socioeconomic class	Middle	90	60
	Lower	200	100

Set this study up in formal statistical terms and draw the appropriate conclusion.

17.8 Construct the sampling distribution of the binomial when $P = 0.20$, $Q = 0.80$, and $n = 6$.

17.9 Out of 300 castings on a given mold, 27 were found to be defective. Another mold produced 31 defective castings in 500. Determine whether there is a significant difference in the proportion of defective castings produced by the two molds.

17.10 Employ the χ^2 test, one-variable case, for the example shown in Section 17.3 of the text. Verify that $\chi^2 = z^2$ in the one-degree-of-freedom situation.

17.11 In a study concerned with preferences of packaging for potato chips, 100 women in a high income group and 200 women in a lower income group were interviewed. The results of their choices are given in the next table.

	Upper income group	Lower income group
Prefer metallic package	36	84
Prefer waxed paper package	39	51
Prefer cellophane package	16	44
Have no preference	9	21

What conclusions would you draw from these data?

17.12 Suppose that 100 random drawings from a deck of cards produced 28 hearts, 19 clubs, 31 diamonds, and 22 spades. Would you consider these results unusual?

17.13 In polling 46 interviewees, we find that 28 favor a certain routing of a highway. Is it likely that this represents a majority vote in the population in the same direction? Use the normal approximation to the binomial and the χ^2, one-variable case. Verify that $z = \sqrt{\chi^2}$.

17.14 Suppose a study was conducted which compared the types of investments made by 115 persons who were considered conservative and 125 individuals who were judged as likely to take risks. What can you conclude from the following results?

	Conservative	Risky
Government bonds	75	45
Stocks	40	80

17.15 Suppose a company manufactured an equal number of three different file cabinets. The amount held by each cabinet was equal, but the designs were different. The first 300 buyers ordered 125 of type I, 100 of type II, 75 of type III. Should the company continue to produce an equal number of each design?

17.16 In Exercise 17.15, if the first 300 buyers ordered 115 cabinets of type I, 95 of type II, and 90 of type III, should the company continue its present production policy?

17.17 Assume you are testing a die to determine if it is biased. With 90 tosses, you obtain 13 ones, 18 twos, 13 threes, 23 fours, 9 fives, and 14 sixes. What would you conclude?

17.18 In Exercise 17.17, suppose you wanted to compare the number of times a toss is odd with the number of times it is even. Using the z-score, can you conclude that the die is unbiased?

17.19 Use the χ^2 test to determine the answers for Exercise 17.18.

17.20 Suppose that a recording company is interested in the type of cover to put on an album. It sends the same record with three different covers to a store. At the end of a month, it is found that the following number of albums have been sold:

Type I cover	Type II cover	Type III cover
41	50	20

Is the company in a position to determine which type of cover it should use?

17.21 Earlier in this text we pointed out that red dye number 2 became the focus of a swirling controversy in 1976 when a report issued by the Food and Drug Administration implicated it in cancer among laboratory rats. The table below shows the findings.

	Low dosage	High dosage	
Contracted cancer	4	14	18
Did not contract cancer	40	30	70
	44	44	88

Set up and test H_0, using $\alpha = 0.01$.

18 Statistical Inference with Ordinally Scaled Variables

18.1 INTRODUCTION

In the previous chapter, we pointed out that the researcher is frequently faced with a choice as to which statistical test is appropriate for his or her data. You will recall that this was not really a problem in relation to categorical variables because nonparametric tests alone are suitable for nominally scaled data.

In this chapter we shall discuss several statistical techniques which are frequently employed as alternatives to parametric tests.

18.2 MANN-WHITNEY *U*-TEST

The Mann-Whitney U-test is one of the most powerful nonparametric statistical tests, since, as we shall see, it utilizes most of the quantitative information that is inherent in the data. It is most commonly employed as an alternative to the Student t-ratio when the measurements fail to achieve interval scaling or when the researcher wishes to avoid the assumptions of the parametric counterpart.

Imagine that we have drawn two independent samples of n_1 and n_2 observations. The null hypothesis is that both samples are drawn from populations with the same distributions. The two-tailed alternative hypothesis, against which we test the null hypothesis, is that the parent populations, from which the samples were drawn, are different. Imagine, further, that we combine the $n_1 + n_2$ observations and assign a rank of 1 to the smallest value, a rank of 2 to the next smallest value, and continue until we have assigned ranks to all the observations. Let us refer to our two groups as E and C, respectively. If we were to count the number of times each C precedes each E in the ranks, we would expect, under the null hypothesis, that it would equal the number of times each E precedes a C. In other words, if there is no difference between the two groups, the order of E's preceding C's, and vice versa, should be random. However, if the null hypothesis is not true, we would expect a bulk of the E-scores or the C-scores to precede their opposite number.

Let us take an example. Suppose you have the hypothesis that leadership is a trainable quality. You set up two groups, one to receive special training in leadership (*E*) and the other to receive no special instruction (*C*). Following the training, independent estimates of the leadership qualities of all the subjects are obtained. The results are:

E-scores	12	18	31	45	47
C-scores	2	8	15	19	38

In employing the Mann-Whitney test, we are concerned with the sampling distribution of the statistic "*U*." To find *U*, we must first rank all the scores from the lowest to the highest, retaining the identity of each score as *E* or *C* (Table 18.1).

Table 18.1

Rank	1	2	3	4	5	6	7	8	9	10
Score	2	8	12	15	18	19	31	38	45	47
Condition	*C*	*C*	*E*	*C*	*E*	*C*	*E*	*C*	*E*	*E*

You note that the number of *E*'s preceding *C*'s is less than the number of *C*'s preceding *E*'s. The next step is to count the number of times each *E* precedes a *C*. Note that the first *E* (score of 12) precedes three *C's* (scores of 15, 19, and 38, respectively). The second *E* (score of 18) precedes two *C*'s (scores of 19 and 38). The third *E* (score of 31) precedes one *C* (score of 38). Finally, the last two *E*'s precede no *C*'s. *U* is the sum of the number of times each *E* precedes a *C*. Thus in our hypothetical problem, $U = 3 + 2 + 1 + 0 + 0 = 6$. Had we concentrated on the number of times *C*'s precede *E*'s we would have obtained a sum of $5 + 5 + 4 + 3 + 2 = 19$. We shall refer to this greater sum as U'. Under the null hypotheses, U and U' should be equal. The question is whether the magnitude of the observed difference is sufficient to warrant the rejection of the null hypothesis.

The sampling distribution of U under the null hypothesis is known. Tables I_1 through I_4 show the values of U and U' which are significant at various α levels. To be significant at a given α-level, the obtained U must be equal to or *less* than the tabled value, or, the obtained U' must be equal to or *greater* than its corresponding critical value. Employing $\alpha = 0.05$, two-tailed test, we find (Table I_3) that for $n_1 = 5$ and $n_2 = 5$, either $U \leq 2$ or $U' \geq 23$ is required to reject H_0. Since our obtained U of $6 > 2$, we may not reject the null hypothesis.

Formula (18.1) may be employed as a check on the calculation of U and U':

$$U = n_1 n_2 - U'. \tag{18.1}$$

The counting technique for arriving at U can become tedious, particularly with large n's, and frequently leads to error. An alternative procedure, which provides identical results, is to assign ranks to the combined groups, as we did before, and then employ either of the following formulas to arrive at U.

$$U = n_1 n_2 + \frac{n_1(n_1 + 1)}{2} - R_1, \tag{18.2}$$

or

$$U = n_1 n_2 + \frac{n_2(n_2 + 1)}{2} - R_2, \tag{18.3}$$

where

R_1 = the sum of ranks assigned to the group with a sample size of n_1,

R_2 = the sum of ranks assigned to the group with a sample size of n_2.

Let us suppose that we conducted a study to determine the effects of a drug on the reaction time to a visual stimulus. Since reaction time (and related measures such as latency, time to traverse a runway, etc.) are commonly skewed to the right because of the restriction on the left of the distribution (i.e., no score can be less than zero) and no restrictions on the right (i.e., the score can take on *any* value greater than zero), the Mann-Whitney U-test is selected in preference to the Student t-ratio. The results of the hypothetical study and the computational procedures are shown in Table 18.2.

Table 18.2

The calculation of the Mann-Whitney U employing formula (18.2) (hypothetical data)

Experimental		Control		
Time, milliseconds	Rank	Time, milliseconds	Rank	
140	4	130	1	$U = n_1 n_2 + \frac{n_1(n_1 + 1)}{2} - 81$
147	6	135	2	
153	8	138	3	
160	10	144	5	$= 56 + \frac{8(9)}{2} - 81$
165	11	148	7	
170	13	155	9	
171	14	168	12	$= 56 + 36 - 81 = 11.$
193	15			
	$R_1 = 81$ $n_1 = 8$		$R_2 = 39$ $n_2 = 7$	

To check the above calculations, we should first obtain the value of U', employing formula (18.3):

$$U' = n_1 n_2 + \frac{n_2(n_2 + 1)}{2} - 39 = 45.$$

We employ formula (18.1) as a check for our calculations:

$$\begin{aligned} U &= n_1 n_2 - U' \\ &= 56 - 45 \\ &= 11. \end{aligned}$$

Employing $\alpha = 0.01$, two-tailed test, for $n_1 = 8$ and $n_2 = 7$, we find (Table I_1) that a $U \leq 6$ is required to reject H_0. Since the obtained U of 11 is greater than this value, we accept H_0.

Tables I_1 through I_4 have been constructed so that it is not necessary to calculate both U and U'. Indeed, it is not even necessary to identify which of these statistics has been calculated. For any given n_1 and n_2, at a specific α level, the tabled values represent the upper and the lower limits of the critical region. The obtained statistic, whether it is actually U or U', must fall *outside* these limits to be significant. Thus you need not be concerned about labeling which of the statistics you have calculated.

18.2.1 Mann-Whitney *U*-Test with Tied Ranks

A problem that often arises with data is that several scores may be exactly the same. Although the underlying dimension on which we base our measures may be continuous, our measures are, for the most part, quite crude. Even though, theoretically, there should be no ties (if we had sufficiently sensitive measuring instruments), we do, as a matter of fact, obtain ties quite often. Although ties within a group do not constitute a problem (U is unaffected), we do face some difficulty when ties occur between two or more observations which involve both groups. There is a formula available which corrects for the effects of ties. Unfortunately, the use of this formula is rather involved and is beyond the scope of this introductory text.* However, the failure to correct for ties results in a test which is more "conservative," i.e., decreases the probability of a type I error (rejecting the null hypothesis when it should not be rejected). Correcting for ties is recommended only when their proportion is high and when the uncorrected U approaches our previously set level of significance.

18.3 NONPARAMETRIC TESTS INVOLVING CORRELATED SAMPLES

In Chapter 15, when discussing the Student t-ratio for correlated samples and the algebraically equivalent Sandler A-statistic, we noted the advantages of employing correlated samples wherever feasible. The same advantages accrue to nonparametric

* See Siegel (1956), pp. 123–125, for corrections when a large number of ties occur.

tests involving matched or correlated samples. In this section we shall discuss two such tests for ordinally scaled variables, i.e., the *sign test* and the *Wilcoxon signed rank test.*

18.4 THE SIGN TEST

Let us suppose that we are repeating the leadership experiment with which we introduced the chapter, employing larger samples. On the expectation that intelligence and leadership ability are correlated variables, we set up two groups, an experimental and a control, which are matched on the basis of intelligence. On completion of the leadership training course, independent observers are asked to rate the leadership qualities of each subject on a 50-point scale. The results are listed in Table 18.3.

Table 18.3

Ratings of two groups of matched subjects on qualities of leadership (hypothetical data)

	Leadership score		
Matched pair	Experimental	Control	Sign of difference $(E - C)$
A	47	40	+
B	43	38	+
C	36	42	−
D	38	25	+
E	30	29	+
F	22	26	−
G	25	16	+
H	21	18	+
I	14	8	+
J	12	4	+
K	5	7	−
L	9	3	+
M	5	5	(0)

The rating scales seem to be extremely crude and we are unwilling to affirm that the scores have any precise quantitative properties. The only assumption we feel justified in making is that any difference which exists between two paired scores is a valid indicator of the direction and not the magnitude of the difference.

There are 13 pairs of observations in Table 18.3. Since pair *M* is tied and there is, consequently, no indication of a difference one way or another, we drop these paired observations. Of the remaining 12 pairs, we would expect, on the basis of the null hypothesis, half of the changes to be in the positive direction and half of the changes to be in the negative direction. In other words, under H_0, the probability of any difference being positive is equal to the probability that it will be negative. Since we are dealing with a two-category population (positive differences and negative differences), H_0 may be expressed in precisely the same fachion as in the binomial test when $P = Q = \frac{1}{2}$. That is, in the present problem, H_0: $P = Q = \frac{1}{2}$. Indeed, the sign test is merely a variation of the binomial test introduced in Section 17.2.

In the present example, out of 12 comparisons showing a difference ($N = 12$), 9 are positive and 3 are negative. Since $P = Q = \frac{1}{2}$, we refer to Table M, under $x = 3$, $N = 12$, and find that the one-tailed probability is 0.073. The two-tailed probability is therefore 0.146. Employing $\alpha = 0.05$ (two-tailed test) we accept H_0 since $p > 0.05$.

The assumptions underlying the use of **the sign test** are that the pairs of measurements must be independent of each other and that these measurements must represent, at least, ordinal scaling.

One of the disadvantages of the sign test is that it completely eliminates any quantitative information that may be inherent in the data, (for example, $-8 = -7 = -6$, etc.). The sign test treats all plus differences as if they were the same and all minus differences as if they were the same.

If this is the only assumption warranted by the scale of measurement employed, we have little choice but to employ the sign test. If, on the other hand, the data *do* permit us to make such quantitative statements as "a difference of $8 > 7 > 6 > \cdots$," we lose power when we employ the sign test.

18.5 WILCOXON MATCHED-PAIRS SIGNED-RANK TEST

We have seen that the sign test simply utilizes information concerning the direction of the differences between pairs. If the *magnitude* as well as the *direction* of these differences may be considered, a more powerful test may be employed. The **Wilcoxon matched-pairs signed-rank test** achieves greater power by utilizing the quantitative information inherent in the ranking of the differences.

For heuristic purposes, let us return to the data in Table 18.3 and make a different assumption about the scale of measurement employed. Suppose that the rating scale is not as crude as we had imagined; i.e., not only do the measurements achieve ordinal scaling but also the differences between measures achieve ordinality. Table 18.4 reproduces these data, with an additional entry indicating the magnitude of the differences.

Note that the difference column represents differences in scores rather than in ranks. The following column represents the ranking of these differences from smallest to largest without regard to the algebraic sign. Now, if the null hypothesis were correct, we would expect the sum of the positive and that of the negative ranks to more or less balance each other. The more the sum of the ranks are preponderantly positive or negative, the more likely we are to reject the null hypothesis.

The statistic T is the sum of the ranks with the smaller sum. In the above problem, T is equal to -13. Table J presents the critical values of T for sample sizes up

Table 18.4

Ratings of two groups of matched subjects on qualities of leadership (hypothetical data)

	Leadership score				
Matched pair	Experimental	Control	Difference	Rank of difference	Ranks with smaller sum
A	47	40	+7	9	
B	43	38	+5	5	
C	36	42	−6	−7	−7
D	38	25	+13	12	
E	30	29	+1	1	
F	22	26	−4	−4	−4
G	25	16	+9	11	
H	21	18	+3	3	
I	14	8	+6	7	
J	12	4	+8	10	
K	5	7	−2	−2	−2
L	9	3	+6	7	
M	5	5	(0)	—	
					$T = -13$

to 50 pairs. All entries are for the absolute value of T. In the present example, we find that a T of 13 or less is required for significance at the 0.05 level (two-tailed test) when $n = 12$. Note that we dropped the M-pair from our calculations since, as with the sign test, a zero difference in scores cannot be considered as either a negative or a positive change. Since our obtained T was 13, we may reject the null hypothesis.

You will recall that the sign test applied to these same data did not lead to the rejection of the null hypothesis. The reason should be apparent, i.e., we were not taking advantage of all the information inherent in our data when we employed the sign test.

18.5.1 Assumptions Underlying Wilcoxon's Matched-pairs Signed-Rank Test

An assumption involved in the use of the Wilcoxon signed-rank test is that the scale of measurement is at least ordinal in nature. In other words, the assumption is that the scores permit the ordering of the data into relationships of greater than and less than. However, the signed-rank test makes one additional assumption which may rule it out of some potential applications, namely, that the differences in scores also constitute an ordinal scale. It is not always clear whether or not this assumption is valid for a given set of data. Take, for example, a personality scale purported to measure "manifest anxiety" in a testing situation. Can we validly claim that a difference between matched pairs of, say, 5 points on one part of the scale is greater than a difference of 4 points on another part of the scale? If we cannot validly make this assumption, we must employ another form of statistical analysis, even if it requires that we move to a less sensitive test of significance. Once again, our basic conservatism as scientists makes us more willing to risk a type II rather than a type I error.

CHAPTER SUMMARY

Let us briefly review what we have learned in this chapter.

We have pointed out that the behavioral scientist does not first collect data and then "shop around" for a statistical test to determine the significance of differences between experimental conditions. *The researcher must specify in advance of the experiment* his null hypothesis, alternative hypothesis, test of significance, and the probability value which he will accept as the basis for rejecting the null hypothesis.

We demonstrated the use of the Mann-Whitney U-test as an alternative to the Student t-ratio when the measurements fail to achieve interval scaling or when the researcher wishes to avoid the assumptions of the parametric counterpart. It is one of the most powerful of the nonparametric tests, since it utilizes most of the quantitative information inherent in the data.

We have seen that by taking into account correlations between subjects on a variable correlated with the criterion measure, we can increase the sensitivity of our statistical test.

The sign test accomplishes this objective by employing before-after measures on the same individuals.

We have also seen that the sign test, although taking advantage of the *direction* of differences involved in ordinal measurement, fails to make use of information concerning *magnitudes* of difference.

The Wilcoxon matched-pairs signed-rank test takes advantage of both *direction* and *magnitude* implicit in ordinal measurement with correlated samples. When the assumptions underlying the test are met, the Wilcoxon paired replicates technique is an extremely sensitive basis for obtaining probability values.

Terms to Remember:

Mann-Whitney U-test

Sign test

Wilcoxon matched-pairs signed-rank test

EXERCISES

18.1 For the data presented below, determine whether there is a significant difference in the number of stolen bases obtained by two leagues, employing

a) the sign test,

b) the Wilcoxon matched-pairs test,
c) the Mann-Whitney U-test.
Which is the best statistical test for these data? Why?

Number of stolen bases		
Team standing	League 1	League 2
1	91	81
2	46	51
3	108	63
4	99	51
5	110	46
6	105	45
7	191	66
8	57	64
9	34	90
10	81	28

18.2 In a study to determine the effect of a drug on aggressiveness, group A received a drug and group B received a placebo. A test of aggressiveness was applied following the drug administration. The scores obtained were as follows (the higher the score, the greater the aggressiveness):

Group A	10	8	12	16	5	9	7	11	6
Group B	12	15	20	18	13	14	9	16	

Set this study up in formal statistical terms and state the conclusion which is warranted by the statistical evidence.

18.3 The personnel director at a large insurance office claims that insurance agents who are trained in personal-social relations make more favorable impressions on prospective clients. To test this hypothesis, 22 individuals are randomly selected from those most recently hired and half are assigned to the personal-social relations course. The remaining 11 individuals constitute the control group. Following the training period, all 22 individuals are observed in a simulated interview with a client, and they are rated on a ten-point scale (0-9) for their ease in establishing relationships. The higher the score, the better the rating. Set up and test H_0 employing the appropriate test statistic. Use $\alpha = 0.01$.

Experimentals:	8	7	9	4	7	9	3	7	8	9	3
Controls:	5	6	2	6	0	2	6	5	1	0	5

18.4 Assume that the subjects in Example 18.3 were matched on a variable known to be correlated with the criterion variable. Employ the appropriate test statistic to test H_0: $\alpha = 0.01$.

18.5 Fifteen married couples were administered an opinion scale to assess their attitudes about a particular political issue. The results were as follows (the higher the score, the more favorable the attitude):

Husband	Wife	Husband	Wife
37	33	32	46
46	44	35	32
59	48	39	29
17	30	37	45
41	56	36	29
36	30	45	48
29	35	40	35
38	38		

What do you conclude?

18.6 Suppose that during last track season, there was no difference in the mean running speeds of the runners from two schools. Assume that the same people are on the teams this year. School A trains as usual for this season. However, the coach at school B introduces bicycle riding in the training classes. During a meet, the following times (in minutes) were recorded for the runners of the two schools:

A	10.2	11.1	10.5	10.0	9.7	12.0	10.7	10.9	11.5	10.4
B	9.9	10.3	11.0	10.1	9.8	9.5	10.8	10.6	9.6	9.4

Test the hypothesis that bicycle riding does not affect running speed.

18.7 Suppose that in Exercise 18.6 the people on each team had been previously matched on running speed for the 50-yard dash. The matches are as listed above. Using the sign test and the Wilcoxon matched-pairs signed-rank test, set up and test the null hypothesis.

18.8 An investigator wants to measure the effectiveness of an advertisement which promotes his brand of toothpaste. He matched subjects (all of whom had never bought his brand of toothpaste) according to the number of tubes of toothpaste they usually buy in six months. He then divided the sample into two groups and showed one group the advertisement. After six months, he found that the number of tubes of his brand of toothpaste the people bought during that time was:

Advertisement group	4	4	3	1	2	0	1	0
No advertisement group	1	2	0	2	0	1	0	1

Was the advertisement effective?

18.9 Suppose a supervisor is interested in increasing the efficiency of her employees. She divides 20 employees into two groups and gives a special training program to one group. Because the job is diversified, there is no scale available to measure efficiency. Therefore, she elicits the help of an efficiency expert who observes all 20 employees and ranks them on job efficiency, with the following results (a rank of one is most efficient).

Training group	1	2	4	5	6	7	8	10	11	12
Control	3	9	13	14	15	16	17	18	19	10

What can the supervisor conclude about the effectiveness of the training program?

Review of Section II Inferential Statistics

PART A. PARAMETRIC TESTS OF SIGNIFICANCE

In Section II, Part A (Chapters 11–16) you learned the elements of probability theory and their application to the problem of drawing inferences from samples presumed to be selected from normally distributed populations. The problems presented below will help you to integrate the material you have learned in this section.

1. Assume that the following is a population of scores: 2, 3, 3, 4, 4, 4, 5, 5, 6.
   a) Determine the mean and standard deviation of the population.
   b) Construct a frequency distribution of means for $N = 2$, employing sampling *with replacement.*
   c) Construct a probability distribution of means for $N = 2$.
   d) Determine $\sigma_{\bar{X}}$ by use of the following formula:

$$\sigma_{\bar{X}} = \sigma/\sqrt{N}.$$

   e) Determine $\sigma_{\bar{X}}$ by direct calculation from the sampling distribution of means.
   f) Determine the probability of randomly selecting samples ($N = 2$) with
      i) a sample mean of 2.0,
      ii) a sample mean of at least 5.0,
      iii) a sample mean as rare as 3.0.

2. Assume a second population of scores: 4, 5, 5, 6, 6, 6, 7, 7, 8.

   Individual A selects samples (with replacement) from population 1 and individual B selects samples from population 2 (with replacement). They test the null hypothesis that the samples were drawn from a common population, employing $\alpha = 0.05$, two-tailed test. In which of the following examples will they make a type II error? Use only the statistics calculated from the samples. Do not use the population values.

   a)

A	2	3	4	5	6
B	4	5	6	6	8

b)

A	2	3	3	4	4	4	4	4	4	6
B	4	4	4	5	5	5	5	5	6	7

3. Using the data above, Problem 2(a) and (b), assume matching of the scores and test for the significance of differences, employing $\alpha = 0.05$, two-tailed test.

4. Each of three universities claims that it has the brightest students. Each sends ten of its best students to compete in a national contest. The results are as follows:

1	2	3	1	2	3
99	94	74	95	93	99
93	98	99	97	96	71
84	97	98	91	96	70
89	92	99	88	91	79
72	92	97	89	90	83

Set up and test the appropriate null hypothesis.

PART B. NONPARAMETRIC TESTS OF SIGNIFICANCE

In the two preceding chapters you have seen several ways of handling data for which parametric tests of significance were not appropriate. Below are some data, based on a hypothetical experiment. Formulate the null hypothesis, the alternative hypothesis, and conduct the statistical analysis appropriate to the assumptions enumerated in each problem.

Experimental	37	35	33	28	26	20	16	14	12	10	7	5
Control	32	24	29	31	15	18	23	5	9	2	4	1

1. Subjects are randomly assigned to both groups. The scale of measurement is ordinal, and the population is *not* normally distributed.

2. Subjects are matched on a variable known to be correlated with the criterion variable. However, differences in scores *cannot* be assumed to represent *magnitudes* of differences but only direction.

3. Subjects are matched. Scores are based on an ordered scale. Differences in scores may be assumed to be ordinal. However, the population of scores is *not* normally distributed.

4. Determine how effective the matching techniques were.

References

Auble, D. (1953), Extended tables for the Mann-Whitney statistic. *Bulletin of the Institute of Educational Research at Indiana University*, **1**, No. 2.

Bingham, W. V. (1937), *Aptitudes and Aptitude Testing.* New York: Harper and Bros.

Dunlap, J. W., and A. K. Kurtz (1932), *Handbook of Statistical Nomographs, Tables, and Formulas.* New York: World Book.

Edwards, A. L. (1950), *Experimental Design in Psychological Research.* New York: Rinehart.

Edwards, A. L. (1969), *Statistical Analysis*, 3rd. ed. New York: Holt, Rinehart and Winston.

Eels, W. C. (1926), The relative merits of circles and bars for representing component parts. *J. Amer. Stat. Assoc.*, **21**, 119–132.

Fisher, R. A. (1935), *The Design of Experiments.* Edinburgh: Oliver and Boyd.

Fisher, R. A. (1950), *Statistical Methods for Research Workers.* Edinburgh: Oliver and Boyd.

Fisher, R. A., and F. Yates (1948), *Statistical Tables for Biological, Agricultural, and Medical Research.* Edinburgh: Oliver and Boyd.

Haber, Audrey, and H. I. Kalish (1963), Prediction of discrimination from generalization after variations in schedule of reinforcement. *Science*, **142**, 3590, 412–413.

Haber, A., R. P. Runyon, and P. Badia (1970), *Readings in Statistics.* Reading, Mass.: Addison-Wesley.

Huff, D. (1954), *How to Lie with Statistics.* New York: W. W. Norton and Co.

Johnston, J. J. (1975), Sticking with first responses on multiple-choice exams: for better or for worse? *Teaching of Psychology*, **2**, 4, 178–179.

Kirk, R. E. (1968), *Experimental Design: Procedures for the Behavioral Sciences.* California: Brooks/Cole.

Mann, H. B., and D. R. Whitney (1947), On a test of whether one of two random variables is stochastically larger than the other. *Ann. Math. Statist.*, **18**, 52–54.

Mathews, C. O. (1926), The grade placement of curriculum materials in the social studies. *Contributions to Education*, 241. New York: Teachers College, Columbia University.

McNemar, Q. (1962), *Psychological Statistics.* New York: John Wiley and Sons.

Merrington, M. and C. M. Thompson. Tables of Percentage Points of the Inverted Beta Distribution. *Biometrika*, **33**, 73.

Olds, E. G. (1949), The 5 percent significance levels of sums of squares of rank differences and a correction. *Ann. Math. Statist.*, **20**, 117–118.

Pearson, E. S. and H. O. Hartley (1958), *Biometrika Tables for Statisticians*, Vol. 1, 2nd ed. New York: Cambridge.

PETERSON, L. V., and W. SCHRAMM (1954). How accurately are different kinds of graphs read? *Audio-Visual Communication Review*, **2**, 178–189.

RAND CORPORATION (1955), *A Million Random Digits*. Glencoe, Ill.: The Free Press of Glencoe.

REED, E. W. and S. C. REED (1965), *Mental Retardation: A Family Study*. Philadelphia: Saunders.

ROCKS, L., and R. P. RUNYON (1972), *The Energy Crisis*. New York: Crown Publishers.

RUNYON, R. P. (1968), Note on use of the A statistic as a substitute for t in the one-sample case. *Psychological Reports*, **22**, 361–362.

RUNYON, R. P., and M. KOSACOFF (1965), *Olfactory Stimuli as Reinforcers of Bar Pressing Behavior*. Paper presented at Eastern Psych. Assn. meeting.

RUNYON, R. P., and W. J. TURNER (1964), *A Study of the Effects of Drugs on the Social Behavior of White Rats*. New York: Long Island University.

SANDLER, J. (1955), A test of the significance of the difference between the means of correlated measures, based on a simplification of Student's t. *Brit. J. Psychol.*, **46**, 225–226.

SEARS, R. R. and G. W. WISE (1950), Relation of cup feeding in infancy to thumbsucking and the oral drive. *American Journal of Orthopsychiatry*, **20**, 123–138.

SIEGEL, S. (1956), *Non-Parametric Statistics*. New York: McGraw-Hill.

SNEDECOR, G. W. (1956), *Statistical Methods*. Iowa: Iowa State University Press.

TUKEY, J. W. (1953), The Problem of Multiple Comparisons. Ditto, Princeton University, 396 pp.

WALKER, H., and J. LEV (1953), *Statistical Inference*. New York: Henry Holt & Co.

WHITING, J. W. M. and I. L. CHILD. (1953), *Child Training and Personality*. New Haven: Yale University Press.

WHITING, J. W. M. (1968), Methods and problems in cross-cultural research (in) *The Handbook of Social Psychology*, Vol. II, (2nd. ed.), edited by Lindzey and Aronson, Reading, Mass.: Addison-Wesley.

WILCOXON, F., S. KATTE, and R. A. WILCOX (1963), *Critical Values and Probability Levels for the Wilcoxon Rank Sum Test and the Wilcoxon Signed Rank Test*. New York: American Cyanamid.

WILCOXON, F., and R. A. WILCOX (1964), *Some Rapid Approximate Statistical Procedures*. New York: Lederle Laboratories.

Appendixes

APPENDIX I

Review of Basic Mathematics

ARITHMETIC OPERATIONS

You already know that addition is indicated by the sign "+," subtraction by the sign "−," multiplication in one of three ways, 2×4, $2(4)$, or $2 \cdot 4$, and division by a slash, "/," a bar, "—," or the symbol "÷." However, it is not unusual to forget the rules concerning addition, subtraction, multiplication, and division, particularly when these operations occur in a single problem.

Addition and Subtraction

When a series of numbers is added together, the order of adding the numbers has no influence on the sum. Thus we may add $2 + 5 + 3$ in any of the following ways:

$$\begin{array}{lll} 2 + 5 + 3, & 5 + 2 + 3, & 2 + 3 + 5, \\ 5 + 3 + 2, & 3 + 2 + 5, & 3 + 5 + 2. \end{array}$$

When a series of numbers containing both positive and negative signs are added, the order of adding the numbers has no influence on the sum. However, it is often desirable to group together the numbers preceded by positive signs, group together the numbers preceded by negative signs, add each group together separately, and subtract the latter sum from the former. Thus

$$-2 + 3 + 5 - 4 + 2 + 1 - 8$$

may best be added by grouping in the following ways:

$$\begin{array}{rr} +3 & -2 \\ +5 & -4 \\ +2 & -8 \\ +1 & \\ \hline +11 & -14 = -3. \end{array}$$

Incidentally, to subtract a larger numerical value from a smaller numerical value, as in the above example $(11 - 14)$, we ignore the signs, subtract the smaller number from the larger, and affix a negative sign to the sum. Thus $-14 + 11 = -3$.

Multiplication

The order in which numbers are multiplied has no effect on the product. In other words,

$$\begin{aligned} 2 \times 3 \times 4 &= 2 \times 4 \times 3 = 3 \times 2 \times 4 \\ &= 3 \times 4 \times 2 = 4 \times 2 \times 3 = 4 \times 3 \times 2 = 24. \end{aligned}$$

However, when addition, subtraction, and multiplication occur in the same expression, the order does affect the result; thus we must develop certain procedures governing which operations are to be performed first.

In the following expression,

$$2 \times 4 + 7 \times 3 - 5$$

multiplication is performed first. Thus the above expression is equal to

$$\begin{aligned} &1)\quad 2 \times 4 = 8, &\qquad 2)\quad 8 + 21 - 5 = 24. \\ &\qquad 7 \times 3 = 21, & \\ &\qquad\quad -5 = -5, & \end{aligned}$$

We may *not* add first and then multiply. Thus $2 \times 4 + 7$ is *not* equal to $2(4 + 7)$ or 22.

If a problem involves finding the product of one term multiplied by a second expression that includes two or more terms which are either added or subtracted, we may multiply first and then add, or add first and then multiply. Thus the solution to the following problem becomes

$$\begin{aligned} 8(6 - 4) &= 8 \times 6 - 8 \times 4 \\ &= 48 - 32 \\ &= 16, \end{aligned}$$

or

$$\begin{aligned} 8(6 - 4) &= 8(2) \\ &= 16. \end{aligned}$$

In most cases, however, it is more convenient to reduce the expression within the parentheses first. Thus generally speaking, the second solution appearing above will be more frequently employed.

Finally, if numbers having like signs are multiplied, the product is always positive; e.g., $(+2) \times (+4) = +8$ and $(-2) \times (-4) = +8$. If numbers bearing unlike signs are multiplied, the product is always negative; e.g.,

$$(+2) \times (-4) = -8$$

and

$$(-2) \times (+4) = -8.$$

The same rule applies also to division: when we obtain the quotient of two numbers of like signs, it is always positive; when the numbers differ in sign, the quotient is always negative.

Multiplication as successive addition. Many students tend to forget that multiplication is a special form of successive addition. Thus

$$15 + 15 + 15 + 15 + 15 = 5(15)$$

and

$$(15 + 15 + 15 + 15 + 15) + (16 + 16 + 16 + 16) = 5(15) + 4(16).$$

This formulation is useful in understanding the advantages of "grouping" scores into what is called a frequency distribution. In obtaining the sum of an array of scores, some of which occur a number of times, it is desirable to multiply each score by the frequency with which it occurs, and then add the products. Thus, if we were to obtain the following distribution of scores,

12, 13, 13, 13, 14, 14, 14, 14, 15, 15, 15, 15,
15, 15, 15, 16, 16, 16, 17, 17, 17, 17, 18,

and wanted the sum of these scores, it would be advantageous to form the following frequency distribution:

X	f	fX
12	1	12
13	3	39
14	4	56
15	7	105
16	3	48
17	4	68
18	1	18
	$N = 23$	$\sum fX = 346$

ALGEBRAIC OPERATIONS

Transposing

To transpose a term from one side of an equation to another, you merely have to change the sign of the transposed term. All the following are equivalent statements.

$$\begin{aligned} a + b &= c, \\ a &= c - b, \\ b &= c - a, \\ 0 &= c - a - b, \\ 0 &= c - (a + b). \end{aligned}$$

Solving Equations Involving Fractions

Much of the difficulty encountered in solving equations which involve fractions can be avoided by remembering one important mathematical principle:

Equals multiplied by equals are equal.

Let us look at a few sample problems.

1. Solve the following equation for x;

$$b = a/x.$$

In solving for x, we want to express the value of x in terms of a and b. In other words, we want our final equation to read:

$$x = \underline{\qquad}.$$

Note that we may multiply both sides of the equation by x/b and obtain the following:

$$\cancel{b} \cdot \frac{x}{\cancel{b}} = \frac{a}{\cancel{x}} \cdot \frac{\cancel{x}}{b}.$$

This reduces to

$$x = \frac{a}{b}.$$

2. Solve the above equation for a.

Similarly, if we wanted to solve the equation in terms of a, we could multiply both sides of the equation by x. Thus

$$b \cdot x = \frac{a}{\cancel{x}} \cdot \cancel{x}$$

becomes $bx = a$, or $a = bx$.

In each of the above solutions, you will note that the net effect of multiplying by a constant has been to rearrange the terms in the numerator and the denominator of the equations. In fact, we may state two general rules which will permit us to solve the above problems without having to employ multiplication by equals (although multiplication by equals is implicit in the arithmetic operations):

a) A term which is in the denominator on one side of the equation may be moved to the other side of the equation by multiplying the term to be moved by the numerator on the side to which it is to be moved. Thus

$$\frac{x}{a} = b \qquad \text{becomes} \qquad x = ab.$$

b) A term in the numerator on one side of an equation may be moved to the other side of the equation by dividing both sides by the term to be moved. Thus

$$ab = x \qquad \text{may become} \qquad a = \frac{x}{b} \quad \text{or} \quad b = \frac{x}{a}.$$

Therefore we see that all of the following are equivalent statements:

$$b = \frac{a}{x}, \qquad a = bx, \qquad x = \frac{a}{b}.$$

Similarly,

$$\frac{\sum X}{N} = \bar{X}, \qquad \sum X = N\bar{X}, \qquad \frac{\sum X}{\bar{X}} = N.$$

Dividing by a sum or a difference. While it is true that

$$\frac{x+y}{z} = \frac{x}{z} + \frac{y}{z} \qquad \text{and} \qquad \frac{x-y}{z} = \frac{x}{z} - \frac{y}{z},$$

the following expressions cannot be simplified as easily:

$$\frac{x}{y+z} \qquad \text{or} \qquad \frac{x}{y-z}.$$

That is,

$$\frac{x}{y+z} \neq \frac{x}{y} + \frac{x}{z},$$

in which $\neq$ means "not equal to."

REDUCING FRACTIONS TO SIMPLEST EXPRESSIONS

A corollary to the rule that equals multiplied by equals are equal is:

Unequals multiplied by equals remain proportional.

Thus if we were to multiply $\frac{1}{4}$ by $\frac{8}{8}$, the product, $\frac{8}{32}$, is in the same proportion as

$\frac{1}{4}$. This corollary is useful in reducing the complex fractions to their simplest expression. Let us look at an example.

Example: Reduce
$$\frac{\frac{a}{b}}{\frac{c}{d}} \quad \text{or} \quad \frac{a}{b} \div \frac{c}{d}$$
to its simplest expression.

Note that if we multiply both the numerator and the denominator by
$$\frac{\frac{bd}{1}}{\frac{bd}{1}}$$
we obtain
$$\frac{\frac{a}{\not b} \cdot \frac{\not b d}{1}}{\frac{c}{\not d} \cdot \frac{b \not d}{1}},$$
which becomes ad/bc.

However, we could obtain the same result if we invert the divisor and multiply. Thus
$$\frac{\frac{a}{b}}{\frac{c}{d}} = \frac{a}{b} \cdot \frac{d}{c} = \frac{ad}{bc}.$$

We may now formulate a general rule for dividing one fraction into another fraction. In dividing fractions, we invert the divisor and multiply. Thus
$$\frac{\frac{x}{y}}{\frac{a^2}{b}} \quad \text{becomes} \quad \frac{x}{y} \cdot \frac{b}{a^2} \quad \text{which equals} \quad \frac{bx}{a^2 y}.$$

To illustrate: If $a = 5$, $b = 2$, $x = 3$, and $y = 4$, the above expressions become
$$\frac{\frac{3}{4}}{\frac{5^2}{2}} = \frac{3}{4} \cdot \frac{2}{5^2} = \frac{2 \cdot 3}{4 \times 5^2} = \frac{6}{100}.$$

A general practice you should follow when substituting numerical values into fractional expressions is to reduce the expression to its simplest form *prior* to substitution.

MULTIPLICATION AND DIVISION OF TERMS HAVING EXPONENTS

An exponent indicates how many times a number is to be multiplied by itself. For example, X^5 means that X is to be multiplied by itself 5 times, or

$$X^5 = X \cdot X \cdot X \cdot X \cdot X.$$

If $X = 3$,

$$X^5 = 3 \cdot 3 \cdot 3 \cdot 3 \cdot 3 = 243 \quad \text{and} \quad \left(\frac{1}{X}\right)^5 = \frac{1^5}{X^5} = \frac{1 \cdot 1 \cdot 1 \cdot 1 \cdot 1}{3 \cdot 3 \cdot 3 \cdot 3 \cdot 3} = \frac{1}{243}.$$

To multiply X raised to the ath power (X^a) times X raised to the bth power, you simply add the exponents and raise X to the $(a + b)$th power. The reason for the addition of exponents may be seen from the following illustration.

If $a = 3$ and $b = 5$, then

$$X^a \cdot X^b = X^3X^5 = (X \cdot X \cdot X)(X \cdot X \cdot X \cdot X \cdot X),$$

which equals X^8.

Now, if $X = 5$, $a = 3$, and $b = 5$, then

$$X^a \cdot X^b = X^{a+b} = X^{3+5} = X^8 = 5^8 = 390{,}625.$$

If $X = \frac{1}{6}$, $a = 2$, and $b = 3$,

$$X^a \cdot X^b = X^{a+b} = \left(\frac{1}{6}\right)^{2+3} = \left(\frac{1}{6}\right)^5 = \frac{1^5}{6^5} = \frac{1}{7776}.$$

To divide X raised to the ath power by X raised to the bth power, you simply subtract the exponent in the denominator from the exponent in the numerator.* The reason for the subtraction is made clear in the following illustration. Substituting $\frac{5}{6}$ for X, we have

$$X^2 = \left(\frac{5}{6}\right)^2 = \frac{5^2}{6^2} = \frac{25}{36}.$$

* This leads to an interesting exception to the rule that an exponent indicates the number of times a number is multiplied by itself; that is,

$$\frac{X^n}{X^n} = X^{n-n} = X^0;$$

however,

$$\frac{X^n}{X^n} = 1; \quad \text{therefore} \quad X^0 = 1.$$

Any number raised to the zero power is equal to 1.

EXTRACTING SQUARE ROOTS

The square root of a number is the value which, when multiplied by itself, equals that number. Appendix IV contains a table of square roots.

The usual difficulty encountered in calculating square roots is the decision as to how many digits precede the decimal, for example $\sqrt{25{,}000{,}000} = 5000$, not 500 or 50,000; i.e., there are four digits before the decimal. In order to calculate the number of digits preceding the decimal, simply count the number of pairs to the left of the the decimal:

$$\text{number of pairs} = \text{number of digits}.$$

However, if there is an odd number of digits, then the number of digits preceding the decimal equals the number of pairs $+1$. The following examples illustrate this point:

a) $\dfrac{50.0}{\sqrt{2500.00}}$, $\dfrac{5.0}{\sqrt{25.00}}$;

b) $\dfrac{15.8}{\sqrt{250.00}}$, $\dfrac{1.58}{\sqrt{2.5000}}$.

APPENDIX II

Glossary of Symbols

Listed below are definitions of the symbols which appear in the text followed by the page number showing the first reference to the symbol.

English letters and Greek letters are listed separately in their approximate alphabetical order. Mathematical operators are also listed separately.

MATHEMATICAL OPERATORS

Symbol	Definition	
$\neq$	Not equal to	23
$a < b$	a is less than b	23
$a > b$	a is greater than b	23
$\leq$	Less than or equal to	208
$\geq$	Greater than or equal to	208
$\sqrt{\ }$	Square root	19
X^a	X raised to the ath power	19
$N!$	Factorial: multiply N by all integers less than it but greater than zero: $(N)(N-1)(N-2)\cdots(2)(1)$	317
$\lvert X \rvert$	Absolute value of X	120
$\sum$	Sum all quantities or scores that follow	19
$\sum_{i=1}^{N} X_i$	Sum all quantities X_1 through X_N: $X_1 + X_2 + \cdots + X_N$	20

GREEK LETTERS

Symbol	Definition	
α	Probability of a type I error, probability of rejecting H_0 when it is true	243
β	Probability of a type II error, probability of accepting H_0 when it is false	243
χ^2	Chi square	320
μ	Population mean	102
μ_0	Value of the population mean under H_0	253
$\mu_{\bar{X}}$	Mean of the distribution of sample means	252
$\mu_{\bar{X}_1-\bar{X}_2}$	Mean of the distribution of the difference between pairs of sample means	272
μ_D	Mean of the difference between paired scores	288
ρ	Population correlation coefficient	263
σ^2	Population variance	121
σ	Population standard deviation	121
$\sigma_{\bar{X}}^2 = \dfrac{\sigma^2}{N}$	Variance of the sampling distribution of the mean	252
$\sigma_{\bar{X}} = \dfrac{\sigma}{\sqrt{N}}$	True standard error of the mean given random samples of a fixed N	252
$\sigma_{\bar{X}_1-\bar{X}_2}$	True standard error of the difference between means	272

ENGLISH LETTERS

Symbol	Definition	
$A = \dfrac{\sum D^2}{(\sum D)^2}$	Statistic employed to test hypotheses for correlated samples	290
a	Constant term in a regression equation	179
b_y	Slope of a line relating values of Y to values of X	178
c	Number of columns in a contingency table	322
cum f	Cumulative frequency	53
cum %	Cumulative percent	53

Symbol	Meaning	Page
D	(1) Rank on X-variable − rank on Y-variable (r_{rho} formula)	162
	(2) Score on X-variable − score on Y- variable ($X - Y$)	288
$\bar{D}$	Mean of the differences between the paired scores	288
d	Deviation of a difference score (D) from $\bar{D}$	288
df	Degrees of freedom: number of values free to vary after certain restrictions have been placed on the data	257
F	A ratio of two variances	276
f	Frequency	47
f_e	Expected number in a given category	320
f_o	Observed number in a given category	320
fX	A score multiplied by its corresponding frequency	102
H_0	The null hypothesis; hypothesis actually tested	241
H_1	The alternative hypothesis; hypothesis entertained if H_0 is rejected	241
i	Width of the class interval	49
k	Number of groups or categories	301
M.D.	Mean deviation	120
N	Total number of scores or quantities	19
n	(1) Number of pairs	156
	(2) Number in either sample	272
p	(1) Proportion	31
	(2) Probability	208
$p(A)$	Probability of event A	207
$p(B\|A)$	Probability of B given that A has occurred	213
P	(1) Probability of the occurrence of an event	208
	(2) Proportion of cases in one class in a two-category population	316
Q	(1) Probability of the nonoccurrence of an event	208
	(2) Proportion of cases in the other class of a two-category population	316
Q_1	First quartile, 25th percentile	119
Q_3	Third quartile, 75th percentile	119
r	(1) Pearson product-moment correlation coefficient	154
	(2) Number of rows in a contingency table	322
r^2	Coefficient of determination	186

Symbol	Meaning	Page
r_{rho}	Spearman rank-order correlation coefficient	162
R_1	Sum of ranks assigned to the group with a sample size of n_1 (Mann-Whitney U-formula)	332
R_2	Sum of ranks assigned to the group with a sample size of n_2 (Mann-Whitney U-formula)	332
$s^2 = \dfrac{\sum x^2}{N - 1}$	Variance of a sample	122
$s = \sqrt{\dfrac{\sum x^2}{N - 1}}$	Standard deviation of a sample	123
$s_{\bar{X}}^2$	Estimated variance of the sampling distribution of the mean	255
$s_{\bar{X}} = \dfrac{s}{\sqrt{N}}$	Estimated standard error of the mean	255
$s_{\bar{X}_1 - \bar{X}_2}$	Estimated standard error of the difference between means	272
s_D	Standard deviation of the difference scores	288
$s_{\bar{D}}$	Estimated standard error of the difference between means, direct-difference method	288
s_B^2	Between-group variance estimate	301
s_W^2	Within-group variance estimate	301
$s_{\text{est } y}$	Standard error of estimate when predictions are made from X to Y	183
$s_{\text{est } x}$	Standard error of estimate when predictions are made from Y to X	183
T	Sum of the ranks with the least frequent sign	335
t	Statistic employed to test hypotheses when σ is unknown	256
U, U'	Statistics in the Mann-Whitney test	331
V	Coefficient of variation	144
X, Y	Variables; quantities or scores of variables	19
X_i, Y_i	Specific quantities indicated by the subscript i	19
$\bar{X}, \bar{Y}$	Arithmetic means	102
$\bar{X}_i$	Mean of the ith group	300
$\bar{X}_{\text{tot}} = \dfrac{\sum X_{\text{tot}}}{N}$	Overall mean	300
$\bar{X}_w$	Weighted mean	107
$x = X - \bar{X}$	Deviation of a score from its mean	122

Symbol	Meaning	Page
x	Number of objects in one category or the number of successes	316
$\sum X^2$	Sum of the squares of the raw scores	122
$(\sum X)^2$	Sum of the raw scores, the quantity squared	123
$\sum x^2 = \sum (X - \bar{X})^2$	Sum squares, sum of the squared deviations from the mean	122
$\sum x_{\text{tot}}^2$	Total sum squares, sum of the squared deviations of each score (X) from the overall mean ($\bar{X}_{\text{tot}}$)	300
$\sum x_w^2$	Within-group sum squares, sum of the squared deviations of each score (X) from the mean of its own group ($\bar{X}_i$)	300
$\sum x_B^2$	Between-group sum squares, sum of the squared deviations of each group mean ($\bar{X}_i$) from the overall mean ($\bar{X}_{\text{tot}}$), multiplied by the n in each group	300
X', Y'	Scores predicted by regression equations	178
z	(1) Deviation of a specific score from the mean expressed in standard deviation units	136
	(2) Statistic employed to test hypotheses when σ is known	253
$z_{0.01} = \pm 2.58$	Critical value of z, minimum z required to reject H_0 at the 0.01 level of significance, two-tailed test	254
$z_{0.05} = \pm 1.96$	Minimum value of z required to reject H_0 at the 0.05 level of significance, two-tailed test	254
z_r	Transformed value of sample r	263
Z_r	Transformed value of population correlation coefficient	263
$z_{y'}$	Y' expressed in terms of a z-score	181

APPENDIX III

Glossary of Terms

Abscissa (*X-axis*). Horizontal axis of a graph.
Absolute value. Value of a number without regard to sign.
Alpha (α) *error.* See *Type I error.*
Alpha (α) *level.* See *Significance level.*
Alternative hypothesis (H_1). A statement specifying that the population parameter is some value other than the one specified under the null hypothesis.
Analysis of variance. A method, described initially by R. A. Fisher, for partitioning the sum of squares for experimental data into known components of variation.
Array. Arrangement of data according to their magnitude from the smallest to the largest value.
Attributive data. Tabulations of the number of occurrences in each class of a given variable.
Average deviation. See *Mean deviation.*
Before–after design. A correlated-samples design in which each individual is measured on the criterion task both before and after the introduction of the experimental conditions.
Beta (β) *error.* See *Type II error.*
Biased estimate. A sample statistic is a biased estimate of a corresponding population parameter if the mean of a large number of sample values, obtained by repeated sampling, does not approach the parameter as a limit.
Categorical data. See *Attributive data.*
Coefficient of determination. Ratio of explained variation to the total variation.
Coefficient of variation (V). A measure which expresses variation relative to the mean.
Conditional probability. The probability of an event given that another event has occurred. Represented symbolically as $p(A|B)$, the probability of A given that B has occurred.
Continuous variables. Variables that can assume an unlimited number of intermediate values.
Correlated samples design. See *Before–after design* and *Matched group design.*

Correlation coefficient. A measure which expresses the extent to which two variables are related.
Critical region. That portion of the area under a curve which includes those values of a statistic which lead to rejection of the null hypothesis.
Data. Numbers or measurements which are collected as a result of observations.
Degrees of freedom. The number of values that are free to vary after we have placed certain restrictions upon our data.
Descriptive statistics. Procedures employed to organize and present data in a convenient, usable, and communicable form.
Directional hypothesis. See *One-tailed p value.*
Discontinuous variables (discrete variables). Variables that can assume only a certain finite number of values.
Distribution ratio. Ratio of a part to a total which includes that part.
Enumerative data. See *Attributive data.*
Exhaustive. Two or more events are said to be exhaustive if they exhaust all possible outcomes. Symbolically, $p(A \text{ or } B \text{ or } \ldots) = 1.00$.
F-ratio. A ratio between sample variances. Named in honor of R. A. Fisher.
Frequency data. See *Attributive data.*
Heterogeneity of variance. The condition that exists when two or more sample variances have been drawn from populations with different variances.
Homogeneity of variance. The condition that exists when two or more sample variances have been drawn from populations with equal variances.
Homoscedasticity. Homogeneous variability within the columns and the rows.
Independence. The condition that exists when the occurrence of a given event will not affect the probability of the occurrence of another event. Symbolically, $p(A|B) = p(A)$ and $p(B|A) = p(B)$.
Inductive statistics. Procedures employed to arrive at broader generalizations or inferences from sample data to populations.
Inferential statistics. See *Inductive statistics.*
Interclass ratio. Ratio of a part in a total to another part in the same total.
Interval estimation. The determination of an interval within which the population parameter is presumed to fall. (Contrast *Point estimation.*)
Joint occurrence. The occurrence of two events simultaneously. Such events cannot be *mutually exclusive.* Symbolically, $p(A \text{ and } B) > 0.00$.
Matched group design. A correlated-samples design in which pairs of subjects are matched in a variable correlated with the criterion measure. Each member of a pair receives different experimental conditions.
Mean. Sum of the scores or values of a variable divided by their number.
Mean deviation (average deviation). Sum of the deviations of each score from the mean, without regard to sign, divided by the number of scores.
Median. Score or potential score in a distribution of scores, above and below which one-half of the frequencies fall.
Mode. Score which occurs with the greatest frequency.

Mutually exclusive. Events A and B are said to be mutually exclusive if both cannot occur simultaneously. Symbolically, for mutually exclusive events, $p(A \text{ and } B) = 0.00$.
Negatively skewed distribution. Distribution which has relatively fewer frequencies at the low end of the horizontal axis.
Nominal numerals. Numerals used to name.
Null hypothesis (H_0). A statement that specifies hypothesized values for one or more of the population parameters. Commonly, although not necessarily, involves the hypothesis of "no difference."
Numbers. Specific types of numerals used to represent amount or quantity.
Numerals. Symbols used to represent numbers but lacking arithmetical properties.
One-tailed p value (one-directional or one-sided test). The probability value representing one tail (or one side) of the distribution.
Ordinal numerals. Numerals used to represent position, or order, in a series.
Ordinate (*Y-axis*). Vertical axis of a graph.
Parameter. Any characteristic of a population which is measurable.
Percentile rank. Number which represents the percent of cases in a comparison group which achieved scores lower than the one cited.
Point estimation. An estimate of a population parameter which involves a single value, selected by the criterion of "best estimate." (Contrast *Interval estimation.*)
Population universe. A complete set of individuals, objects, or measurements having some common observable characteristic.
Positively skewed distribution. Distribution which has relatively fewer frequencies at the high end of the horizontal axis.
Random sample. A sample drawn in such a way that every member of a population is equally likely to be included.
Range. Measure of dispersion; the scale distance between the largest and the smallest score.
Regression line. Straight line which represents the relationship between two variables.
Sample. A subset of a population or universe.
Sampling distribution. A theoretical probability distribution of a statistic which would result from drawing all possible samples of a given size from some population.
Scatter diagram. Graphic device employed to represent the variation in two variables.
Significance level. A probability value that is considered so rare in the sampling distribution specified under the null hypothesis that one is willing to assert the operation of nonchance factors. Common significance levels are 0.05 and 0.01.
Skewed distribution. Distribution which departs from symmetry and tails off at one end.
Standard error of the mean. A theoretical standard deviation of sample means, of a given sample size, drawn from some specified population. When based upon a known population standard deviation, $\sigma_{\bar{X}} = \sigma/\sqrt{N}$; when estimated from a single sample, $s_{\bar{X}} = s/\sqrt{N}$.

Standard error of the mean difference. Standard error based upon a correlated-samples design; is employed in the Student t-ratio for correlated samples.
Standard score (z). A score which represents the deviation of a specific score from the mean expressed in standard deviation units.
Statistic. A number resulting from the manipulation of raw data according to certain specified procedures.
Statistics. Collection of numerical facts which are expressed in summarizing statements; method of dealing with data: a tool for collecting, organizing, and analyzing numerical facts or observations.
Student t-ratio. See *t-ratio.*
Sum of squares. Deviations from the mean squared and summed.
t-distributions. Theoretical symmetrical sampling distributions with a mean of zero and a standard deviation which becomes smaller as degrees of freedom (df) increase. Employed in relation to the Student t-ratio.
Time ratio. Measure which expresses the change in a series of values arranged in a time sequence; it is typically shown as a percentage.
t-ratio. A test statistic for determining the significance of a difference between means (two-sample case) or for testing the hypothesis that a given sample mean was drawn from a population with the mean specified under the null hypothesis (one-sample case) which is employed when population standard deviation (or standard deviations) is (are) not known.
Two-tailed p value (non-directional or two-sided test). The probability value which represents two tails (or two sides) of the distribution.
Type I error (type α error). The rejection of H_0 when it is actually true. The probability of a type I error is given by the α level.
Type II error (type β error). The probability of accepting H_0 when it is actually false. The probability of a type II error is given by β.
Variable. A characteristic or phenomenon which may take on different values.
X-axis. See *Abscissa.*
Y-axis. See *Ordinate.*

APPENDIX IV

List of Formulas

Listed below are those formulas appearing in the text which the authors have found to be useful for computational purposes. Following each formula is the page number on which it first appears.

Numbers	Formula	Page
(5.1)	Percentile rank $= \frac{\text{cum } f}{N} \times 100$	89
(5.2)	$\text{Cum } f = \frac{\text{percentile rank} \times N}{100}$	91
(6.1)	$\bar{X} = \frac{\sum X}{N}$	102
(6.2)	$\bar{X} = \frac{\sum fX}{N}$	103
(6.4)	$\bar{X}_w = \frac{\sum (n \cdot \bar{X})}{\sum n}$	107
(7.3)	$x = X - \bar{X}$	122
(7.6)	$s = \sqrt{\frac{\sum x^2}{N - 1}}$	122
(7.7)	$\sum x^2 = \sum X^2 - \frac{(\sum X)^2}{N}$	123
(7.8)	$s = \sqrt{\frac{N\sum X^2 - (\sum X)^2}{N(N - 1)}}$	124

(8.1) $z = \dfrac{X - \bar{X}}{s}$ 136

(8.2) $z = \dfrac{x}{s}$ 137

(8.3) $V = \dfrac{100s}{\bar{X}}$ 144

(9.2) $r = \dfrac{\sum xy}{\sqrt{\sum x^2 \sum y^2}}$ 156

(9.3) $\sum xy = \sum XY - \dfrac{(\sum X)(\sum Y)}{n}$ 157

(9.4) $r = \dfrac{\sum XY - \dfrac{(\sum X)(\sum Y)}{n}}{\sqrt{\left[\sum X^2 - \dfrac{(\sum X)^2}{n}\right]\left[\sum Y^2 - \dfrac{(\sum Y)^2}{n}\right]}}$ 158

(9.5) $r_{\text{rho}} = 1 - \dfrac{6\sum D^2}{n(n^2 - 1)}$ 162

(10.7) $Y' = \bar{Y} + r\dfrac{s_y}{s_x}(X - \bar{X})$ 179

(10.9) $z_{y'} = rz_x$ 181

(10.14) $s_{\text{est } y} = s_y\sqrt{1 - r^2}$ 184

(11.3) $p(A \text{ or } B) = p(A) + p(B)$ when A and B are *mutually exclusive* 210

(11.5) $P + Q = 1.00$ when the events are *mutually exclusive* and *exhaustive* 210

(11.8) $p(A \text{ and } B) = p(A)p(B)$ when events are *independent* 211

(12.1) $(P + Q)^N = P^N + \dfrac{N}{1}P^{N-1}Q + \dfrac{N(N - 1)}{(1)(2)}P^{N-2}Q^2 + \dfrac{N(N - 1)(N - 2)}{(1)(2)(3)}P^{N-3}Q^3 + \cdots + Q^N$ 238

(13.5) $s_{\bar{X}} = \dfrac{s}{\sqrt{N}}$ 255

(13.6) $t = \dfrac{\bar{X} - \mu_0}{s_{\bar{X}}}$ 256

(13.7) Upper limit $\mu_0 = \bar{X} + t_{0.05}(s_{\bar{X}})$ 261

(13.8) Lower limit $\mu_0 = \bar{X} - t_{0.05}(s_{\bar{X}})$ 261

(13.9) $z = \dfrac{z_r - Z_r}{\sqrt{\dfrac{1}{n - 3}}}$ 263

(14.1) $s_{\bar{X}_1 - \bar{X}_2} = \sqrt{s_{\bar{X}_1}^2 + s_{\bar{X}_2}^2}$ 272

(14.2) $s_{\bar{X}_1 - \bar{X}_2} = \sqrt{\dfrac{s_1^2}{n_1} + \dfrac{s_2^2}{n_2}}, \qquad n_1 \neq n_2$ 272

(14.3) $s_{\bar{X}_1 - \bar{X}_2} = \sqrt{\dfrac{s_1^2 + s_2^2}{n}}, \qquad n_1 = n_2 = n$ 273

(14.4) $s_{\bar{X}_1 - \bar{X}_2} = \sqrt{\left(\dfrac{\sum x_1^2 + \sum x_2^2}{n_1 + n_2 - 2}\right)\left(\dfrac{1}{n_1} + \dfrac{1}{n_2}\right)}$

$= \sqrt{\dfrac{(\sum X_1^2 + \sum X_2^2) - (n_1\bar{X}_1^2 + n_2\bar{X}_2)}{n_1 + n_2 - 2}\left(\dfrac{1}{n_1} + \dfrac{1}{n_2}\right)}, \qquad n_1 \neq n_2$ 273

(14.5) $s_{\bar{X}_1 - \bar{X}_2} = \sqrt{\dfrac{\sum x_1^2 + \sum x_2^2}{n(n - 1)}}$

$= \sqrt{\dfrac{\sum X_1^2 + \sum X_2^2 - n(\bar{X}_1^2 + \bar{X}_2^2)}{n(n - 1)}}, \qquad n_1 = n_2 = n$ 273

(14.6) $t = \dfrac{(\bar{X}_1 - \bar{X}_2) - (\mu_1 - \mu_2)}{s_{\bar{X}_1 - \bar{X}_2}}$ 273

(15.1) $s_{\bar{X}_1 - \bar{X}_2} = \sqrt{s_{\bar{X}_1}^1 + s_{\bar{X}_2}^2 - 2r s_{\bar{X}_1} s_{\bar{X}_2}}$ 285

(15.2) $t = \dfrac{\bar{D} - \mu_D}{s_{\bar{D}}} = \dfrac{\bar{D}}{s_{\bar{D}}}$ 288

(15.3) $\sum d^2 = \sum D^2 - (\sum D)^2/n$ 288

(15.5) $s_{\bar{D}} = \sqrt{\dfrac{\sum d^2}{n(n - 1)}}$ 288

(15.7) $A = \dfrac{\sum D^2}{(\sum D)^2}$ 290

(16.2) $\sum x_{\text{tot}}^2 = \sum X_{\text{tot}}^2 - (\sum X_{\text{tot}})^2/N$ 300

(16.5) $\sum x_B^2 = \sum \dfrac{(\sum X_i)^2}{n_i} - \dfrac{(\sum X_{\text{tot}})^2}{N}$ 300

(16.6) $\sum x_{\text{tot}}^2 = \sum x_W^2 + \sum x_B^2$ 301

(16.7) $\mathrm{df}_B = k - 1$ 301

(16.8) $s_B^2 = \dfrac{\sum x_B^2}{\mathrm{df}_B}$ 301

(16.9) $\mathrm{df}_W = N - k$ 301

(16.10) $s_W^2 = \dfrac{\sum x_W^2}{\mathrm{df}_W}$ 301

(16.11) $F = s_B^2/s_W^2$ 301

(16.12) $\mathrm{df}_{\mathrm{tot}} = N - 1$ 301

(16.13) $\mathrm{HSD} = q_\alpha \sqrt{\dfrac{s_W^2}{n}}$ 305

(17.1) $p(x) = \dfrac{N!}{x!(N - x)!} P^x Q^{N-x}$ 316

(17.3) $z = \dfrac{|x - NP| - 0.5}{\sqrt{NPQ}}$ 319

(17.5) $\chi^2 = \sum_{r=1}^{r} \sum_{c=1}^{c} \dfrac{(f_o - f_e)^2}{f_e}$ 322

(17.6) $\chi^2 = \sum_{r=1}^{2} \sum_{c=1}^{2} \dfrac{(|f_o - f_e| - 0.5)^2}{f_e}$ when df = 1 323

(18.1) $U = n_1 n_2 - U'$ 331

(18.2) $U = n_1 n_2 + \dfrac{n_1(n_1 + 1)}{2} - R_1$ 332

APPENDIX V

Tables

List of Tables **Page**

ACKNOWLEDGMENTS

The authors are grateful to the authors and publishers listed below for permission to adapt from the following tables.

Table B — Table III of R. A. FISHER and F. YATES (1950), *Methods for Research Workers.* Edinburgh: Oliver and Boyd.

Table C — Table III of R. A. FISHER and F. YATES (1948), *Statistical Tables for Biological, Agricultural, and Medical Research.* Edinburgh: Oliver and Boyd.

Table D — G. W. SNEDECOR and WILLIAM G. COCHRAN, © 1967, *Statistical Methods*, 6th ed., Ames, Iowa: Iowa State University Press.

Table D_1 — M. MERRINGTON and C. M. THOMPSON (1943), "Tables of Percentage Points of the Inverted Beta Distribution." *Biometrika*, **33**, 73.

Table E — J. SANDLER (1955), A test of the significance of the difference between the means of correlated measures, based on a simplification of Student's *t*. *Brit. J. Psychol.*, **46**, 225–226.

Table F — Q. MCNEMAR (1962), Table B of *Psychological Statistics.* New York: John Wiley and Sons.

Table G — E. G. OLDS (1949), The 5 percent significance levels of sums of squares of rank differences and a correction. *Ann. Math. Statist.*, **20**, 117–118. E. G. OLDS (1938), Distribution of sums of squares of rank differences for small numbers of individuals. *Ann. Math. Statist*, **9**, 133–148.

Table H — W. V. BINGHAM (1937), Table XVII of *Aptitudes and Aptitude Testing.* New York: Harper and Bros.

Table I — H. B. MANN and D. R. WHITNEY (1947), On a test of whether one of two random variables is stochastically larger than the other. *Ann. Math. Statist.*, **18**, 52–54.

D. AUBLE (1953), Extended tables for the Mann-Whitney statistic. *Bulletin of the Institute of Educational Research at Indiana University*, **1**, 2.

Table J — F. WILCOXON, S. KATTI, and R. A. WILCOX (1963), *Critical Values and Probability Levels for the Wilcoxon Rank Sum Test and the Wilcoxon Signed Rank Test.* New York: American Cyanamid.

F. WILCOXON and R. A. WILCOX (1964), *Some Rapid Approximate Statistical Procedures.* New York: Lederle Laboratories.

Table L — S. SIEGEL (1956), *Non-Parametric Statistics.* New York: McGraw-Hill.

Table M — Table IV B of H. WALKER and J. LEV (1953), *Statistical Inference.* New York: Henry Holt & Co.

Table P — E. S. PEARSON and H. O. HARTLEY (1958), *Biometrika Tables for Statisticians*, vol. 1, 2nd ed. New York: Cambridge.

Table Q — A. L. EDWARDS (1969), *Statistical Analysis*, 3rd ed. New York: Holt, Rinehart and Winston.

J. W. DUNLAP and A. K. KURTZ (1932), *Handbook of Statistical Nomographs, Tables, and Formulas.* New York: World Book.

Table R — RAND Corporation (1955), *A Million Random Digits.* Glencoe, Ill.: Free Press of Glencoe.

THE USE OF TABLE A

The use of Table A requires that the raw score be transformed into a z-score and that the variable be normally distributed.

The values in Table A represent the proportion of area in the standard normal curve which has a mean of 0, a standard deviation of 1.00, and a total area also equal to 1.00.

Since the normal curve is symmetrical, it is sufficient to indicate only the areas corresponding to positive z-values. Negative z-values will have precisely the same proportions of area as their positive counterparts.

Column B represents the proportion of area between the mean and a given z.

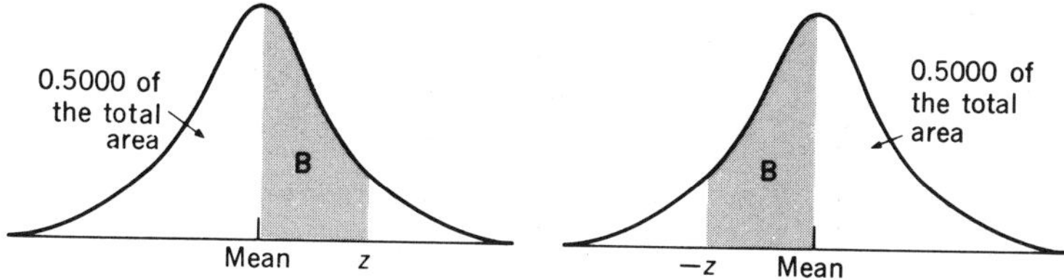

Column C represents the proportion of area beyond a given z.

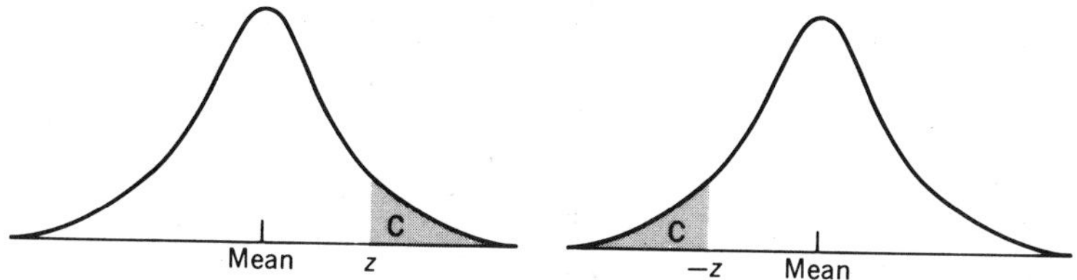

Table A

Proportions of area under the normal curve

(A) z	(B) area between mean and z	(C) area beyond z
0.00	.0000	.5000
0.01	.0040	.4960
0.02	.0080	.4920
0.03	.0120	.4880
0.04	.0160	.4840
0.05	.0199	.4801
0.06	.0239	.4761
0.07	.0279	.4721
0.08	.0319	.4681
0.09	.0359	.4641
0.10	.0398	.4602
0.11	.0438	.4562
0.12	.0478	.4522
0.13	.0517	.4483
0.14	.0557	.4443
0.15	.0596	.4404
0.16	.0636	.4364
0.17	.0675	.4325
0.18	.0714	.4286
0.19	.0753	.4247
0.20	.0793	.4207
0.21	.0832	.4168
0.22	.0871	.4129
0.23	.0910	.4090
0.24	.0948	.4052
0.25	.0987	.4013
0.26	.1026	.3974
0.27	.1064	.3936
0.28	.1103	.3897
0.29	.1141	.3859
0.30	.1179	.3821
0.31	.1217	.3783
0.32	.1255	.3745
0.33	.1293	.3707
0.34	.1331	.3669
0.35	.1368	.3632
0.36	.1406	.3594
0.37	.1443	.3557
0.38	.1480	.3520
0.39	.1517	.3483
0.40	.1554	.3446
0.41	.1591	.3409
0.42	.1628	.3372
0.43	.1664	.3336
0.44	.1700	.3300
0.45	.1736	.3264
0.46	.1772	.3228
0.47	.1808	.3192
0.48	.1844	.3156
0.49	.1879	.3121
0.50	.1915	.3085
0.51	.1950	.3050
0.52	.1985	.3015
0.53	.2019	.2981
0.54	.2054	.2946
0.55	.2088	.2912
0.56	.2123	.2877
0.57	.2157	.2843
0.58	.2190	.2810
0.59	.2224	.2776
0.60	.2257	.2743
0.61	.2291	.2709
0.62	.2324	.2676
0.63	.2357	.2643
0.64	.2389	.2611
0.65	.2422	.2578
0.66	.2454	.2546
0.67	.2486	.2514
0.68	.2517	.2483
0.69	.2549	.2451
0.70	.2580	.2420
0.71	.2611	.2389
0.72	.2642	.2358
0.73	.2673	.2327
0.74	.2704	.2296
0.75	.2734	.2266
0.76	.2764	.2236
0.77	.2794	.2206
0.78	.2823	.2177
0.79	.2852	.2148
0.80	.2881	.2119
0.81	.2910	.2090
0.82	.2939	.2061
0.83	.2967	.2033
0.84	.2995	.2005
0.85	.3023	.1977
0.86	.3051	.1949
0.87	.3078	.1922
0.88	.3106	.1894
0.89	.3133	.1867
0.90	.3159	.1841
0.91	.3186	.1814
0.92	.3212	.1788
0.93	.3238	.1762
0.94	.3264	.1736
0.95	.3289	.1711
0.96	.3315	.1685
0.97	.3340	.1660
0.98	.3365	.1635
0.99	.3389	.1611
1.00	.3413	.1587
1.01	.3438	.1562
1.02	.3461	.1539
1.03	.3485	.1515
1.04	.3508	.1492
1.05	.3531	.1469
1.06	.3554	.1446
1.07	.3577	.1423
1.08	.3599	.1401
1.09	.3621	.1379
1.10	.3643	.1357
1.11	.3665	.1335
1.12	.3686	.1314
1.13	.3708	.1292
1.14	.3729	.1271
1.15	.3749	.1251
1.16	.3770	.1230
1.17	.3790	.1210
1.18	.3810	.1190
1.19	.3830	.1170
1.20	.3849	.1151
1.21	.3869	.1131
1.22	.3888	.1112
1.23	.3907	.1093
1.24	.3925	.1075
1.25	.3944	.1056
1.26	.3962	.1038
1.27	.3980	.1020
1.28	.3997	.1003
1.29	.4015	.0985
1.30	.4032	.0968
1.31	.4049	.0951
1.32	.4066	.0934
1.33	.4082	.0918
1.34	.4099	.0901
1.35	.4115	.0885
1.36	.4131	.0869
1.37	.4147	.0853
1.38	.4162	.0838
1.39	.4177	.0823
1.40	.4192	.0808
1.41	.4207	.0793
1.42	.4222	.0778
1.43	.4236	.0764
1.44	.4251	.0749
1.45	.4265	.0735
1.46	.4279	.0721
1.47	.4292	.0708
1.48	.4306	.0694
1.49	.4319	.0681
1.50	.4332	.0668
1.51	.4345	.0655
1.52	.4357	.0643
1.53	.4370	.0630
1.54	.4382	.0618
1.55	.4394	.0606
1.56	.4406	.0594
1.57	.4418	.0582
1.58	.4429	.0571
1.59	.4441	.0559
1.60	.4452	.0548
1.61	.4463	.0537
1.62	.4474	.0526
1.63	.4484	.0516
1.64	.4495	.0505

Table A. *(continued)*

(A) z	(B) area between mean and z	(C) area beyond z
1.65	.4505	.0495
1.66	.4515	.0485
1.67	.4525	.0475
1.68	.4535	.0465
1.69	.4545	.0455
1.70	.4554	.0446
1.71	.4564	.0436
1.72	.4573	.0427
1.73	.4582	.0418
1.74	.4591	.0409
1.75	.4599	.0401
1.76	.4608	.0392
1.77	.4616	.0384
1.78	.4625	.0375
1.79	.4633	.0367
1.80	.4641	.0359
1.81	.4649	.0351
1.82	.4656	.0344
1.83	.4664	.0336
1.84	.4671	.0329
1.85	.4678	.0322
1.86	.4686	.0314
1.87	.4693	.0307
1.88	.4699	.0301
1.89	.4706	.0294
1.90	.4713	.0287
1.91	.4719	.0281
1.92	.4726	.0274
1.93	.4732	.0268
1.94	.4738	.0262
1.95	.4744	.0256
1.96	.4750	.0250
1.97	.4756	.0244
1.98	.4761	.0239
1.99	.4767	.0233
2.00	.4772	.0228
2.01	.4778	.0222
2.02	.4783	.0217
2.03	.4788	.0212
2.04	.4793	.0207
2.05	.4798	.0202
2.06	.4803	.0197
2.07	.4808	.0192
2.08	.4812	.0188
2.09	.4817	.0183
2.10	.4821	.0179
2.11	.4826	.0174
2.12	.4830	.0170
2.13	.4834	.0166
2.14	.4838	.0162
2.15	.4842	.0158
2.16	.4846	.0154
2.17	.4850	.0150
2.18	.4854	.0146
2.19	.4857	.0143
2.20	.4861	.0139
2.21	.4864	.0136

(A) z	(B) area between mean and z	(C) area beyond z
2.22	.4868	.0132
2.23	.4871	.0129
2.24	.4875	.0125
2.25	.4878	.0122
2.26	.4881	.0119
2.27	.4884	.0116
2.28	.4887	.0113
2.29	.4890	.0110
2.30	.4893	.0107
2.31	.4896	.0104
2.32	.4898	.0102
2.33	.4901	.0099
2.34	.4904	.0096
2.35	.4906	.0094
2.36	.4909	.0091
2.37	.4911	.0089
2.38	.4913	.0087
2.39	.4916	.0084
2.40	.4918	.0082
2.41	.4920	.0080
2.42	.4922	.0078
2.43	.4925	.0075
2.44	.4927	.0073
2.45	.4929	.0071
2.46	.4931	.0069
2.47	.4932	.0068
2.48	.4934	.0066
2.49	.4936	.0064
2.50	.4938	.0062
2.51	.4940	.0060
2.52	.4941	.0059
2.53	.4943	.0057
2.54	.4945	.0055
2.55	.4946	.0054
2.56	.4948	.0052
2.57	.4949	.0051
2.58	.4951	.0049
2.59	.4952	.0048
2.60	.4953	.0047
2.61	.4955	.0045
2.62	.4956	.0044
2.63	.4957	.0043
2.64	.4959	.0041
2.65	.4960	.0040
2.66	.4961	.0039
2.67	.4962	.0038
2.68	.4963	.0037
2.69	.4964	.0036
2.70	.4965	.0035
2.71	.4966	.0034
2.72	.4967	.0033
2.73	.4968	.0032
2.74	.4969	.0031
2.75	.4970	.0030
2.76	.4971	.0029
2.77	.4972	.0028
2.78	.4973	.0027

(A) z	(B) area between mean and z	(C) area beyond z
2.79	.4974	.0026
2.80	.4974	.0026
2.81	.4975	.0025
2.82	.4976	.0024
2.83	.4977	.0023
2.84	.4977	.0023
2.85	.4978	.0022
2.86	.4979	.0021
2.87	.4979	.0021
2.88	.4980	.0020
2.89	.4981	.0019
2.90	.4981	.0019
2.91	.4982	.0018
2.92	.4982	.0018
2.93	.4983	.0017
2.94	.4984	.0016
2.95	.4984	.0016
2.96	.4985	.0015
2.97	.4985	.0015
2.98	.4986	.0014
2.99	.4986	.0014
3.00	.4987	.0013
3.01	.4987	.0013
3.02	.4987	.0013
3.03	.4988	.0012
3.04	.4988	.0012
3.05	.4989	.0011
3.06	.4989	.0011
3.07	.4989	.0011
3.08	.4990	.0010
3.09	.4990	.0010
3.10	.4990	.0010
3.11	.4991	.0009
3.12	.4991	.0009
3.13	.4991	.0009
3.14	.4992	.0008
3.15	.4992	.0008
3.16	.4992	.0008
3.17	.4992	.0008
3.18	.4993	.0007
3.19	.4993	.0007
3.20	.4993	.0007
3.21	.4993	.0007
3.22	.4994	.0006
3.23	.4994	.0006
3.24	.4994	.0006
3.25	.4994	.0006
3.30	.4995	.0005
3.35	.4996	.0004
3.40	.4997	.0003
3.45	.4997	.0003
3.50	.4998	.0002
3.60	.4998	.0002
3.70	.4999	.0001
3.80	.4999	.0001
3.90	.49995	.00005
4.00	.49997	.00003

Table B

Table of χ^2

Degrees of freedom df	P = .99	.98	.95	.90	.80	.70	.50	.30	.20	.10	.05	.02	.01
1	.000157	.000628	.00393	.0158	.0642	.148	.455	1.074	1.642	2.706	3.841	5.412	6.635
2	.0201	.0404	.103	.211	.446	.713	1.386	2.408	3.219	4.605	5.991	7.824	9.210
3	.115	.185	.352	.584	1.005	1.424	2.366	3.665	4.642	6.251	7.815	9.837	11.341
4	.297	.429	.711	1.064	1.649	2.195	3.357	4.878	5.989	7.779	9.488	11.668	13.277
5	.554	.752	1.145	1.610	2.343	3.000	4.351	6.064	7.289	9.236	11.070	13.388	15.086
6	.872	1.134	1.635	2.204	3.070	3.828	5.348	7.231	8.558	10.645	12.592	15.033	16.812
7	1.239	1.564	2.167	2.833	3.822	4.671	6.346	8.383	9.803	12.017	14.067	16.622	18.475
8	1.646	2.032	2.733	3.490	4.594	5.527	7.344	9.524	11.030	13.362	15.507	18.168	20.090
9	2.088	2.532	3.325	4.168	5.380	6.393	8.343	10.656	12.242	14.684	16.919	19.679	21.666
10	2.558	3.059	3.940	4.865	6.179	7.267	9.342	11.781	13.442	15.987	18.307	21.161	23.209
11	3.053	3.609	4.575	5.578	6.989	8.148	10.341	12.899	14.631	17.275	19.675	22.618	24.725
12	3.571	4.178	5.226	6.304	7.807	9.034	11.340	14.011	15.812	18.549	21.026	24.054	26.217
13	4.107	4.765	5.892	7.042	8.634	9.926	12.340	15.119	16.985	19.812	22.362	25.472	27.688
14	4.660	5.368	6.571	7.790	9.467	10.821	13.339	16.222	18.151	21.064	23.685	26.873	29.141
15	5.229	5.985	7.261	8.547	10.307	11.721	14.339	17.322	19.311	22.307	24.996	28.259	30.578
16	5.812	6.614	7.962	9.312	11.152	12.624	15.338	18.418	20.465	23.542	26.296	29.633	32.000
17	6.408	7.255	8.672	10.085	12.002	13.531	16.338	19.511	21.615	24.769	27.587	30.995	33.409
18	7.015	7.906	9.390	10.865	12.857	14.440	17.338	20.601	22.760	25.989	28.869	32.346	34.805
19	7.633	8.567	10.117	11.651	13.716	15.352	18.338	21.689	23.900	27.204	30.144	33.687	36.191
20	8.260	9.237	10.851	12.443	14.578	16.266	19.337	22.775	25.038	28.412	31.410	35.020	37.566
21	8.897	9.915	11.591	13.240	15.445	17.182	20.337	23.858	26.171	29.615	32.671	36.343	38.932
22	9.542	10.600	12.338	14.041	16.314	18.101	21.337	24.939	27.301	30.813	33.924	37.659	40.289
23	10.196	11.293	13.091	14.848	17.187	19.021	22.337	26.018	28.429	32.007	35.172	38.968	41.638
24	10.856	11.992	13.848	15.659	18.062	19.943	23.337	27.096	29.553	33.196	36.415	40.270	42.980
25	11.524	12.697	14.611	16.473	18.940	20.867	24.337	28.172	30.675	34.382	37.652	41.566	44.314
26	12.198	13.409	15.379	17.292	19.820	21.792	25.336	29.246	31.795	35.563	38.885	42.856	45.642
27	12.879	14.125	16.151	18.114	20.703	22.719	26.336	30.319	32.912	36.741	40.113	44.140	46.963
28	13.565	14.847	16.928	18.939	21.588	23.647	27.336	31.391	34.027	37.916	41.337	45.419	48.278
29	14.256	15.574	17.708	19.768	22.475	24.577	28.336	32.461	35.139	39.087	42.557	46.693	49.588
30	14.953	16.306	18.493	20.599	23.364	25.508	29.336	33.530	36.250	40.256	43.773	47.962	50.892

Table C

Critical values of *t*

For any given df, the table shows the values of t corresponding to various levels of probability. Obtained t is significant at a given level if it is equal to or greater than the value shown in the table.

	Level of significance for one-tailed test					
	.10	.05	.025	.01	.005	.0005
	Level of significance for two-tailed test					
df	.20	.10	.05	.02	.01	.001
1	3.078	6.314	12.706	31.821	63.657	636.619
2	1.886	2.920	4.303	6.965	9.925	31.598
3	1.638	2.353	3.182	4.541	5.841	12.941
4	1.533	2.132	2.776	3.747	4.604	8.610
5	1.476	2.015	2.571	3.365	4.032	6.859
6	1.440	1.943	2.447	3.143	3.707	5.959
7	1.415	1.895	2.365	2.998	3.499	5.405
8	1.397	1.860	2.306	2.896	3.355	5.041
9	1.383	1.833	2.262	2.821	3.250	4.781
10	1.372	1.812	2.228	2.764	3.169	4.587
11	1.363	1.796	2.201	2.718	3.106	4.437
12	1.356	1.782	2.179	2.681	3.055	4.318
13	1.350	1.771	2.160	2.650	3.012	4.221
14	1.345	1.761	2.145	2.624	2.977	4.140
15	1.341	1.753	2.131	2.602	2.947	4.073
16	1.337	1.746	2.120	2.583	2.921	4.015
17	1.333	1.740	2.110	2.567	2.898	3.965
18	1.330	1.734	2.101	2.552	2.878	3.922
19	1.328	1.729	2.093	2.539	2.861	3.883
20	1.325	1.725	2.086	2.528	2.845	3.850
21	1.323	1.721	2.080	2.518	2.831	3.819
22	1.321	1.717	2.074	2.508	2.819	3.792
23	1.319	1.714	2.069	2.500	2.807	3.767
24	1.318	1.711	2.064	2.492	2.797	3.745
25	1.316	1.708	2.060	2.485	2.787	3.725
26	1.315	1.706	2.056	2.479	2.779	3.707
27	1.314	1.703	2.052	2.473	2.771	3.690
28	1.313	1.701	2.048	2.467	2.763	3.674
29	1.311	1.699	2.045	2.462	2.756	3.659
30	1.310	1.697	2.042	2.457	2.750	3.646
40	1.303	1.684	2.021	2.423	2.704	3.551
60	1.296	1.671	2.000	2.390	2.660	3.460
120	1.289	1.658	1.980	2.358	2.617	3.373
∞	1.282	1.645	1.960	2.326	2.576	3.291

Table D

Critical values of *F*

The obtained F is significant at a given level if it is equal to or greater than the value shown in the table.
0.05 (light row) and 0.01 (dark row) points for the distribution of F

Degrees of freedom for lesser mean square	Degrees of freedom for greater mean square: 1	2	3	4	5	6	7	8	9	10	11	12	14	16	20	24	30	40	50	75	100	200	500	∞
1	161 **4052**	200 **4999**	216 **5403**	225 **5625**	230 **5764**	234 **5859**	237 **5928**	239 **5981**	241 **6022**	242 **6056**	243 **6082**	244 **6106**	245 **6142**	246 **6169**	248 **6208**	249 **6234**	250 **6258**	251 **6286**	252 **6302**	253 **6323**	253 **6334**	254 **6352**	254 **6361**	254 **6366**
2	18.51 **98.49**	19.00 **99.01**	19.16 **99.17**	19.25 **99.25**	19.30 **99.30**	19.33 **99.33**	19.36 **99.34**	19.37 **99.36**	19.38 **99.38**	19.39 **99.40**	19.40 **99.41**	19.41 **99.42**	19.42 **99.43**	19.43 **99.44**	19.44 **99.45**	19.45 **99.46**	19.46 **99.47**	19.47 **99.48**	19.47 **99.48**	19.48 **99.49**	19.49 **99.49**	19.49 **99.49**	19.50 **99.50**	19.50 **99.50**
3	10.13 **34.12**	9.55 **30.81**	9.28 **29.46**	9.12 **28.71**	9.01 **28.24**	8.94 **27.91**	8.88 **27.67**	8.84 **27.49**	8.81 **27.34**	8.78 **27.23**	8.76 **27.13**	8.74 **27.05**	8.71 **26.92**	8.69 **26.83**	8.66 **26.69**	8.64 **26.60**	8.62 **26.50**	8.60 **26.41**	8.58 **26.30**	8.57 **26.27**	8.56 **26.23**	8.54 **26.18**	8.54 **26.14**	8.53 **26.12**
4	7.71 **21.20**	6.94 **18.00**	6.59 **16.69**	6.39 **15.98**	6.26 **15.52**	6.16 **15.21**	6.09 **14.98**	6.04 **14.80**	6.00 **14.66**	5.96 **14.54**	5.93 **14.45**	5.91 **14.37**	5.87 **14.24**	5.84 **14.15**	5.80 **14.02**	5.77 **13.93**	5.74 **13.83**	5.71 **13.74**	5.70 **13.69**	5.68 **13.61**	5.66 **13.57**	5.65 **13.52**	5.64 **13.48**	5.63 **13.46**
5	6.61 **16.26**	5.79 **13.27**	5.41 **12.06**	5.19 **11.39**	5.05 **10.97**	4.95 **10.67**	4.88 **10.45**	4.82 **10.27**	4.78 **10.15**	4.74 **10.05**	4.70 **9.96**	4.68 **9.89**	4.64 **9.77**	4.60 **9.68**	4.56 **9.55**	4.53 **9.47**	4.50 **9.38**	4.46 **9.29**	4.44 **9.24**	4.42 **9.17**	4.40 **9.13**	4.38 **9.07**	4.37 **9.04**	4.36 **9.02**
6	5.99 **13.74**	5.14 **10.92**	4.76 **9.78**	4.53 **9.15**	4.39 **8.75**	4.28 **8.47**	4.21 **8.26**	4.15 **8.10**	4.10 **7.98**	4.06 **7.87**	4.03 **7.79**	4.00 **7.72**	3.96 **7.60**	3.92 **7.52**	3.87 **7.39**	3.84 **7.31**	3.81 **7.23**	3.77 **7.14**	3.75 **7.09**	3.72 **7.02**	3.71 **6.99**	3.69 **6.94**	3.68 **6.90**	3.67 **6.88**
7	5.59 **12.25**	4.74 **9.55**	4.35 **8.45**	4.12 **7.85**	3.97 **7.46**	3.87 **7.19**	3.79 **7.00**	3.73 **6.84**	3.68 **6.71**	3.63 **6.62**	3.60 **6.54**	3.57 **6.47**	3.52 **6.35**	3.49 **6.27**	3.44 **6.15**	3.41 **6.07**	3.38 **5.98**	3.34 **5.90**	3.32 **5.85**	3.29 **5.78**	3.28 **5.75**	3.25 **5.70**	3.24 **5.67**	3.23 **5.65**
8	5.32 **11.26**	4.46 **8.65**	4.07 **7.59**	3.84 **7.01**	3.69 **6.63**	3.58 **6.37**	3.50 **6.19**	3.44 **6.03**	3.39 **5.91**	3.34 **5.82**	3.31 **5.74**	3.28 **5.67**	3.23 **5.56**	3.20 **5.48**	3.15 **5.36**	3.12 **5.28**	3.08 **5.20**	3.05 **5.11**	3.03 **5.06**	3.00 **5.00**	2.98 **4.96**	2.96 **4.91**	2.94 **4.88**	2.93 **4.86**
9	5.12 **10.56**	4.26 **8.02**	3.86 **6.99**	3.63 **6.42**	3.48 **6.06**	3.37 **5.80**	3.29 **5.62**	3.23 **5.47**	3.18 **5.35**	3.13 **5.26**	3.10 **5.18**	3.07 **5.11**	3.02 **5.00**	2.98 **4.92**	2.93 **4.80**	2.90 **4.73**	2.86 **4.64**	2.82 **4.56**	2.80 **4.51**	2.77 **4.45**	2.76 **4.41**	2.73 **4.36**	2.72 **4.33**	2.71 **4.31**
10	4.96 **10.04**	4.10 **7.56**	3.71 **6.55**	3.48 **5.99**	3.33 **5.64**	3.22 **5.39**	3.14 **5.21**	3.07 **5.06**	3.02 **4.95**	2.97 **4.85**	2.94 **4.78**	2.91 **4.71**	2.86 **4.60**	2.82 **4.52**	2.77 **4.41**	2.74 **4.33**	2.70 **4.25**	2.67 **4.17**	2.64 **4.12**	2.61 **4.05**	2.59 **4.01**	2.56 **3.96**	2.55 **3.93**	2.54 **3.91**
11	4.84 **9.65**	3.98 **7.20**	3.59 **6.22**	3.36 **5.67**	3.20 **5.32**	3.09 **5.07**	3.01 **4.88**	2.95 **4.74**	2.90 **4.63**	2.86 **4.54**	2.82 **4.46**	2.79 **4.40**	2.74 **4.29**	2.70 **4.21**	2.65 **4.10**	2.61 **4.02**	2.57 **3.94**	2.53 **3.86**	2.50 **3.80**	2.47 **3.74**	2.45 **3.70**	2.42 **3.66**	2.41 **3.62**	2.40 **3.60**
12	4.75 **9.33**	3.88 **6.93**	3.49 **5.95**	3.26 **5.41**	3.11 **5.06**	3.00 **4.82**	2.92 **4.65**	2.85 **4.50**	2.80 **4.39**	2.76 **4.30**	2.72 **4.22**	2.69 **4.16**	2.64 **4.05**	2.60 **3.98**	2.54 **3.86**	2.50 **3.78**	2.46 **3.70**	2.42 **3.61**	2.40 **3.56**	2.36 **3.49**	2.35 **3.46**	2.32 **3.41**	2.31 **3.38**	2.30 **3.36**
13	4.67 **9.07**	3.80 **6.70**	3.41 **5.74**	3.18 **5.20**	3.02 **4.86**	2.92 **4.62**	2.84 **4.44**	2.77 **4.30**	2.72 **4.19**	2.67 **4.10**	2.63 **4.02**	2.60 **3.96**	2.55 **3.85**	2.51 **3.78**	2.46 **3.67**	2.42 **3.59**	2.38 **3.51**	2.34 **3.42**	2.32 **3.37**	2.28 **3.30**	2.26 **3.27**	2.24 **3.21**	2.22 **3.18**	2.21 **3.16**
14	4.60 **8.86**	3.74 **6.51**	3.34 **5.56**	3.11 **5.03**	2.96 **4.69**	2.85 **4.46**	2.77 **4.28**	2.70 **4.14**	2.65 **4.03**	2.60 **3.94**	2.56 **3.86**	2.53 **3.80**	2.48 **3.70**	2.44 **3.62**	2.39 **3.51**	2.35 **3.43**	2.31 **3.34**	2.27 **3.26**	2.24 **3.21**	2.21 **3.14**	2.19 **3.11**	2.16 **3.06**	2.14 **3.02**	2.13 **3.00**
15	4.54 **8.68**	3.68 **6.36**	3.29 **5.42**	3.06 **4.89**	2.90 **4.56**	2.79 **4.32**	2.70 **4.14**	2.64 **4.00**	2.59 **3.89**	2.55 **3.80**	2.51 **3.73**	2.48 **3.67**	2.43 **3.56**	2.39 **3.48**	2.33 **3.36**	2.29 **3.29**	2.25 **3.20**	2.21 **3.12**	2.18 **3.07**	2.15 **3.00**	2.12 **2.97**	2.10 **2.92**	2.08 **2.89**	2.07 **2.87**

Table D. *(continued)*

Degrees of freedom for lesser mean square																								
16	4.49 **8.53**	3.63 **6.23**	3.24 **5.29**	3.01 **4.77**	2.85 **4.44**	2.74 **4.20**	2.66 **4.03**	2.59 **3.89**	2.54 **3.78**	2.49 **3.69**	2.45 **3.61**	2.42 **3.55**	2.37 **3.45**	2.33 **3.37**	2.28 **3.25**	2.24 **3.18**	2.20 **3.10**	2.16 **3.01**	2.13 **2.96**	2.09 **2.89**	2.07 **2.86**	2.04 **2.80**	2.02 **2.77**	2.01 **2.75**
17	4.45 **8.40**	3.59 **6.11**	3.20 **5.18**	2.96 **4.67**	2.81 **4.34**	2.70 **4.10**	2.62 **3.93**	2.55 **3.79**	2.50 **3.68**	2.45 **3.59**	2.41 **3.52**	2.38 **3.45**	2.33 **3.35**	2.29 **3.27**	2.23 **3.16**	2.19 **3.08**	2.15 **3.00**	2.11 **2.92**	2.08 **2.86**	2.04 **2.79**	2.02 **2.76**	1.99 **2.70**	1.97 **2.67**	1.96 **2.65**
18	4.41 **8.28**	3.55 **6.01**	3.16 **5.09**	2.93 **4.58**	2.77 **4.25**	2.66 **4.01**	2.58 **3.85**	2.51 **3.71**	2.46 **3.60**	2.41 **3.51**	2.37 **3.44**	2.34 **3.37**	2.29 **3.27**	2.25 **3.19**	2.19 **3.07**	2.15 **3.00**	2.11 **2.91**	2.07 **2.83**	2.04 **2.78**	2.00 **2.71**	1.98 **2.68**	1.95 **2.62**	1.93 **2.59**	1.92 **2.57**
19	4.38 **8.18**	3.52 **5.93**	3.13 **5.01**	2.90 **4.50**	2.74 **4.17**	2.63 **3.94**	2.55 **3.77**	2.48 **3.63**	2.43 **3.52**	2.38 **3.43**	2.34 **3.36**	2.31 **3.30**	2.26 **3.19**	2.21 **3.12**	2.15 **3.00**	2.11 **2.92**	2.07 **2.84**	2.02 **2.76**	2.00 **2.70**	1.96 **2.63**	1.94 **2.60**	1.91 **2.54**	1.90 **2.51**	1.88 **2.49**
20	4.35 **8.10**	3.49 **5.85**	3.10 **4.94**	2.87 **4.43**	2.71 **4.10**	2.60 **3.87**	2.52 **3.71**	2.45 **3.56**	2.40 **3.45**	2.35 **3.37**	2.31 **3.30**	2.28 **3.23**	2.23 **3.13**	2.18 **3.05**	2.12 **2.94**	2.08 **2.86**	2.04 **2.77**	1.99 **2.69**	1.96 **2.63**	1.92 **2.56**	1.90 **2.53**	1.87 **2.47**	1.85 **2.44**	1.84 **2.42**
21	4.32 **8.02**	3.47 **5.78**	3.07 **4.87**	2.84 **4.37**	2.68 **4.04**	2.57 **3.81**	2.49 **3.65**	2.42 **3.51**	2.37 **3.40**	2.32 **3.31**	2.28 **3.24**	2.25 **3.17**	2.20 **3.07**	2.15 **2.99**	2.09 **2.88**	2.05 **2.80**	2.00 **2.72**	1.96 **2.63**	1.93 **2.58**	1.80 **2.51**	1.87 **2.47**	1.84 **2.42**	1.82 **2.38**	1.81 **2.36**
22	4.30 **7.94**	3.44 **5.72**	3.05 **4.82**	2.82 **4.31**	2.66 **3.99**	2.55 **3.76**	2.47 **3.59**	2.40 **3.45**	2.35 **3.35**	2.30 **3.26**	2.26 **3.18**	2.23 **3.12**	2.18 **3.02**	2.13 **2.94**	2.07 **2.83**	2.03 **2.75**	1.98 **2.67**	1.93 **2.58**	1.91 **2.53**	1.87 **2.46**	1.84 **2.42**	1.81 **2.37**	1.80 **2.33**	1.78 **2.31**
23	4.28 **7.88**	3.42 **5.66**	3.03 **4.76**	2.80 **4.26**	2.64 **3.94**	2.53 **3.71**	2.45 **3.54**	2.38 **3.41**	2.32 **3.30**	2.28 **3.21**	2.24 **3.14**	2.20 **3.07**	2.14 **2.97**	2.10 **2.89**	2.04 **2.78**	2.00 **2.70**	1.96 **2,62**	1.91 **2.53**	1.88 **2.48**	1.84 **2.41**	1.82 **2.37**	1.79 **2.32**	1.77 **2.28**	1.76 **2.26**
24	4.26 **7.82**	3.40 **5.61**	3.01 **4.72**	2.78 **4.22**	2.62 **3.90**	2.51 **3.67**	2.43 **3.50**	2.36 **3.36**	2.30 **3.25**	2.26 **3.17**	2.22 **3.09**	2.18 **3.03**	2.13 **2.93**	2.09 **2.85**	2.02 **2.74**	1.98 **2.66**	1.94 **2.58**	1.89 **2.49**	1.86 **2.44**	1.82 **2.36**	1.80 **2.33**	1.76 **2.27**	1.74 **2.23**	1.73 **2.21**
25	4.24 **7.77**	3.38 **5.57**	2.99 **4.68**	2.76 **4.18**	2.60 **3.86**	2.49 **3.63**	2.41 **3.46**	2.34 **3.32**	2.28 **3.21**	2.24 **3.13**	2.20 **3.05**	2.16 **2.99**	2.11 **2.89**	2.06 **2.81**	2.00 **2.70**	1.96 **2.62**	1.92 **2.54**	1.87 **2.45**	1.84 **2.40**	1.80 **2.32**	1.77 **2.29**	1.74 **2.23**	1.72 **2.19**	1.71 **2.17**
26	4.22 **7.72**	3.37 **5.53**	2.89 **4.64**	2.74 **4.14**	2.59 **3.82**	2.47 **3.59**	2.39 **3.42**	2.32 **3.29**	2.27 **3.17**	2.22 **3.09**	2.18 **3.02**	2.15 **2.96**	2.10 **2.86**	2.05 **2.77**	1.99 **2.66**	1.95 **2.58**	1.90 **2.50**	1.85 **2.41**	1.82 **2.36**	1.78 **2.28**	1.76 **2.25**	1.72 **2.19**	1.70 **2.15**	1.69 **2.13**
27	4.21 **7.68**	3.35 **5.49**	2.96 **4.60**	2.73 **4.11**	2.57 **3.79**	2.46 **3.56**	2.37 **3.39**	2.30 **3.26**	2.25 **3.14**	2.20 **3.06**	2.16 **2.98**	2.13 **2.93**	2.08 **2.83**	2.03 **2.74**	1.97 **2.63**	1.93 **2.55**	1.88 **2.47**	1.84 **2.38**	1.80 **2.33**	1.76 **2.25**	1.74 **2.21**	1.71 **2.16**	1.68 **2.12**	1.67 **2.10**
28	4.20 **7.64**	3.34 **5.45**	2.95 **4.57**	2.71 **4.07**	2.56 **3.76**	2.44 **3.53**	2.36 **3.36**	2.29 **3.23**	3.24 **3.11**	2.19 **3.03**	2.15 **2.95**	2.12 **2.90**	2.06 **2.80**	2.02 **2.71**	1.96 **2.60**	1.91 **2.52**	1.87 **2.44**	1.81 **2.35**	1.78 **2.30**	1.75 **2.22**	1.72 **2.18**	1.69 **2.13**	1.67 **2.09**	1.65 **2.06**
29	4.18 **7.60**	3.33 **5.52**	2.93 **4.54**	2.70 **4.04**	2.54 **3.73**	2.43 **3.50**	2.35 **3.32**	2.28 **3.20**	2.22 **3.08**	2.18 **3.00**	2.14 **2.92**	2.10 **2.87**	2.05 **2.77**	2.00 **2.68**	1.94 **2.57**	1.90 **2.49**	1.85 **2.41**	1.80 **2.32**	1.77 **2.27**	1.73 **2.19**	1.71 **2.15**	1.68 **2.10**	1.65 **2.06**	1.64 **2.03**
30	4.17 **7.56**	3.32 **5.39**	2.92 **4.51**	2.69 **4.02**	2.53 **3.70**	2.42 **3.47**	2.34 **3.30**	2.27 **3.17**	2.21 **3.06**	2.16 **2.98**	2.12 **2.90**	2.09 **2.84**	2.04 **2.74**	1.99 **2.66**	1.93 **2.55**	1.89 **2.47**	1.84 **2.38**	1.79 **2.29**	1.76 **2.24**	1.72 **2.16**	1.69 **2.13**	1.66 **2.07**	1.64 **2.03**	1.62 **2.01**

Table D. *(continued)*

0.05 (light row) and 0.01 (dark row) points for the distribution of F

Degrees of freedom for lesser mean square	Degrees of freedom for greater mean square																						
	1	2	3	4	5	6	7	8	9	10	11	12	14	16	20	24	30	40	50	75	100	200	500	∞
32	4.15	3.30	2.90	2.67	2.51	2.40	2.32	2.25	2.19	2.14	2.10	2.07	2.02	1.97	1.91	1.86	1.82	1.76	1.74	1.69	1.67	1.64	1.61	1.59
	7.50	**5.34**	**4.46**	**3.97**	**3.66**	**3.42**	**3.25**	**3.12**	**3.01**	**2.94**	**2.86**	**2.80**	**2.70**	**2.62**	**2.51**	**2.42**	**2.34**	**2.25**	**2.20**	**2.12**	**2.08**	**2.02**	**1.98**	**1.96**
34	4.13	3.28	2.88	2.65	2.49	2.38	2.30	2.23	2.17	2.12	2.08	2.05	2.00	1.95	1.89	1.84	1.80	1.74	1.71	1.67	1.64	1.61	1.59	1.57
	7.44	**5.29**	**4.42**	**3.93**	**3.61**	**3.38**	**3.21**	**3.08**	**2.97**	**2.89**	**2.82**	**2.76**	**2.66**	**2.58**	**2.47**	**2.38**	**2.30**	**2.21**	**2.15**	**2.08**	**2.04**	**1.98**	**1.94**	**1.91**
36	4.11	3.26	2.86	2.63	2.48	2.36	2.28	2.21	2.15	2.10	2.06	2.03	1.89	1.93	1.87	1.82	1.78	1.72	1.69	1.65	1.62	1.59	1.56	1.55
	7.39	**5.25**	**4.38**	**3.89**	**3.58**	**3.35**	**3.18**	**3.04**	**2.94**	**2.86**	**2.78**	**2.72**	**2.62**	**2.54**	**2.43**	**2.35**	**2.26**	**2.17**	**2.12**	**2.04**	**2.00**	**1.94**	**1.90**	**1.87**
38	4.10	3.25	2.85	2.62	2.46	2.35	2.26	2.19	2.14	2.09	2.05	2.02	1.96	1.92	1.85	1.80	1.76	1.71	1.67	1.63	1.60	1.57	1.54	1.53
	7.35	**5.21**	**4.34**	**3.86**	**3.54**	**3.32**	**3.15**	**3.02**	**2.91**	**2.82**	**2.75**	**2.69**	**2.59**	**2.51**	**2.40**	**2.32**	**2.22**	**2.14**	**2.08**	**2.00**	**1.97**	**1.90**	**1.86**	**1.84**
40	4.08	3.23	2.84	2.61	2.45	2.34	2.25	2.18	2.12	2.07	2.04	2.00	1.95	1.90	1.84	1.79	1.74	1.69	1.66	1.61	1.59	1.55	1.53	1.51
	7.31	**5.18**	**4.31**	**3.83**	**3.51**	**3.29**	**3.12**	**2.99**	**2.88**	**2.80**	**2.73**	**2.66**	**2.56**	**2.49**	**2.37**	**2.29**	**2.20**	**2.11**	**2.05**	**1.97**	**1.94**	**1.88**	**1.84**	**1.81**
42	4.07	3.22	2.83	2.59	2.44	2.32	2.24	2.17	2.11	2.06	2.02	1.90	1.94	1.89	1.82	1.78	1.73	1.68	1.64	1.60	1.57	1.54	1.51	1.49
	7.27	**5.15**	**4.29**	**3.80**	**3.49**	**3.26**	**3.10**	**2.96**	**2.86**	**2.77**	**2.70**	**2.64**	**2.54**	**2.46**	**2.35**	**2.26**	**2.17**	**2.08**	**2.02**	**1.94**	**1.91**	**1.85**	**1.80**	**1.78**
44	4.06	3.21	2.82	2.58	2.43	2.31	2.23	2.16	2.10	2.05	2.01	1.98	1.92	1.88	1.81	1.76	1.72	1.66	1.63	1.58	1.56	1.52	1.50	1.48
	7.24	**5.12**	**4.26**	**3.78**	**3.46**	**3.24**	**3.07**	**2.94**	**2.84**	**2.75**	**2.68**	**2.62**	**2.52**	**2.44**	**2.32**	**2.24**	**2.15**	**2.06**	**2.09**	**1.92**	**1.88**	**1.82**	**1.78**	**1.75**
46	4.05	3.20	2.81	2.57	2.42	2.30	2.22	2.14	2.09	2.04	2.00	1.97	1.91	1.87	1.80	1.75	1.71	1.65	1.62	1.57	1.54	1.51	1.48	1.46
	7.21	**5.10**	**4.24**	**3.76**	**3.44**	**3.22**	**3.05**	**2.92**	**2.82**	**2.73**	**2.66**	**2.60**	**2.50**	**2.42**	**2.30**	**2.22**	**2.13**	**2.04**	**1.98**	**1.90**	**1.86**	**1.80**	**1.76**	**1.72**
48	4.04	3.19	2.80	2.56	2.41	2.30	2.21	2.14	2.08	2.03	1.99	1.96	1.90	1.86	1.79	1.74	1.70	1.64	1.61	1.56	1.53	1.50	1.47	1.45
	7.19	**5.08**	**4.22**	**3.74**	**3.42**	**3.20**	**3.04**	**2.90**	**2.80**	**2.71**	**2.64**	**2.58**	**2.48**	**2.40**	**2.28**	**2.20**	**2.11**	**2.02**	**1.96**	**1.88**	**1.84**	**1.78**	**1.73**	**1.70**
50	4.03	3.18	2.79	2.56	2.40	2.29	2.20	2.13	2.07	2.02	1.98	1.95	1.90	1.85	1.78	1.74	1.69	1.63	1.60	1.55	1.52	1.48	1.46	1.44
	7.17	**5.06**	**4.20**	**3.72**	**3.41**	**3.18**	**3.02**	**2.88**	**2.78**	**2.70**	**2.62**	**2.56**	**2.46**	**2.39**	**2.26**	**2.18**	**2.10**	**2.00**	**1.94**	**1.86**	**1.82**	**1.76**	**1.71**	**1.68**
55	4.02	3.17	2.78	2.54	2.38	2.27	2.18	2.11	2.05	2.00	1.97	1.93	1.88	1.83	1.76	1.72	1.67	1.61	1.58	1.52	1.50	1.46	1.43	1.41
	7.12	**5.01**	**4.16**	**3.68**	**3.37**	**3.15**	**2.98**	**2.85**	**2.75**	**2.66**	**2.59**	**2.53**	**2.43**	**2.35**	**2.23**	**2.15**	**2.06**	**1.96**	**1.90**	**1.82**	**1.78**	**1.71**	**1.66**	**1.64**
60	4.00	3.15	2.76	2.52	2.37	2.25	2.17	2.10	2.04	1.99	1.95	1.92	1.86	1.81	1.75	1.70	1.65	1.59	1.56	1.50	1.48	1.44	1.41	1.39
	7.08	**4.98**	**4.13**	**3.65**	**3.34**	**3.12**	**2.95**	**2.82**	**2.72**	**2.63**	**2.56**	**2.50**	**2.40**	**2.32**	**2.20**	**2.12**	**2.03**	**1.93**	**1.87**	**1.79**	**1.74**	**1.68**	**1.63**	**1.60**
65	3.99	3.14	2.75	2.51	2.36	2.24	2.15	2.08	2.02	1.98	1.94	1.90	1.85	1.80	1.73	1.68	1.63	1.57	1.54	1.49	1.46	1.42	1.39	1.37
	7.04	**4.95**	**4.10**	**3.62**	**3.31**	**3.09**	**2.93**	**2.79**	**2.70**	**2.61**	**2.54**	**2.47**	**2.37**	**2.30**	**2.18**	**2.09**	**2.00**	**1.90**	**1.84**	**1.76**	**1.71**	**1.64**	**1.60**	**1.56**
70	3.98	3.13	2.74	2.50	2.35	2.32	2.14	2.07	2.01	1.97	1.93	1.89	1.84	1.79	1.72	1.67	1.62	1.56	1.53	1.47	1.45	1.40	1.37	1.35
	7.01	**4.92**	**4.08**	**3.60**	**3.29**	**3.07**	**2.91**	**2.77**	**2.67**	**2.59**	**2.51**	**2.45**	**2.35**	**2.28**	**2.15**	**2.07**	**1.98**	**1.88**	**1.82**	**1.74**	**1.69**	**1.62**	**1.56**	**1.53**
80	3.96	3.11	2.72	2.48	2.33	2.21	2.12	2.05	1.99	1.95	1.91	1.88	1.82	1.77	1.70	1.65	1.60	1.54	1.51	1.45	1.42	1.38	1.35	1.32
	6.96	**4.88**	**4.04**	**3.56**	**3.25**	**3.04**	**2.87**	**2.74**	**2.64**	**2.55**	**2.48**	**2.41**	**2.32**	**2.24**	**2.11**	**2.03**	**1.94**	**1.84**	**1.78**	**1.70**	**1.65**	**1.57**	**1.52**	**1.49**

Degrees of freedom for lesser mean square																									
	100	3.94 **6.90**	3.09 **4.82**	2.70 **3.98**	2.46 **3.51**	2.30 **3.20**	2.19 **2.99**	2.10 **2.82**	2.03 **2.69**	1.97 **2.59**	1.92 **2.51**	1.88 **2.43**	1.85 **2.36**	1.79 **2.26**	1.75 **2.19**	1.68 **2.06**	1.63 **1.98**	1.57 **1.89**	1.51 **1.79**	1.48 **1.73**	1.42 **1.64**	1.39 **1.59**	1.34 **1.51**	1.30 **1.46**	1.28 **1.43**
	125	3.92 **6.84**	3.07 **4.78**	2.68 **3.94**	2.44 **3.47**	2.29 **3.17**	2.17 **2.95**	2.08 **2.79**	2.01 **2.65**	1.95 **2.56**	1.90 **2.47**	1.86 **2.40**	1.83 **2.33**	1.77 **2.23**	1.72 **2.15**	1.65 **2.03**	1.60 **1.94**	1.55 **1.85**	1.49 **1.75**	1.45 **1.68**	1.39 **1.59**	1.36 **1.54**	1.31 **1.46**	1.27 **1.40**	1.25 **1.37**
	150	3.91 **6.81**	3.06 **4.75**	2.67 **3.91**	2.43 **3.44**	2.27 **3.13**	2.16 **2.92**	2.07 **2.76**	2.00 **2.62**	1.94 **2.53**	1.89 **2.44**	1.85 **2.37**	1.82 **2.30**	1.76 **2.20**	1.71 **2.12**	1.64 **2.00**	1.59 **1.91**	1.54 **1.83**	1.47 **1.72**	1.44 **1.66**	1.37 **1.56**	1.34 **1.51**	1.29 **1.43**	1.25 **1.37**	1.22 **1.33**
	200	3.89 **6.76**	3.04 **4.71**	2.65 **3.38**	2.41 **3.41**	2.26 **3.11**	2.14 **2.90**	2.05 **2.73**	1.98 **2.60**	1.92 **2.50**	1.87 **2.41**	1.83 **2.34**	1.80 **2.28**	1.74 **1.17**	1.69 **2.09**	1.62 **1.97**	1.57 **1.88**	1.52 **1.79**	1.45 **1.69**	1.42 **1.62**	1.35 **1.53**	1.32 **1.48**	1.26 **1.39**	1.22 **1.33**	1.19 **1.28**
	400	3.86 **6.70**	3.02 **4.66**	2.62 **3.83**	2.39 **3.36**	2.23 **3.06**	2.12 **2.85**	2.03 **2.69**	1.96 **2.55**	1.90 **2.46**	1.85 **2.37**	1.81 **2.29**	1.78 **2.23**	1.72 **2.12**	1.67 **2.04**	1.60 **1.92**	1.54 **1.84**	1.49 **1.74**	1.42 **1.64**	1.38 **1.57**	1.32 **1.47**	1.28 **1.42**	1.22 **1.32**	1.16 **1.24**	1.13 **1.19**
	1000	3.85 **6.66**	3.00 **4.62**	2.61 **3.80**	2.38 **3.34**	2.22 **3.04**	2.10 **2.82**	2.02 **2.66**	1.95 **2.53**	1.89 **2.43**	1.84 **2.34**	1.80 **2.26**	1.76 **2.20**	1.70 **2.09**	1.65 **2.01**	1.58 **1.89**	1.53 **1.81**	1.47 **1.71**	1.41 **1.61**	1.36 **1.54**	1.30 **1.44**	1.26 **1.38**	1.19 **1.28**	1.13 **1.19**	1.08 **1.11**
	∞	3.84 **6.64**	2.99 **4.60**	2.60 **3.78**	2.37 **3.32**	2.21 **3.02**	2.09 **2.80**	2.01 **2.64**	1.94 **2.51**	1.88 **2.41**	1.83 **2.32**	1.79 **2.24**	1.75 **2.18**	1.69 **2.07**	1.64 **1.99**	1.57 **1.87**	1.52 **1.79**	1.46 **1.69**	1.40 **1.59**	1.35 **1.52**	1.28 **1.41**	1.24 **1.36**	1.17 **1.25**	1.11 **1.15**	1.00 **1.00**

Table D_1

Values of F exceeded by 0.025 of the values in the sampling distribution

If, in testing the homogeneity of two sample variances, the larger variance is placed over the smaller, the number of ratios greater than any given value, equal to or greater than unity, is doubled. Therefore use of the values tabulated below, in comparing two sample variances, will provide a 0.05 level of significance.

df for larger variance (numerator)

		4	5	6	7	8	9	10	12	15	20
	4	9.60	9.36	9.20	9.07	8.98	8.90	8.84	8.75	8.66	8.56
	5	7.39	7.15	6.98	6.85	6.76	6.68	6.62	6.52	6.43	6.33
df	6	6.23	5.99	5.82	5.70	5.60	5.52	5.46	5.37	5.27	5.17
for	7	5.52	5.29	5.12	4.99	4.90	4.82	4.76	4.67	4.57	4.47
smaller	8	5.05	4.82	4.65	4.53	4.43	4.36	4.30	4.20	4.10	4.00
variance	9	4.72	4.48	4.32	4.20	4.10	4.03	3.96	3.87	3.77	3.67
(denomi-	10	4.47	4.24	4.07	3.95	3.85	3.78	3.72	3.62	3.52	3.42
nator)	12	4.12	3.89	3.73	3.61	3.51	3.44	3.37	3.28	3.18	3.07
	15	3.80	3.58	3.41	3.29	3.20	3.12	3.06	2.96	2.86	2.76
	20	3.51	3.29	3.13	3.01	2.91	2.84	2.77	2.68	2.57	2.46

Interpolation may be performed using reciprocals of the degrees of freedom.

Table E

Critical values of *A*

For any given value of n − 1, the table shows the values of A corresponding to various levels of probability. A is significant at a given level if it is equal to or less than the value shown in the table.

	Level of significance for one-tailed test					
	.05	.025	.01	.005	.0005	
	Level of significance for two-tailed test					
n − 1*	.10	.05	.02	.01	.001	n − 1*
1	0.5125	0.5031	0.50049	0.50012	0.5000012	1
2	0.412	0.369	0.347	0.340	0.334	2
3	0.385	0.324	0.286	0.272	0.254	3
4	0.376	0.304	0.257	0.238	0.211	4
5	0.372	0.293	0.240	0.218	0.184	5
6	0.370	0.286	0.230	0.205	0.167	6
7	0.369	0.281	0.222	0.196	0.155	7
8	0.368	0.278	0.217	0.190	0.146	8
9	0.368	0.276	0.213	0.185	0.139	9
10	0.368	0.274	0.210	0.181	0.134	10
11	0.368	0.273	0.207	0.178	0.130	11
12	0.368	0.271	0.205	0.176	0.126	12
13	0.368	0.270	0.204	0.174	0.124	13
14	0.368	0.270	0.202	0.172	0.121	14
15	0.368	0.269	0.201	0.170	0.119	15
16	0.368	0.268	0.200	0.169	0.117	16
17	0.368	0.268	0.199	0.168	0.116	17
18	0.368	0.267	0.198	0.167	0.114	18
19	0.368	0.267	0.197	0.166	0.113	19
20	0.368	0.266	0.197	0.165	0.112	20
21	0.368	0.266	0.196	0.165	0.111	21
22	0.368	0.266	0.196	0.164	0.110	22
23	0.368	0.266	0.195	0.163	0.109	23
24	0.368	0.265	0.195	0.163	0.108	24
25	0.368	0.265	0.194	0.162	0.108	25
26	0.368	0.265	0.194	0.162	0.107	26
27	0.368	0.265	0.193	0.161	0.107	27
28	0.368	0.265	0.193	0.161	0.106	28
29	0.368	0.264	0.193	0.161	0.106	29
30	0.368	0.264	0.193	0.160	0.105	30
40	0.368	0.263	0.191	0.158	0.102	40
60	0.369	0.262	0.189	0.155	0.099	60
120	0.369	0.261	0.187	0.153	0.095	120
∞	0.370	0.260	0.185	0.151	0.092	∞

*n = number of pairs

Table F
Transformation of r to z_r

r	z_r	r	z_r	r	z_r
.01	.010	.34	.354	.67	.811
.02	.020	.35	.366	.68	.829
.03	.030	.36	.377	.69	.848
.04	.040	.37	.389	.70	.867
.05	.050	.38	.400	.71	.887
.06	.060	.39	.412	.72	.908
.07	.070	.40	.424	.73	.929
.08	.080	.41	.436	.74	.950
.09	.090	.42	.448	.75	.973
.10	.100	.43	.460	.76	.996
.11	.110	.44	.472	.77	1.020
.12	.121	.45	.485	.78	1.045
.13	.131	.46	.497	.79	1.071
.14	.141	.47	.510	.80	1.099
.15	.151	.48	.523	.81	1.127
.16	.161	.49	.536	.82	1.157
.17	.172	.50	.549	.83	1.188
.18	.181	.51	.563	.84	1.221
.19	.192	.52	.577	.85	1.256
.20	.203	.53	.590	.86	1.293
.21	.214	.54	.604	.87	1.333
.22	.224	.55	.618	.88	1.376
.23	.234	.56	.633	.89	1.422
.24	.245	.57	.648	.90	1.472
.25	.256	.58	.663	.91	1.528
.26	.266	.59	.678	.92	1.589
.27	.277	.60	.693	.93	1.658
.28	.288	.61	.709	.94	1.738
.29	.299	.62	.725	.95	1.832
.30	.309	.63	.741	.96	1.946
.31	.321	.64	.758	.97	2.092
.32	.332	.65	.775	.98	2.298
.33	.343	.66	.793	.99	2.647

Table G

Critical values of r_{rho} (Rank-Order Correlation Coefficient)

	Level of significance for one-tailed test			
	.05	.025	.01	.005
	Level of significance for two-tailed test			
n*	.10	.05	.02	.01
5	.900	1.000	1.000	--
6	.829	.886	.943	1.000
7	.714	.786	.893	.929
8	.643	.738	.833	.881
9	.600	.683	.783	.833
10	.564	.648	.746	.794
12	.506	.591	.712	.777
14	.456	.544	.645	.715
16	.425	.506	.601	.665
18	.399	.475	.564	.625
20	.377	.450	.534	.591
22	.359	.428	.508	.562
24	.343	.409	.485	.537
26	.329	.392	.465	.515
28	.317	.377	.448	.496
30	.306	.364	.432	.478

*n = number of pairs

Table H

Functions of *r*

r	$\sqrt{r}$	r^2	$\sqrt{r - r^2}$	$\sqrt{1 - r}$	$1 - r^2$	$\sqrt{1 - r^2}$	100(1 − k)	r
						k	% Eff.	
1.00	1.0000	1.0000	0.0000	0.0000	0.0000	0.0000	100.00	1.00
.99	.9950	.9801	.0995	.1000	.0199	.1411	85.89	.99
.98	.9899	.9604	.1400	.1414	.0396	.1990	80.10	.98
.97	.9849	.9409	.1706	.1732	.0591	.2431	75.69	.97
.96	.9798	.9216	.1960	.2000	.0784	.2800	72.00	.96
.95	.9747	.9025	.2179	.2236	.0975	.3122	68.78	.95
.94	.9695	.8836	.2375	.2449	.1164	.3412	65.88	.94
.93	.9644	.8649	.2551	.2646	.1351	.3676	63.24	.93
.92	.9592	.8464	.2713	.2828	.1536	.3919	60.81	.92
.91	.9539	.8281	.2862	.3000	.1719	.4146	58.54	.91
.90	.9487	.8100	.3000	.3162	.1900	.4359	56.41	.90
.89	.9434	.7921	.3129	.3317	.2079	.4560	54.40	.89
.88	.9381	.7744	.3250	.3464	.2256	.4750	52.50	.88
.87	.9327	.7569	.3363	.3606	.2431	.4931	50.69	.87
.86	.9274	.7396	.3470	.3742	.2604	.5103	48.97	.86
.85	.9220	.7225	.3571	.3873	.2775	.5268	47.32	.85
.84	.9165	.7056	.3666	.4000	.2944	.5426	45.74	.84
.83	.9110	.6889	.3756	.4123	.3111	.5578	44.22	.83
.82	.9055	.6724	.3842	.4243	.3276	.5724	42.76	.82
.81	.9000	.6561	.3923	.4359	.3439	.5864	41.36	.81
.80	.8944	.6400	.4000	.4472	.3600	.6000	40.00	.80
.79	.8888	.6241	.4073	.4583	.3759	.6131	38.69	.79
.78	.8832	.6084	.4142	.4690	.3916	.6258	37.42	.78
.77	.8775	.5929	.4208	.4796	.4071	.6380	36.20	.77
.76	.8718	.5776	.4271	.4899	.4224	.6499	35.01	.76
.75	.8660	.5625	.4330	.5000	.4375	.6614	33.86	.75
.74	.8602	.5476	.4386	.5099	.4524	.6726	32.74	.74
.73	.8544	.5329	.4440	.5196	.4671	.6834	31.66	.73
.72	.8485	.5184	.4490	.5292	.4816	.6940	30.60	.72
.71	.8426	.5041	.4538	.5385	.4959	.7042	29.58	.71
.70	.8367	.4900	.4583	.5477	.5100	.7141	28.59	.70
.69	.8307	.4761	.4625	.5568	.5239	.7238	27.62	.69
.68	.8246	.4624	.4665	.5657	.5376	.7332	26.68	.68
.67	.8185	.4489	.4702	.5745	.5511	.7424	25.76	.67
.66	.8124	.4356	.4737	.5831	.5644	.7513	24.87	.66
.65	.8062	.4225	.4770	.5916	.5775	.7599	24.01	.65
.64	.8000	.4096	.4800	.6000	.5904	.7684	23.16	.64
.63	.7937	.3969	.4828	.6083	.6031	.7766	22.34	.63
.62	.7874	.3844	.4854	.6164	.6156	.7846	21.54	.62
.61	.7810	.3721	.4877	.6245	.6279	.7924	20.76	.61
.60	.7746	.3600	.4899	.6325	.6400	.8000	20.00	.60
.59	.7681	.3481	.4918	.6403	.6519	.8074	19.26	.59
.58	.7616	.3364	.4936	.6481	.6636	.8146	18.54	.58
.57	.7550	.3249	.4951	.6557	.6751	.8216	17.84	.57
.56	.7483	.3136	.4964	.6633	.6864	.8285	17.15	.56
.55	.7416	.3025	.4975	.6708	.6975	.8352	16.48	.55
.54	.7348	.2916	.4984	.6782	.7084	.8417	15.83	.54
.53	.7280	.2809	.4991	.6856	.7191	.8480	15.20	.53
.52	.7211	.2704	.4996	.6928	.7296	.8542	14.58	.52
.51	.7141	.2601	.4999	.7000	.7399	.8602	13.98	.51
.50	.7071	.2500	.5000	.7071	.7500	.8660	13.40	.50

Table H. (*continued*)

r	$\sqrt{r}$	r^2	$\sqrt{r-r^2}$	$\sqrt{1-r}$	$1-r^2$	$\sqrt{1-r^2}$	$100(1-k)$	r
						k	% Eff.	
.50	.7071	.2500	.5000	.7071	.7500	.8660	13.40	.50
.49	.7000	.2401	.4999	.7141	.7599	.8717	12.83	.49
.48	.6928	.2304	.4996	.7211	.7696	.8773	12.27	.48
.47	.6856	.2209	.4991	.7280	.7791	.8827	11.73	.47
.46	.6782	.2116	.4984	.7348	.7884	.8879	11.21	.46
.45	.6708	.2025	.4975	.7416	.7975	.8930	10.70	.45
.44	.6633	.1936	.4964	.7483	.8064	.8980	10.20	.44
.43	.6557	.1849	.4951	.7550	.8151	.9028	9.72	.43
.42	.6481	.1764	.4936	.7616	.8236	.9075	9.25	.42
.41	.6403	.1681	.4918	.7681	.8319	.9121	8.79	.41
.40	.6325	.1600	.4899	.7746	.8400	.9165	8.35	.40
.39	.6245	.1521	.4877	.7810	.8479	.9208	7.92	.39
.38	.6164	.1444	.4854	.7874	.8556	.9250	7.50	.38
.37	.6083	.1369	.4828	.7937	.8631	.9290	7.10	.37
.36	.6000	.1296	.4800	.8000	.8704	.9330	6.70	.36
.35	.5916	.1225	.4770	.8062	.8775	.9367	6.33	.35
.34	.5831	.1156	.4737	.8124	.8844	.9404	5.96	.34
.33	.5745	.1089	.4702	.8185	.8911	.9440	5.60	.33
.32	.5657	.1024	.4665	.8246	.8976	.9474	5.25	.32
.31	.5568	.0961	.4625	.8307	.9039	.9507	4.93	.31
.30	.5477	.0900	.4583	.8367	.9100	.9539	4.61	.30
.29	.5385	.0841	.4538	.8426	.9159	.9570	4.30	.29
.28	.5292	.0784	.4490	.8485	.9216	.9600	4.00	.28
.27	.5196	.0729	.4440	.8544	.9271	.9629	3.71	.27
.26	.5099	.0676	.4386	.8602	.9324	.9656	3.44	.26
.25	.5000	.0625	.4330	.8660	.9375	.9682	3.18	.25
.24	.4899	.0576	.4271	.8718	.9424	.9708	2.92	.24
.23	.4796	.0529	.4208	.8775	.9471	.9732	2.68	.23
.22	.4690	.0484	.4142	.8832	.9516	.9755	2.45	.22
.21	.4583	.0441	.4073	.8888	.9559	.9777	2.23	.21
.20	.4472	.0400	.4000	.8944	.9600	.9798	2.02	.20
.19	.4359	.0361	.3923	.9000	.9639	.9818	1.82	.19
.18	.4243	.0324	.3842	.9055	.9676	.9837	1.63	.18
.17	.4123	.0289	.3756	.9110	.9711	.9854	1.46	.17
.16	.4000	.0256	.3666	.9165	.9744	.9871	1.29	.16
.15	.3873	.0225	.3571	.9220	.9775	.9887	1.13	.15
.14	.3742	.0196	.3470	.9274	.9804	.9902	.98	.14
.13	.3606	.0169	.3363	.9327	.9831	.9915	.85	.13
.12	.3464	.0144	.3250	.9381	.9856	.9928	.72	.12
.11	.3317	.0121	.3129	.9434	.9879	.9939	.61	.11
.10	.3162	.0100	.3000	.9487	.9900	.9950	.50	.10
.09	.3000	.0081	.2862	.9539	.9919	.9959	.41	.09
.08	.2828	.0064	.2713	.9592	.9936	.9968	.32	.08
.07	.2646	.0049	.2551	.9644	.9951	.9975	.25	.07
.06	.2449	.0036	.2375	.9695	.9964	.9982	.18	.06
.05	.2236	.0025	.2179	.9747	.9975	.9987	.13	.05
.04	.2000	.0016	.1960	.9798	.9984	.9992	.08	.04
.03	.1732	.0009	.1706	.9849	.9991	.9995	.05	.03
.02	.1414	.0004	.1400	.9899	.9996	.9998	.02	.02
.01	.1000	.0001	.0995	.9950	.9999	.9999	.01	.01
.00	.0000	.0000	.0000	1.0000	1.0000	1.0000	.00	.00

Table I₁

Critical values of U and U' for a one-tailed test at $\alpha = 0.005$ or a two-tailed test at $\alpha = 0.01$

To be significant for any given n_1 and n_2: Obtained U must be equal to or less than the value shown in the table. Obtained U′ must be equal to or greater than the value shown in the table.

n_2 \ n_1	1	2	3	4	5	6	7	8	9	10	11	12	13	14	15	16	17	18	19	20
1	--	--	--	--	--	--	--	--	--	--	--	--	--	--	--	--	--	--	--	--
2	--	--	--	--	--	--	--	--	--	--	--	--	--	--	--	--	--	--	0 38	0 40
3	--	--	--	--	--	--	--	--	0 27	0 30	0 33	1 35	1 38	1 41	2 43	2 46	2 49	2 52	3 54	3 57
4	--	--	--	--	--	0 24	0 28	1 31	1 35	2 38	2 42	3 45	3 49	4 52	5 55	5 59	6 62	6 66	7 69	8 72
5	--	--	--	--	0 25	1 29	1 34	2 38	3 42	4 46	5 50	6 54	7 58	7 63	8 67	9 71	10 75	11 79	12 83	13 87
6	--	--	--	0 24	1 29	2 34	3 39	4 44	5 49	6 54	7 59	9 63	10 68	11 73	12 78	13 83	15 87	16 92	17 97	18 102
7	--	--	--	0 28	1 34	3 39	4 45	6 50	7 56	9 61	10 67	12 72	13 78	15 83	16 89	18 94	19 100	21 105	22 111	24 116
8	--	--	--	1 31	2 38	4 44	6 50	7 57	9 63	11 69	13 75	15 81	17 87	18 94	20 100	22 106	24 112	26 118	28 124	30 130
9	--	--	0 27	1 35	3 42	5 49	7 56	9 63	11 70	13 77	16 83	18 90	20 97	22 104	24 111	27 117	29 124	31 131	33 138	36 144
10	--	--	0 30	2 38	4 46	6 54	9 61	11 69	13 77	16 84	18 92	21 99	24 106	26 114	29 121	31 129	34 136	37 143	39 151	42 158
11	--	--	0 33	2 42	5 50	7 59	10 67	13 75	16 83	18 92	21 100	24 108	27 116	30 124	33 132	36 140	39 148	42 156	45 164	48 172
12	--	--	1 35	3 45	6 54	9 63	12 72	15 81	18 90	21 99	24 108	27 117	31 125	34 134	37 143	41 151	44 160	47 169	51 177	54 186
13	--	--	1 38	3 49	7 58	10 68	13 78	17 87	20 97	24 106	27 116	31 125	34 125	38 144	42 153	45 163	49 172	53 181	56 191	60 200
14	--	--	1 41	4 52	7 63	11 73	15 83	18 94	22 104	26 114	30 124	34 134	38 144	42 154	46 164	50 174	54 184	58 194	63 203	67 213
15	--	--	2 43	5 55	8 67	12 78	16 89	20 100	24 111	29 121	33 132	37 143	42 153	46 164	51 174	55 185	60 195	64 206	69 216	73 227
16	--	--	2 46	5 59	9 71	13 83	18 94	22 106	27 117	31 129	36 140	41 151	45 163	50 174	55 185	60 196	65 207	70 218	74 230	79 241
17	--	--	2 49	6 62	10 75	15 87	19 100	24 112	29 124	34 148	39 148	44 160	49 172	54 184	60 195	65 207	70 219	75 231	81 242	86 254
18	--	--	2 52	6 66	11 79	16 92	21 105	26 118	31 131	37 143	42 156	47 169	53 181	58 194	64 206	70 218	75 231	81 243	87 255	92 268
19	--	0 38	3 54	7 69	12 83	17 97	22 111	28 124	33 138	39 151	45 164	51 177	56 191	63 203	69 216	74 230	81 242	87 255	93 268	99 281
20	--	0 40	3 57	8 72	13 87	18 102	24 116	30 130	36 144	42 158	48 172	54 186	60 200	67 213	73 227	79 241	86 254	92 268	99 281	105 295

(Dashes in the body of the table indicate that no decision is possible at the stated level of significance.)

Table I_2

Critical values of U and U' for a one-tailed test at $\alpha = 0.01$ or a two-tailed test at $\alpha = 0.02$

To be significant for any given n_1 and n_2: Obtained U must be equal to or less than the value shown in the table. Obtained U′ must be equal to or greater than the value shown in the table.

n_2 \ n_1	1	2	3	4	5	6	7	8	9	10	11	12	13	14	15	16	17	18	19	20
1	--	--	--	--	--	--	--	--	--	--	--	--	--	--	--	--	--	--	--	--
2	--	--	--	--	--	--	--	--	--	--	--	--	0 26	0 28	0 30	0 32	0 34	0 36	1 37	1 39
3	--	--	--	--	--	--	0 21	0 24	1 26	1 29	1 32	2 34	2 37	2 40	3 42	3 45	4 47	4 50	4 52	5 55
4	--	--	--	--	0 20	1 23	1 27	2 30	3 33	3 37	4 40	5 43	5 47	6 50	7 53	7 57	8 60	9 63	9 67	10 70
5	--	--	--	0 20	1 24	2 28	3 32	4 36	5 40	6 44	7 48	8 52	9 56	10 60	11 64	12 68	13 72	14 76	15 80	16 84
6	--	--	--	1 23	2 28	3 33	4 38	6 42	7 47	8 52	9 57	11 61	12 66	13 71	15 75	16 80	18 84	19 89	20 94	22 98
7	--	--	0 21	1 27	3 32	4 38	6 43	7 49	9 54	11 59	12 65	14 70	16 75	17 81	19 86	21 91	23 96	24 102	26 107	28 112
8	--	--	0 24	2 30	4 36	6 42	7 49	9 55	11 61	13 67	15 73	17 79	20 84	22 90	24 96	26 102	28 108	30 114	32 120	34 126
9	--	--	1 26	3 33	5 40	7 47	9 54	11 61	14 67	16 74	18 81	21 87	23 94	26 100	28 107	31 113	33 120	36 126	38 133	40 140
10	--	--	1 29	3 37	6 44	8 52	11 59	13 67	16 74	19 81	22 88	24 96	27 103	30 110	33 117	36 124	38 132	41 139	44 146	47 153
11	--	--	1 32	4 40	7 48	9 57	12 65	15 73	18 81	22 88	25 96	28 104	31 112	34 120	37 128	41 135	44 143	47 151	50 159	53 167
12	--	--	2 34	5 43	8 52	11 61	14 70	17 79	21 87	24 96	28 104	31 113	35 121	38 130	42 138	46 146	49 155	53 163	56 172	60 180
13	--	0 26	2 37	5 47	9 56	12 66	16 75	20 84	23 94	27 103	31 112	35 121	39 130	43 139	47 148	51 157	55 166	59 175	63 184	67 193
14	--	0 28	2 40	6 50	10 60	13 71	17 81	22 90	26 100	30 110	34 120	38 130	43 139	47 149	51 159	56 168	60 178	65 187	69 197	73 207
15	--	0 30	3 42	7 53	11 64	15 75	19 86	24 96	28 107	33 117	37 128	42 138	47 148	51 159	56 169	61 179	66 189	70 200	75 210	80 220
16	--	0 32	3 45	7 57	12 68	16 80	21 91	26 102	31 113	36 124	41 135	46 146	51 157	56 168	61 179	66 190	71 201	76 212	82 222	87 233
17	--	0 34	4 47	8 60	13 72	18 84	23 96	28 108	33 120	38 132	44 143	49 155	55 166	60 178	66 189	71 201	77 212	82 224	88 234	93 247
18	--	0 36	4 50	9 63	14 76	19 89	24 102	30 114	36 126	41 139	47 151	53 163	59 175	65 187	70 200	76 212	82 224	88 236	94 248	100 260
19	--	1 37	4 53	9 67	15 80	20 94	26 107	32 120	38 133	44 146	50 159	56 172	63 184	69 197	75 210	82 222	88 235	94 248	101 260	107 273
20	--	1 39	5 55	10 70	16 84	22 98	28 112	34 126	40 140	47 153	53 167	60 180	67 193	73 207	80 220	87 233	93 247	100 260	107 273	114 286

(Dashes in the body of the table indicate that no decision is possible at the stated level of significance.)

Table I_3

Critical values of U and U' for a one-tailed test at $\alpha = 0.025$ or a two-tailed test at $\alpha = 0.05$

To be significant for any given n_1 and n_2: Obtained U must be equal to or <u>less than</u> the value shown in the table. Obtained U′ must be equal to or <u>greater than</u> the value shown in the table.

n_2 \ n_1	1	2	3	4	5	6	7	8	9	10	11	12	13	14	15	16	17	18	19	20
1	--	--	--	--	--	--	--	--	--	--	--	--	--	--	--	--	--	--	--	--
2	--	--	--	--	--	--	--	0 16	0 18	0 20	0 22	1 23	1 25	1 27	1 29	1 31	2 32	2 34	2 36	2 38
3	--	--	--	--	0 15	1 17	1 20	2 22	2 25	3 27	3 30	4 32	4 35	5 37	5 40	6 42	6 45	7 47	7 50	8 52
4	--	--	--	0 16	1 19	2 22	3 25	4 28	4 32	5 35	6 38	7 41	8 44	9 47	10 50	11 53	11 57	12 60	13 63	13 67
5	--	--	0 15	1 19	2 23	3 27	5 30	6 34	7 38	8 42	9 46	11 49	12 53	13 57	14 61	15 65	17 68	18 72	19 76	20 80
6	--	--	1 17	2 22	3 27	5 31	6 36	8 40	10 44	11 49	13 53	14 58	16 62	17 67	19 71	21 75	22 80	24 84	25 89	27 93
7	--	--	1 20	3 25	5 30	6 36	8 41	10 46	12 51	14 56	16 61	18 66	20 71	22 76	24 81	26 86	28 91	30 96	32 101	34 106
8	--	0 16	2 22	4 28	6 34	8 40	10 46	13 51	15 57	17 63	19 69	22 74	24 80	26 86	29 91	31 97	34 102	36 108	38 111	41 119
9	--	0 18	2 25	4 32	7 38	10 44	12 51	15 57	17 64	20 70	23 76	26 82	28 89	31 95	34 101	37 107	39 114	42 120	45 126	48 132
10	--	0 20	3 27	5 35	8 42	11 49	14 56	17 63	20 70	23 77	26 84	29 91	33 97	36 104	39 111	42 118	45 125	48 132	52 138	55 145
11	--	0 22	3 30	6 38	9 46	13 53	16 61	19 69	23 76	26 84	30 91	33 99	37 106	40 114	44 121	47 129	51 136	55 143	58 151	62 158
12	--	1 23	4 32	7 41	11 49	14 58	18 66	22 74	26 82	29 91	33 99	37 107	41 115	45 123	49 131	53 139	57 147	61 155	65 163	69 171
13	--	1 25	4 35	8 44	12 53	16 62	20 71	24 80	28 89	33 97	37 106	41 115	45 124	50 132	54 141	59 149	63 158	67 167	72 175	76 184
14	--	1 27	5 37	9 47	13 51	17 67	22 76	26 86	31 95	36 104	40 114	45 123	50 132	55 141	59 151	64 160	67 171	74 178	78 188	83 197
15	--	1 29	5 40	10 50	14 61	19 71	24 81	29 91	34 101	39 111	44 121	49 131	54 141	59 151	64 161	70 170	75 180	80 190	85 200	90 210
16	--	1 31	6 42	11 53	15 65	21 75	26 86	31 97	37 107	42 118	47 129	53 139	59 149	64 160	70 170	75 181	81 191	86 202	92 212	98 222
17	--	2 32	6 45	11 57	17 68	22 80	28 91	34 102	39 114	45 125	51 136	57 147	63 158	67 171	75 180	81 191	87 202	93 213	99 224	105 235
18	--	2 34	7 47	12 60	18 72	24 84	30 96	36 108	42 120	48 132	55 143	61 155	67 167	74 178	80 190	86 202	93 213	99 225	106 236	112 248
19	--	2 36	7 50	13 63	19 76	25 89	32 101	38 114	45 126	52 138	58 151	65 163	72 175	78 188	85 200	92 212	99 224	106 236	113 248	119 261
20	--	2 38	8 52	13 67	20 80	27 93	34 106	41 119	48 132	55 145	62 158	69 171	76 184	83 197	90 210	98 222	105 235	112 248	119 261	127 273

(Dashes in the body of the table indicate that no decision is possible at the stated level of significance.)

Table I₄

Critical values of *U* and *U′* for a one-tailed test at $\alpha = 0.05$ or a two-tailed test at $\alpha = 0.10$

To be significant for any given n_1 and n_2: Obtained U must be equal to or less than the value shown in the table. Obtained U′ must be equal to or greater than the value shown in the table.

n_2 \ n_1	1	2	3	4	5	6	7	8	9	10	11	12	13	14	15	16	17	18	19	20
1	--	--	--	--	--	--	--	--	--	--	--	--	--	--	--	--	--	--	0 19	0 20
2	--	--	--	--	0 10	0 12	0 14	1 15	1 17	1 19	1 21	2 22	2 24	2 26	3 27	3 29	3 31	4 32	4 34	4 36
3	--	--	0 9	0 12	1 14	2 16	2 19	3 21	3 24	4 26	5 28	5 31	6 33	7 35	7 38	8 40	9 42	9 45	10 47	11 49
4	--	--	0 12	1 15	2 18	3 21	4 24	5 27	6 30	7 33	8 36	9 39	10 42	11 45	12 48	14 50	15 53	16 56	17 59	18 62
5	--	0 10	1 14	2 18	4 21	5 25	6 29	8 32	9 36	11 39	12 43	13 47	15 50	16 54	18 57	19 61	20 65	22 68	23 72	25 75
6	--	0 12	2 16	3 21	5 25	7 29	8 34	10 38	12 42	14 46	16 50	17 55	19 59	21 63	23 67	25 71	26 76	28 80	30 84	32 88
7	--	0 14	2 19	4 24	6 29	8 34	11 38	13 43	15 48	17 53	19 58	21 63	24 67	26 72	28 77	30 82	33 86	35 91	37 96	39 101
8	--	1 15	3 21	5 27	8 32	10 38	13 43	15 49	18 54	20 60	23 65	26 70	28 76	31 81	33 87	36 92	39 97	41 103	44 108	47 113
9	--	1 17	3 24	6 30	9 36	12 42	15 48	18 54	21 60	24 66	27 72	30 78	33 84	36 90	39 96	42 102	45 108	48 114	51 120	54 126
10	--	1 19	4 26	7 33	11 39	14 46	17 53	20 60	24 66	27 73	31 79	34 86	37 93	41 99	44 106	48 112	51 119	55 125	58 132	62 138
11	--	1 21	5 28	8 36	12 43	16 50	19 58	23 65	27 72	31 79	34 87	38 94	42 101	46 108	50 115	54 122	57 130	61 137	65 144	69 151
12	--	2 22	5 31	9 39	13 47	17 55	21 63	26 70	30 78	34 86	38 94	42 102	47 109	51 117	55 125	60 132	64 140	68 148	72 156	77 163
13	--	2 24	6 33	10 42	15 50	19 59	24 67	28 76	33 84	37 93	42 101	47 109	51 118	56 126	61 134	65 143	70 151	75 159	80 167	84 176
14	--	2 26	7 35	11 45	16 54	21 63	26 72	31 81	36 90	41 99	46 108	51 117	56 126	61 135	66 144	71 153	77 161	82 170	87 179	92 188
15	--	3 27	7 38	12 48	18 57	23 67	28 77	33 87	39 96	44 106	50 115	55 125	61 134	66 144	72 153	77 163	83 172	88 182	94 191	100 200
16	--	3 29	8 40	14 50	19 61	25 71	30 82	36 92	42 102	48 112	54 122	60 132	65 143	71 153	77 163	83 173	89 183	95 193	101 203	107 213
17	--	3 31	9 42	15 53	20 65	26 76	33 86	39 97	45 108	51 119	57 130	64 140	70 151	77 161	83 172	89 183	96 193	102 204	109 214	115 225
18	--	4 32	9 45	16 56	22 68	28 80	35 91	41 103	48 114	55 123	61 137	68 148	75 159	82 170	88 182	95 193	102 204	109 215	116 226	123 237
19	0 19	4 34	10 47	17 59	23 72	30 84	37 96	44 108	51 120	58 132	65 144	72 156	80 167	87 179	94 191	101 203	109 214	116 226	123 238	130 250
20	0 20	4 36	11 49	18 62	25 75	32 88	39 101	47 113	54 126	62 138	69 151	77 163	84 176	92 188	100 200	107 213	115 225	123 237	130 250	138 262

(Dashes in the body of the table indicate that no decision is possible at the stated level of significance.)

Table J

Critical values of *T* at various levels of probability

The symbol T denotes the smaller sum of ranks associated with differences that are all of the same sign. For any given N (number of ranked differences), the obtained T is significant at a given level if it is equal to or less than the value shown in the table.

	Level of significance for one-tailed test			
	.05	.025	.01	.005
	Level of significance for two-tailed test			
N	.10	.05	.02	.01
5	0	--	--	--
6	2	0	--	--
7	3	2	0	--
8	5	3	1	0
9	8	5	3	1
10	10	8	5	3
11	13	10	7	5
12	17	13	9	7
13	21	17	12	9
14	25	21	15	12
15	30	25	19	15
16	35	29	23	19
17	41	34	27	23
18	47	40	32	27
19	53	46	37	32
20	60	52	43	37
21	67	58	49	42
22	75	65	55	48
23	83	73	62	54
24	91	81	69	61
25	100	89	76	68
26	110	98	84	75
27	119	107	92	83

	Level of significance for one-tailed test			
	.05	.025	.01	.005
	Level of significance for two-tailed test			
N	.10	.05	.02	.01
28	130	116	101	91
29	140	126	110	100
30	151	137	120	109
31	163	147	130	118
32	175	159	140	128
33	187	170	151	138
34	200	182	162	148
35	213	195	173	159
36	227	208	185	171
37	241	221	198	182
38	256	235	211	194
39	271	249	224	207
40	286	264	238	220
41	302	279	252	233
42	319	294	266	247
43	336	310	281	261
44	353	327	296	276
45	371	343	312	291
46	389	361	328	307
47	407	378	345	322
48	426	396	362	339
49	446	415	379	355
50	466	434	397	373

(Slight discrepancies will be found between the critical values appearing in the table above and in Table 2 of the 1964 revision of F. Wilcoxon, and R.A. Wilcox, Some Rapid Approximate Statistical Procedures, New York, Lederle Laboratories, 1964. The disparity reflects the latter's policy of selecting the critical value nearest a given significance level, occasionally overstepping that level. For example, for N = 8,

the probability of a T of 3 = 0.0390 (two-tail)

and

the probability of a T of 4 = 0.0546 (two-tail).

Wilcoxon and Wilcox select a T of 4 as the critical value at the 0.05 level of significance (two-tail), whereas Table J reflects a more conservative policy by setting a T of 3 as the critical value at this level.)

Table K

Factorials of numbers 1 to 20

N	N!
0	1
1	1
2	2
3	6
4	24
5	120
6	720
7	5040
8	40320
9	362880
10	3628800
11	39916800
12	479001600
13	6227020800
14	87178291200
15	1307674368000
16	20922789888000
17	355687428096000
18	6402373705728000
19	121645100408832000
20	2432902008176640000

Table L

Binomial coefficients

N	$\binom{N}{0}$	$\binom{N}{1}$	$\binom{N}{2}$	$\binom{N}{3}$	$\binom{N}{4}$	$\binom{N}{5}$	$\binom{N}{6}$	$\binom{N}{7}$	$\binom{N}{8}$	$\binom{N}{9}$	$\binom{N}{10}$
0	1										
1	1	1									
2	1	2	1								
3	1	3	3	1							
4	1	4	6	4	1						
5	1	5	10	10	5	1					
6	1	6	15	20	15	6	1				
7	1	7	21	35	35	21	7	1			
8	1	8	28	56	70	56	28	8	1		
9	1	9	36	84	126	126	84	36	9	1	
10	1	10	45	120	210	252	210	120	45	10	1
11	1	11	55	165	330	462	462	330	165	55	11
12	1	12	66	220	495	792	924	792	495	220	66
13	1	13	78	286	715	1287	1716	1716	1287	715	286
14	1	14	91	364	1001	2002	3003	3432	3003	2002	1001
15	1	15	105	455	1365	3003	5005	6435	6435	5005	3003
16	1	16	120	560	1820	4368	8008	11440	12870	11440	8008
17	1	17	136	680	2380	6188	12376	19448	24310	24310	19448
18	1	18	153	816	3060	8568	18564	31824	43758	48620	43758
19	1	19	171	969	3876	11628	27132	50388	75582	92378	92378
20	1	20	190	1140	4845	15504	38760	77520	125970	167960	184756

Table M

Cumulative probabilities associated with values as small as observed values of *x* in the binomial test (*x* is the smaller of the observed frequencies)

Given in the body of this table are one-tailed probabilities under H_0 for the binomial test when $P = Q = 1/2$. To save space, decimal points are omitted in the p's. (x = smaller of the observed frequencies)

N \ x	0	1	2	3	4	5	6	7	8	9	10	11	12	13	14	15
5	031	188	500	812	969	†										
6	016	109	344	656	891	984	†									
7	008	062	227	500	773	938	992	†								
8	004	035	145	363	637	855	965	996	†							
9	002	020	090	254	500	746	910	980	998	†						
10	001	011	055	172	377	623	828	945	989	999	†					
11		006	033	113	274	500	726	887	967	994	†	†				
12		003	019	073	194	387	613	806	927	981	997	†	†			
13		002	011	046	133	291	500	709	867	954	989	998	†	†		
14		001	006	029	090	212	395	605	788	910	971	994	999	†	†	
15			004	018	059	151	304	500	696	849	941	982	996	†	†	†
16			002	011	038	105	227	402	598	773	895	962	989	998	†	†
17			001	006	025	072	166	315	500	685	834	928	975	994	999	†
18			001	004	015	048	119	240	407	593	760	881	952	985	996	999
19				002	010	032	084	180	324	500	676	820	916	968	990	998
20				001	006	021	058	132	252	412	588	748	868	942	979	994
21				001	004	013	039	095	192	332	500	668	808	905	961	987
22					002	008	026	067	143	262	416	584	738	857	933	974
23					001	005	017	047	105	202	339	500	661	798	895	953
24					001	003	011	032	076	154	271	419	581	729	846	924
25						002	007	022	054	115	212	345	500	655	788	885

†1.0 or approximately 1.0.

Table N

Critical values of x at the 0.05 and 0.01 levels of significance for various values of P and Q when $N \leq 20$

x = frequency in the category with P probability of occurrence. The obtained x must be equal to or *greater than* the value shown in the table for significance at the chosen level.

N	P .01 Q **.99**	.05 **.95**	.10 **.90**	.20 **.80**	.25 **.75**	.30 **.70**	1/3 **2/3**	.40 **.60**
2	1 **2**	2 **2**	2 **2**	2 —	— —	— —	— —	— —
3	1 **2**	2 **2**	2 **3**	3 **3**	3 —	3 —	3 —	— —
4	1 **2**	2 **3**	2 **3**	3 **4**	4 **4**	4 **4**	4 —	4 —
5	1 **2**	2 **3**	3 **3**	4 **4**	4 **5**	4 **5**	4 **5**	5 —
6	2 **2**	2 **3**	3 **4**	4 **5**	4 **5**	5 **6**	5 **6**	5 **6**
7	2 **2**	2 **3**	3 **4**	4 **5**	5 **6**	5 **6**	5 **6**	6 **7**
8	2 **2**	3 **3**	3 **4**	5 **6**	5 **6**	6 **7**	6 **7**	6 **7**
9	2 **2**	3 **3**	4 **4**	5 **6**	5 **6**	6 **7**	6 **7**	7 **8**
10	2 **2**	3 **4**	4 **5**	5 **6**	6 **7**	6 **8**	7 **8**	8 **9**
11	2 **2**	3 **4**	4 **5**	6 **7**	6 **7**	7 **8**	7 **8**	8 **9**
12	2 **2**	3 **4**	4 **5**	6 **7**	7 **8**	7 **8**	8 **9**	9 **10**
13	2 **2**	3 **4**	4 **5**	6 **7**	7 **8**	8 **9**	8 **9**	9 **10**
14	2 **2**	3 **4**	5 **6**	7 **8**	8 **9**	8 **9**	9 **10**	10 **11**
15	2 **2**	3 **4**	5 **6**	7 **8**	8 **9**	9 **10**	9 **10**	10 **11**
16	2 **3**	3 **4**	5 **6**	7 **8**	8 **9**	9 **10**	9 **11**	11 **12**
17	2 **3**	4 **4**	5 **6**	7 **9**	8 **10**	9 **11**	10 **11**	11 **13**
18	2 **3**	4 **5**	5 **6**	8 **9**	9 **10**	10 **11**	10 **12**	12 **13**
19	2 **3**	4 **5**	5 **6**	8 **9**	9 **10**	10 **12**	11 **12**	12 **14**
20	2 **3**	4 **5**	5 **7**	8 **9**	9 **11**	10 **12**	11 **13**	13 **14**

(Dashes in the body of the table indicate that no decision is possible at the stated level of significance.)

Table O

Some approximately normal populations for sampling experiments

X	f	f	f	f
2	4	2	0	0
3	54	27	6	3
4	242	121	24	12
5	400	200	40	20
6	242	121	24	12
7	54	27	6	3
8	4	2	0	0
N	1000	500	100	50
μ	5.00	5.00	5.00	5.00
σ	.99	.99	.98	.98

Table P

Percentage points of the Studentized range

Error df	α	k = *number of means or number of steps between ordered means*									
		2	3	4	5	6	7	8	9	10	11
5	.05	3.64	4.60	5.22	5.67	6.03	6.33	6.58	6.80	6.99	7.17
	.01	5.70	6.98	7.80	8.42	8.91	9.32	9.67	9.97	10.24	10.48
6	.05	3.46	4.34	4.90	5.30	5.63	5.90	6.12	6.32	6.49	6.65
	.01	5.24	6.33	7.03	7.56	7.97	8.32	8.61	8.87	9.10	9.30
7	.05	3.34	4.16	4.68	5.06	5.36	5.61	5.82	6.00	6.16	6.30
	.01	4.95	5.92	6.54	7.01	7.37	7.68	7.94	8.17	8.37	8.55
8	.05	3.26	4.04	4.53	4.89	5.17	5.40	5.60	5.77	5.92	6.05
	.01	4.75	5.64	6.20	6.62	6.96	7.24	7.47	7.68	7.86	8.03
9	.05	3.20	3.95	4.41	4.76	5.02	5.24	5.43	5.59	5.74	5.87
	.01	4.60	5.43	5.96	6.35	6.66	6.91	7.13	7.33	7.49	7.65
10	.05	3.15	3.88	4.33	4.65	4.91	5.12	5.30	5.46	5.60	5.72
	.01	4.48	5.27	5.77	6.14	6.43	6.67	6.87	7.05	7.21	7.36
11	.05	3.11	3.82	4.26	4.57	4.82	5.03	5.20	5.35	5.49	5.61
	.01	4.39	5.15	5.62	5.97	6.25	6.48	6.67	6.84	6.99	7.13
12	.05	3.08	3.77	4.20	4.51	4.75	4.95	5.12	5.27	5.39	5.51
	.01	4.32	5.05	5.50	5.84	6.10	6.32	6.51	6.67	6.81	6.94
13	.05	3.06	3.73	4.15	4.45	4.69	4.88	5.05	5.19	5.32	5.43
	.01	4.26	4.96	5.40	5.73	5.98	6.19	6.37	6.53	6.67	6.79
14	.05	3.03	3.70	4.11	4.41	4.64	4.83	4.99	5.13	5.25	5.36
	.01	4.21	4.89	5.32	5.63	5.88	6.08	6.26	6.41	6.54	6.66
15	.05	3.01	3.67	4.08	4.37	4.59	4.78	4.94	5.08	5.20	5.31
	.01	4.17	4.84	5.25	5.56	5.80	5.99	6.16	6.31	6.44	6.55
16	.05	3.00	3.65	4.05	4.33	4.56	4.74	4.90	5.03	5.15	5.26
	.01	4.13	4.79	5.19	5.49	5.72	5.92	6.08	6.22	6.35	6.46
17	.05	2.98	3.63	4.02	4.30	4.52	4.70	4.86	4.99	5.11	5.21
	.01	4.10	4.74	5.14	5.43	5.66	5.85	6.01	6.15	6.27	6.38
18	.05	2.97	3.61	4.00	4.28	4.49	4.67	4.82	4.96	5.07	5.17
	.01	4.07	4.70	5.09	5.38	5.60	5.79	5.94	6.08	6.20	6.31
19	.05	2.96	3.59	3.98	4.25	4.47	4.65	4.79	4.92	5.04	5.14
	.01	4.05	4.67	5.05	5.33	5.55	5.73	5.89	6.02	6.14	6.25
20	.05	2.95	3.58	3.96	4.23	4.45	4.62	4.77	4.90	5.01	5.11
	.01	4.02	4.64	5.02	5.29	5.51	5.69	5.84	5.97	6.09	6.19
24	.05	2.92	3.53	3.90	4.17	4.37	4.54	4.68	4.81	4.92	5.01
	.01	3.96	4.55	4.91	5.17	5.37	5.54	5.69	5.81	5.92	6.02
30	.05	2.89	3.49	3.85	4.10	4.30	4.46	4.60	4.72	4.82	4.92
	.01	3.89	4.45	4.80	5.05	5.24	5.40	5.54	5.65	5.76	5.85
40	.05	2.86	3.44	3.79	4.04	4.23	4.39	4.52	4.63	4.73	4.82
	.01	3.82	4.37	4.70	4.93	5.11	5.26	5.39	5.50	5.60	5.69
60	.05	2.83	3.40	3.74	3.98	4.16	4.31	4.44	4.55	4.65	4.73
	.01	3.76	4.28	4.59	4.82	4.99	5.13	5.25	5.36	5.45	5.53
120	.05	2.80	3.36	3.68	3.92	4.10	4.24	4.36	4.47	4.56	4.64
	.01	3.70	4.20	4.50	4.71	4.87	5.01	5.12	5.21	5.30	5.37
∞	.05	2.77	3.31	3.63	3.86	4.03	4.17	4.29	4.39	4.47	4.55
	.01	3.64	4.12	4.40	4.60	4.76	4.88	4.99	5.08	5.16	5.23

Table Q

Squares, square roots, and reciprocals of numbers from 1 to 1000

N	N^2	$\sqrt{N}$	1/N	N	N^2	$\sqrt{N}$	1/N	N	N^2	$\sqrt{N}$	1/N
1	1	1.0000	1.000000	61	3721	7.8102	.016393	121	14641	11.0000	.00826446
2	4	1.4142	.500000	62	3844	7.8740	.016129	122	14884	11.0454	.00819672
3	9	1.7321	.333333	63	3969	7.9373	.015873	123	15129	11.0905	.00813008
4	16	2.0000	.250000	64	4096	8.0000	.015625	124	15376	11.1355	.00800452
5	25	2.2361	.200000	65	4225	8.0623	.015385	125	15625	11.1803	.00800000
6	36	2.4495	.166667	66	4356	8.1240	.015152	126	15876	11.2250	.00793651
7	49	2.6458	.142857	67	4489	8.1854	.014925	127	16129	11.2694	.00787402
8	64	2.8284	.125000	68	4624	8.2462	.014706	128	16384	11.3137	.00781250
9	81	3.0000	.111111	69	4761	8.3066	.014493	129	16641	11.3578	.00775194
10	100	3.1623	.100000	70	4900	8.3666	.014286	130	16900	11.4018	.00769231
11	121	3.3166	.090909	71	5041	8.4261	.014085	131	17161	11.4455	.00763359
12	144	3.4641	.083333	72	5184	8.4853	.013889	132	17424	11.4891	.00757576
13	169	3.6056	.076923	73	5329	8.5440	.013699	133	17689	11.5326	.00751880
14	196	3.7417	.071429	74	5476	8.6023	.013514	134	17956	11.5758	.00746269
15	225	3.8730	.066667	75	5625	8.6603	.013333	135	18225	11.6190	.00740741
16	256	4.0000	.062500	76	5776	8.7178	.013158	136	18496	11.6619	.00735294
17	289	4.1231	.058824	77	5929	8.7750	.012987	137	18769	11.7047	.00729927
18	324	4.2426	.055556	78	6084	8.8318	.012821	138	19044	11.7473	.00724638
19	361	4.3589	.052632	79	6241	8.8882	.012658	139	19321	11.7898	.00719424
20	400	4.4721	.050000	80	6400	8.9443	.012500	140	19600	11.8322	.00714286
21	441	4.5826	.047619	81	6561	9.0000	.012346	141	19881	11.8743	.00709220
22	484	4.6904	.045455	82	6724	9.0554	.012195	142	20164	11.9164	.00704225
23	529	4.7958	.043478	83	6889	9.1104	.012048	143	20449	11.9583	.00699301
24	576	4.8990	.041667	84	7056	9.1652	.011905	144	20736	12.0000	.00694444
25	625	5.0000	.040000	85	7225	9.2195	.011765	145	21025	12.0416	.00689655
26	676	5.0990	.038462	86	7396	9.2736	.011628	146	21316	12.0830	.00684932
27	729	5.1962	.037037	87	7569	9.3274	.011494	147	21609	12.1244	.00680272
28	784	5.2915	.035714	88	7744	9.3808	.011364	148	21904	12.1655	.00675676
29	841	5.3852	.034483	89	7921	9.4340	.011236	149	22201	12.2066	.00671141
30	900	5.4772	.033333	90	8100	9.4868	.011111	150	22500	12.2474	.00666667
31	961	5.5678	.032258	91	8281	9.5394	.010989	151	22801	12.2882	.00662252
32	1024	5.6569	.031250	92	8464	9.5917	.010870	152	23104	12.3288	.00657895
33	1089	5.7446	.030303	93	8649	9.6437	.010753	153	23409	12.3693	.00653595
34	1156	5.8310	.029412	94	8836	9.6954	.010638	154	23716	12.4097	.00649351
35	1225	5.9161	.028571	95	9025	9.7468	.010526	155	24025	12.4499	.00645161
36	1296	6.0000	.027778	96	9216	9.7980	.010417	156	24336	12.4900	.00641026
37	1369	6.0828	.027027	97	9409	9.8489	.010309	157	24649	12.5300	.00636943
38	1444	6.1644	.026316	98	9604	9.8995	.010204	158	24964	12.5698	.00632911
39	1521	6.2450	.025641	99	9801	9.9499	.010101	159	25281	12.6095	.00628931
40	1600	6.3246	.025000	100	10000	10.0000	.010000	160	25600	12.6491	.00625000
41	1681	6.4031	.024390	101	10201	10.0499	.00990099	161	25921	12.6886	.00621118
42	1764	6.4807	.023810	102	10404	10.0995	.00980392	162	26244	12.7279	.00617284
43	1849	6.5574	.023256	103	10609	10.1489	.00970874	163	26569	12.7671	.00613497
44	1936	6.6332	.022727	104	10816	10.1980	.00961538	164	26896	12.8062	.00609756
45	2025	6.7082	.022222	105	11025	10.2470	.00952381	165	27225	12.8452	.00606061
46	2116	6.7823	.021739	106	11236	10.2956	.00943396	166	27556	12.8841	.00602410
47	2209	6.8557	.021277	107	11449	10.3441	.00934579	167	27889	12.9228	.00598802
48	2304	6.9282	.020833	108	11664	10.3923	.00925926	168	28224	12.9615	.00595238
49	2401	7.0000	.020408	109	11881	10.4403	.00917431	169	28561	13.0000	.00591716
50	2500	7.0711	.020000	110	12100	10.4881	.00909091	170	28900	13.0384	.00588235
51	2601	7.1414	.019608	111	12321	10.5357	.00900901	171	29241	13.0767	.00584795
52	2704	7.2111	.019231	112	12544	10.5830	.00892857	172	29584	13.1149	.00581395
53	2809	7.2801	.018868	113	12769	10.6301	.00884956	173	29929	13.1529	.00578035
54	2916	7.3485	.018519	114	12996	10.6771	.00877193	174	30276	13.1909	.00574713
55	3025	7.4162	.018182	115	13225	10.7238	.00869565	175	30625	13.2288	.00571429
56	3136	7.4833	.017857	116	13456	10.7703	.00862069	176	30976	13.2665	.00568182
57	3249	7.5498	.017544	117	13689	10.8167	.00854701	177	31329	13.3041	.00564972
58	3364	7.6158	.017241	118	13924	10.8628	.00847458	178	31684	13.3417	.00561798
59	3481	7.6811	.016949	119	14161	10.9087	.00840336	179	32041	13.3791	.00558659
60	3600	7.7460	.016667	120	14400	10.9545	.00833333	180	32400	13.4164	.00555556

Table Q. *(continued)*

N	N^2	$\sqrt{N}$	1/N	N	N^2	$\sqrt{N}$	1/N	N	N^2	$\sqrt{N}$	1/N
181	32761	13.4536	.00552486	241	58081	15.5242	.00414938	301	90601	17.3494	.00332226
182	33124	13.4907	.00549451	242	58564	15.5563	.00413223	302	91204	17.3781	.00331126
183	33489	13.5277	.00546448	243	59049	15.5885	.00411523	303	91809	17.4069	.00330033
184	33856	13.5647	.00543478	244	59536	15.6205	.00409836	304	92416	17.4356	.00328047
185	34225	13.6015	.00540541	245	60025	15.6525	.00408163	305	93025	17.4642	.00328947
186	34596	13.6382	.00537634	246	60516	15.6844	.00406504	306	93636	17.4929	.00326797
187	34969	13.6748	.00534759	247	61009	15.7162	.00404858	307	94249	17.5214	.00325733
188	35344	13.7113	.00531915	248	61504	15.7480	.00403226	308	94864	17.5499	.00321675
189	35721	13.7477	.00529101	249	62001	15.7797	.00401606	309	95481	17.5784	.00323625
190	36100	13.7840	.00526316	250	62500	15.8114	.00400000	310	96100	17.6068	.00322581
191	36481	13.8203	.00523560	251	63001	15.8430	.00398406	311	96721	17.6352	.00321543
192	36864	13.8564	.00520833	252	63504	15.8745	.00396825	312	97344	17.6635	.00320513
193	37249	13.8924	.00518135	253	64009	15.9060	.00395257	313	97969	17.6918	.00319489
194	37636	13.9284	.00515464	254	64516	15.9374	.00393701	314	98596	17.7200	.00318471
195	38025	13.9642	.00512821	255	65025	15.9687	.00392157	315	99225	17.7482	.00317460
196	38416	14.0000	.00510204	256	65536	16.0000	.00390625	316	99856	17.7764	.00316456
197	38809	14.0357	.00507614	257	66049	16.0312	.00389105	317	100489	17.8045	.00315457
198	39204	14.0712	.00505051	258	66564	16.0624	.00387597	318	101124	17.8326	.00314465
199	39601	14.1067	.00502513	259	67081	16.0935	.00386100	319	101761	17.8606	.00313480
200	40000	14.1421	.00500000	260	67600	16.1245	.00384615	320	102400	17.8885	.00312500
201	40401	14.1774	.00497512	261	68121	16.1555	.00383142	321	103041	17.9165	.00311526
202	40804	14.2127	.00495050	262	68644	16.1864	.00381679	322	103684	17.9444	.00310559
203	41209	14.2478	.00492611	263	69169	16.2173	.00380228	323	104329	17.9722	.00309598
204	41616	14.2829	.00490196	264	69696	16.2481	.00378788	324	104976	18.0000	.00308642
205	42025	14.3178	.00487805	265	70225	16.2788	.00377358	325	105625	18.0278	.00307692
206	42436	14.3527	.00485437	266	70756	16.3095	.00375940	326	106276	18.0555	.00306748
207	42849	14.3875	.00483092	267	71289	16.3401	.00374532	327	106929	18.0831	.00305810
208	43264	14.4222	.00480769	268	71824	16.3707	.00373134	328	107584	18.1108	.00304878
209	43681	14.4568	.00478469	269	72361	16.4012	.00371747	329	108241	18.1384	.00303951
210	44100	14.4914	.00476190	270	72900	16.4317	.00370370	330	108900	18.1659	.00303030
211	44521	14.5258	.00473934	271	73441	16.4621	.00369004	331	109561	18.1934	.00302115
212	44944	14.5602	.00471698	272	73984	16.4924	.00367647	332	110224	18.2209	.00301205
213	45369	14.5945	.00469484	273	74529	16.5227	.00366300	333	110889	18.2483	.00300300
214	45796	14.6287	.00467290	274	75076	16.5529	.00364964	334	111556	18.2757	.00299401
215	46225	14.6629	.00465116	275	75625	16.5831	.00363636	335	112225	18.3030	.00298507
216	46656	14.6969	.00462963	276	76176	16.6132	.00362319	336	112896	18.3303	.00297619
217	47089	14.7309	.00460829	277	76729	16.6433	.00361011	337	113569	18.3576	.00296736
218	47524	14.7648	.00458716	278	77284	16.6733	.00359712	338	114244	18.3848	.00295858
219	47961	14.7986	.00456621	279	77841	16.7033	.00358423	339	114921	18.4120	.00294985
220	48400	14.8324	.00454545	280	78400	16.7332	.00357143	340	115600	18.4391	.00294118
221	48841	14.8661	.00452489	281	78961	16.7631	.00355872	341	116281	18.4662	.00293255
222	49284	14.8997	.00450450	282	79524	16.7929	.00354610	342	116964	18.4932	.00292398
223	49729	14.9332	.00448430	283	80089	16.8226	.00353357	343	117649	18.5203	.00291545
224	50176	14.9666	.00446429	284	80656	16.8523	.00352113	344	118336	18.5472	.00290698
225	50625	15.0000	.00444444	285	81225	16.8819	.00350877	345	119025	18.5742	.00289855
226	51076	15.0333	.00442478	286	81796	16.9115	.00349650	346	119716	18.6011	.00289017
227	51529	15.0665	.00440529	287	82369	16.9411	.00348432	347	120409	18.6279	.00288184
228	51984	15.0997	.00438596	288	82944	16.9706	.00347222	348	121104	18.6548	.00287356
229	52441	15.1327	.00436681	289	83521	17.0000	.00346021	349	121801	18.6815	.00286533
230	52900	15.1658	.00434783	290	84100	17.0294	.00344828	350	122500	18.7083	.00285714
231	53361	15.1987	.00432900	291	84681	17.0587	.00343643	351	123201	18.7350	.00284900
232	53824	15.2315	.00431034	292	85264	17.0880	.00342466	352	123904	18.7617	.00284091
233	54289	15.2643	.00429185	293	85849	17.1172	.00341297	353	124609	18.7883	.00283286
234	54756	15.2971	.00427350	294	86436	17.1464	.00340136	354	125316	18.8149	.00282486
235	55225	15.3297	.00425532	295	87025	17.1756	.00338983	355	126025	18.8414	.00281690
236	55696	15.3623	.00423729	296	87616	17.2047	.00337838	356	126736	18.8680	.00280899
237	56169	15.3948	.00421941	297	88209	17.2337	.00336700	357	127449	18.8944	.00280112
238	56644	15.4272	.00420168	298	88804	17.2627	.00335570	358	128164	18.9209	.00279330
239	57121	15.4596	.00418410	299	89401	17.2916	.00334448	359	128881	18.9473	.00278552
240	57600	15.4919	.00416667	300	90000	17.3205	.00333333	360	129600	18.9737	.00277778

Table Q. *(continued)*

N	N^2	$\sqrt{N}$	1/N	N	N^2	$\sqrt{N}$	1/N	N	N^2	$\sqrt{N}$	1/N
361	130321	19.0000	.00277008	421	177241	20.5183	.00237530	481	231361	21.9317	.00207900
362	131044	19.0263	.00276243	422	178084	20.5426	.00236967	482	232324	21.9545	.00207469
363	131769	19.0526	.00275482	423	178929	20.5670	.00236407	483	233289	21.9773	.00207039
364	132496	19.0788	.00274725	424	179776	20.5913	.00235849	484	234256	22.0000	.00206612
365	133225	19.1050	.00273973	425	180625	20.6155	.00235294	485	235225	22.0227	.00206186
366	133956	19.1311	.00273224	426	181476	20.6398	.00234742	486	236196	22.0454	.00205761
367	134689	19.1572	.00272480	427	182329	20.6640	.00234192	487	237169	22.0681	.00205339
368	135424	19.1833	.00271739	428	183184	20.6882	.00233645	488	238144	22.0907	.00204918
369	136161	19.2094	.00271003	429	184041	20.7123	.00233100	489	239121	22.1133	.00204499
370	136900	19.2354	.00270270	430	184900	20.7364	.00232558	490	240100	22.1359	.00204082
371	137641	19.2614	.00269542	431	185761	20.7605	.00232019	491	241081	22.1585	.00203666
372	138384	19.2873	.00268817	432	186624	20.7846	.00231481	492	242064	22.1811	.00203252
373	139129	19.3132	.00268097	433	187489	20.8087	.00230947	493	243049	22.2036	.00202840
374	139876	19.3391	.00267380	434	188356	20.8327	.00230415	494	244036	22.2261	.00202429
375	140625	19.3649	.00266667	435	189225	20.8567	.00229885	495	245025	22.2486	.00202020
376	141376	19.3907	.00265957	436	190096	20.8806	.00229358	496	246016	22.2711	.00201613
377	142129	19.4165	.00265252	437	190969	20.9045	.00228833	497	247009	22.2935	.00201207
378	142884	19.4422	.00264550	438	191844	20.9284	.00228311	498	248004	22.3159	.00200803
379	143641	19.4679	.00263852	439	192721	20.9523	.00227790	499	249001	22.3383	.00200401
380	144400	19.4936	.00263158	440	193600	20.9762	.00227273	500	250000	22.3607	.00200000
381	145161	19.5192	.00262467	441	194481	21.0000	.00226757	501	251001	22.3830	.00199601
382	145924	19.5448	.00261780	442	195364	21.0238	.00226244	502	252004	22.4054	.00199203
383	146689	19.5704	.00261097	443	196249	21.0476	.00225734	503	253009	22.4277	.00198807
384	147456	19.5959	.00260417	444	197136	21.0713	.00225225	504	254016	22.4499	.00198413
385	148225	19.6214	.00259740	445	198025	21.0950	.00224719	505	255025	22.4722	.00198020
386	148996	19.6469	.00259067	446	198916	21.1187	.00224215	506	256036	22.4944	.00197628
387	149769	19.6723	.00258398	447	199809	21.1424	.00223714	507	257049	22.5167	.00197239
388	150544	19.6977	.00257732	448	200704	21.1660	.00223214	508	258064	22.5389	.00196850
389	151321	19.7231	.00257069	449	201601	21.1896	.00222717	509	259081	22.5610	.00196464
390	152100	19.7484	.00256410	450	202500	21.2132	.00222222	510	260100	22.5832	.00196078
391	152881	19.7737	.00255754	451	203401	21.2368	.00221729	511	261121	22.6053	.00195695
392	153664	19.7990	.00255102	452	204304	21.2603	.00221239	512	262144	22.6274	.00195312
393	154449	19.8242	.00254453	453	205209	21.2838	.00220751	513	263169	22.6495	.00194932
394	155236	19.8494	.00253807	454	206116	21.3073	.00220264	514	264196	22.6716	.00194553
395	156025	19.8746	.00253165	455	207025	21.3307	.00219870	515	265225	22.6936	.00194175
396	156816	19.8997	.00252525	456	207936	21.3542	.00219298	516	266256	22.7156	.00193798
397	157609	19.9249	.00251889	457	208849	21.3776	.00218818	517	267289	22.7376	.00193424
398	158404	19.9499	.00251256	458	209764	21.4009	.00218341	518	268324	22.7596	.00193050
399	159201	19.9750	.00250627	459	210681	21.4243	.00217865	519	269361	22.7816	.00192678
400	160000	20.0000	.00250000	460	211600	21.4476	.00217391	520	270400	22.8035	.00192308
401	160801	20.0250	.00249377	461	212521	21.4709	.00216920	521	271441	22.8254	.00191939
402	161604	20.0499	.00248756	462	213444	21.4942	.00216450	522	272484	22.8473	.00191571
403	162409	20.0749	.00248139	463	214369	21.5174	.00215983	523	273529	22.8692	.00191205
404	163216	20.0998	.00247525	464	215296	21.5407	.00215517	524	274576	22.8910	.00190840
405	164025	20.1246	.00246914	465	216225	21.5639	.00215054	525	275625	22.9129	.00190476
406	164836	20.1494	.00246305	466	217156	21.5870	.00214592	526	276676	22.9347	.00190114
407	165649	20.1742	.00245700	467	218089	21.6102	.00214133	527	277729	22.9565	.00189753
408	166464	20.1990	.00245098	468	219024	21.6333	.00213675	528	278784	22.9783	.00189394
409	167281	20.2237	.00244499	469	219961	21.6564	.00213220	529	279841	23.0000	.00189036
410	168100	20.2485	.00243902	470	220900	21.6795	.00212766	530	280900	23.0217	.00188679
411	168921	20.2731	.00243309	471	221841	21.7025	.00212314	531	281961	23.0434	.00188324
412	169744	20.2978	.00242718	472	222784	21.7256	.00211864	532	283024	23.0651	.00187970
413	170569	20.3224	.00242131	473	223729	21.7486	.00211416	533	284089	23.0868	.00187617
414	171396	20.3470	.00241546	474	224676	21.7715	.00210970	534	285156	23.1084	.00187266
415	172225	20.3715	.00240964	475	225625	21.7945	.00210526	535	286225	23.1301	.00186916
416	173056	20.3961	.00240385	476	226576	21.8174	.00210084	536	287296	23.1517	.00186567
417	173889	20.4206	.00239808	477	227529	21.8403	.00209644	537	288369	23.1733	.00186220
418	174724	20.4450	.00239234	478	228484	21.8632	.00209205	538	289444	23.1948	.00185874
419	175561	20.4695	.00238663	479	229441	21.8861	.00208768	539	290521	23.2164	.00185529
420	176400	20.4939	.00238095	480	230400	21.9089	.00208333	540	291600	23.2379	.00185185

Table Q. *(continued)*

N	N^2	$\sqrt{N}$	1/N	N	N^2	$\sqrt{N}$	1/N	N	N^2	$\sqrt{N}$	1/N
541	292681	23.2594	.00184843	601	361201	24.5153	.00166389	661	436921	25.7099	.00151286
542	293764	23.2809	.00184502	602	302404	24.5357	.00166113	662	438244	25.7294	.00151057
543	294849	23.3024	.00184162	603	363609	24.5561	.00165837	663	439569	25.7488	.00150830
544	295936	23.3238	.00183824	604	364816	24.5764	.00165563	664	440896	25.7682	.00150602
545	297025	23.3452	.00183486	605	366025	24.5967	.00165289	665	442225	25.7876	.00150376
566	298116	23.3666	.00183150	606	367236	24.6171	.00165017	666	443556	25.8070	.00150150
547	299209	23.3880	.00182815	607	368449	24.6374	.00164745	667	444889	25.8263	.00149925
548	300304	23.4094	.00182482	608	369664	24.6577	.00164474	668	446224	25.8457	.00149701
549	301401	23.4307	.00182149	609	370881	24.6779	.00164204	669	447561	25.8650	.00149477
550	302500	23.4521	.00181818	610	372100	24.6982	.00163934	670	448900	25.8844	.00149254
551	303601	23.4734	.00181488	611	373321	24.7184	.00163666	671	450241	25.9037	.00149031
552	304704	23.4947	.00181159	612	374544	24.7386	.00163399	672	451584	25.9230	.00148810
553	305809	23.5160	.00180832	613	375769	24.7588	.00163132	673	452929	25.9422	.00148588
554	306916	23.5372	.00180505	614	376996	24.7790	.00162866	674	454276	25.9615	.00148368
555	308025	23.5584	.00180180	615	378225	24.7992	.00162602	675	455625	25.9808	.00148148
556	309136	23.5797	.00179856	616	379456	24.8193	.00162338	676	456976	26.0000	.00147929
557	310249	23.6008	.00179533	617	380689	24.8395	.00162075	677	458329	26.0192	.00147710
558	311364	23.6220	.00179211	618	381924	24.8596	.00161812	678	459684	26.0384	.00147493
559	312481	23.6432	.00178891	619	383161	24.8797	.00161551	679	461041	26.0576	.00147275
560	313600	23.6643	.00178571	620	384400	24.8998	.00161290	680	462400	26.0768	.00147059
561	314721	23.6854	.00178253	621	385641	24.9199	.00161031	681	463761	26.0960	.00146843
562	315844	23.7065	.00177936	622	386884	24.9399	.00160772	682	465124	26.1151	.00146628
563	316969	23.7276	.00177620	623	388129	24.9600	.00160514	683	466489	26.1343	.00146413
564	318096	23.7487	.00177305	624	389376	24.9800	.00160256	684	467856	26.1534	.00146199
565	319225	23.7697	.00176991	625	390625	25.0000	.00160000	685	469225	26.1725	.00145985
566	320356	23.7908	.00176678	626	391876	25.0200	.00159744	686	470596	26.1916	.00145773
567	321489	23.8118	.00176367	627	393129	25.0400	.00159490	687	471969	26.2107	.00145560
568	322624	23.8328	.00176056	628	394384	25.0599	.00159236	688	473344	26.2298	.00145349
569	323761	23.8537	.00175747	629	395641	25.0799	.00158983	689	474721	26.2488	.00145138
570	324900	23.8747	.00175439	630	396900	25.0998	.00158730	690	476100	26.2679	.00144928
571	326041	23.8956	.00175131	631	398161	25.1197	.00158479	691	477481	26.2869	.00144718
572	327184	23.9165	.00164825	632	399424	25.1396	.00158228	692	478864	26.3059	.00144509
573	328329	23.9374	.00174520	633	400689	25.1595	.00157978	693	480249	26.3249	.00144300
574	329476	23.9583	.00174216	634	401956	25.1794	.00157729	694	481636	26.3439	.00144092
575	330625	23.9792	.00173913	635	403225	25.1992	.00157480	695	483025	26.3629	.00143885
576	331776	24.0000	.00173611	636	404496	25.2190	.00157233	696	484416	26.3818	.00143678
577	332929	24.0208	.00173310	637	405769	25.2389	.00156986	697	485809	26.4008	.00143472
578	334084	24.0416	.00173010	638	407044	25.2587	.00156740	698	487204	26.4197	.00143266
579	335241	24.0624	.00172712	639	408321	25.2784	.00156495	699	488601	26.4386	.00143062
580	336400	24.0832	.00172414	640	409600	25.2982	.00156250	700	490000	26.4575	.00142857
581	337561	24.1039	.00172117	641	410881	25.3180	.00156006	701	491401	26.4764	.00142653
582	338724	24.1247	.00171821	642	412164	25.3377	.00155763	702	492804	26.4953	.00142450
583	339889	24.1454	.00171527	643	413449	25.3574	.00155521	703	494209	26.5141	.00142248
584	341056	24.1661	.00171233	644	414736	25.3772	.00155280	704	495616	26.5330	.00142045
585	342225	24.1868	.00170940	645	416025	25.3969	.00155039	705	497025	26.5518	.00141844
586	343396	24.2074	.00170648	646	417316	25.4165	.00154799	706	498436	26.5707	.00141643
587	344569	24.2281	.00170358	647	418609	25.4362	.00154560	707	499849	26.5895	.00141443
588	345744	24.2487	.00170068	648	419904	25.4558	.00154321	708	501264	26.6083	.00141243
589	346921	24.2693	.00169779	649	421201	25.4755	.00154083	709	502681	26.6271	.00141044
590	348100	24.2899	.00169492	650	422500	25.4951	.00153846	710	504100	26.6458	.00140845
591	349281	24.3105	.00169205	651	423801	25.5147	.00153610	711	505521	26.6646	.00140647
592	350464	24.3311	.00168919	652	425104	25.5343	.00153374	712	506944	26.6833	.00140449
593	351649	24.3516	.00168634	653	426409	25.5539	.00153139	713	508369	26.7021	.00140252
594	352836	24.3721	.00168350	654	427716	25.5734	.00152905	714	509796	26.7208	.00140056
595	354025	24.3926	.00168067	655	429025	25.5930	.00152672	715	511225	26.7395	.00139860
596	355216	24.4131	.00167785	656	430336	25.6125	.00152439	716	512656	26.7582	.00139665
597	356409	24.4336	.00167504	657	431649	25.6320	.00152207	717	514089	26.7769	.00139470
598	357604	24.4540	.00167224	658	432964	25.6515	.00151976	718	515524	26.7955	.00139276
599	358801	24.4745	.00166945	659	434281	25.6710	.00151745	719	516961	26.8142	.00139082
600	360000	24.4949	.00166667	660	435600	25.6905	.00151515	720	518400	26.8328	.00138889

Table Q. (*continued*)

N	N^2	$\sqrt{N}$	1/N	N	N^2	$\sqrt{N}$	1/N	N	N^2	$\sqrt{N}$	1/N
721	519841	26.8514	.00138696	781	609961	27.9464	.00128041	841	707281	29.0000	.00118906
722	521284	26.8701	.00138504	782	611524	27.9643	.00127877	842	708964	29.0172	.00118765
723	522729	26.8887	.00138313	783	613089	27.9821	.00127714	843	710649	29.0345	.00118624
724	524176	26.9072	.00138122	784	614656	28.0000	.00127551	844	712336	29.0517	.00118483
725	525625	26.9258	.00137931	785	616225	28.0179	.00127389	845	714025	29.0689	.00118343
726	527076	26.9444	.00137741	786	617796	28.0357	.00127226	846	715716	29.0861	.00118203
727	528529	26.9629	.00137552	787	619369	28.0535	.00127065	847	717409	29.1033	.00118064
728	529984	26.9815	.00137363	788	620944	28.0713	.00126904	848	719104	29.1204	.00117925
729	531441	27.0000	.00137174	789	622521	28.0891	.00126743	849	720801	29.1376	.00117786
730	532900	27.0185	.00136986	790	624100	28.1069	.00126582	850	722500	29.1548	.00117647
731	534361	27.0370	.00136799	791	625681	28.1247	.00126422	851	724201	29.1719	.00117509
732	535824	27.0555	.00136612	792	627264	28.1425	.00126263	852	725904	29.1890	.00117371
733	537289	27.0740	.00136426	793	628849	28.1603	.00126103	853	727609	29.2062	.00117233
734	538756	27.0924	.00136240	794	630436	28.1780	.00125945	854	729316	29.2233	.00117096
735	540225	27.1109	.00136054	795	632025	28.1957	.00125786	855	731025	29.2404	.00116959
736	541696	27.1293	.00135870	796	633616	28.2135	.00125628	856	732736	29.2575	.00116822
737	543169	27.1477	.00135685	797	635209	28.2312	.00125471	857	734449	29.2746	.00116686
738	544644	27.1662	.00135501	798	636804	28.2489	.00125313	858	736164	29.2916	.00116550
739	546121	27.1846	.00135318	799	638401	28.2666	.00125156	859	737881	29.3087	.00116414
740	547600	27.2029	.00135135	800	640000	28.2843	.00125000	860	739600	29.3258	.00116279
741	549081	27.2213	.00134953	801	641601	28.3019	.00124844	861	741321	29.3428	.00116144
742	550564	27.2397	.00134771	802	643204	28.3196	.00124688	862	743044	29.3598	.00116009
743	552049	27.2580	.00134590	803	644809	28.3373	.00124533	863	744769	29.3769	.00115875
744	553536	27.2764	.00134409	804	646416	28.3549	.00124378	864	746496	29.3939	.00115741
745	555025	27.2947	.00134228	805	648025	28.3725	.00124224	865	748225	29.4109	.00115607
746	556516	27.3130	.00134048	806	649636	28.3901	.00124069	866	749956	29.4279	.00115473
747	558009	27.3313	.00133869	807	651249	28.4077	.00123916	867	751689	29.4449	.00115340
748	559504	27.3496	.00133690	808	625864	28.4253	.00123762	868	753424	29.4618	.00115207
749	561001	27.3679	.00133511	809	654481	28.4429	.00123609	869	755161	29.4788	.00115075
750	562500	27.3861	.00133333	810	656100	28.4605	.00123457	870	756900	29.4958	.00114943
751	564001	27.4044	.00133156	811	657721	28.4781	.00123305	871	758641	29.5127	.00114811
752	565504	27.4226	.00132979	812	659344	28.4956	.00123153	872	760384	29.5296	.00114679
753	567009	27.4408	.00132802	813	660969	28.5132	.00123001	873	762129	29.5466	.00114548
754	568516	27.4591	.00132626	814	662596	28.5307	.00122850	874	763876	29.5635	.00114416
755	570025	27.4773	.00132450	815	664225	28.5482	.00122699	875	765625	29.5804	.00114286
756	571536	27.4955	.00132275	816	665856	28.5657	.00122549	876	767376	29.5973	.00114155
757	573049	27.5136	.00132100	817	667489	28.5832	.00122399	877	769129	29.6142	.00114025
758	574564	27.5318	.00131926	818	669124	28.6007	.00122249	878	770884	29.6311	.00113895
759	576081	27.5500	.00131752	819	670761	28.6182	.00122100	879	772641	29.6479	.00113766
760	577600	27.5681	.00131579	820	672400	28.6356	.00121951	880	774400	29.6848	.00113636
761	579121	27.5862	.00131406	821	674041	28.6531	.00121803	881	776161	29.6816	.00113507
762	580644	27.6043	.00131234	822	675684	28.6705	.00121655	882	777924	29.6985	.00113379
763	582169	27.6225	.00131062	823	677329	28.6880	.00121507	883	779689	29.7153	.00113250
764	583696	27.6405	.00130890	824	678976	28.7054	.00121359	884	781456	29.7321	.00113122
765	585225	27.6586	.00130719	825	680625	28.7228	.00121212	885	783225	29.7489	.00112994
766	586756	27.6767	.00130548	826	682276	28.7402	.00121065	886	784996	29.7658	.00112867
767	588289	27.6948	.00130378	827	683929	28.7576	.00120919	887	786769	29.7825	.00112740
768	589824	27.7128	.00130208	828	685584	28.7750	.00120773	888	788544	29.7993	.00112613
769	591361	27.7308	.00130039	829	687241	28.7924	.00120627	889	790321	29.8161	.00112486
770	592900	27.7489	.00129870	830	688900	28.8097	.00120482	890	792100	29.8329	.00112360
771	594441	27.7669	.00129702	831	690561	28.8271	.00120337	891	793881	29.8496	.00112233
772	595984	27.7849	.00129534	832	692224	28.8444	.00120192	892	795664	29.8664	.00112108
773	597529	27.8029	.00129366	833	693889	28.8617	.00120048	893	797449	29.8831	.00111982
774	599076	27.8209	.00129199	834	695556	28.8791	.00119904	894	799236	29.8998	.00111857
775	600625	27.8388	.00129032	835	697225	28.8964	.00119760	895	801025	29.9166	.00111732
776	602176	27.8568	.00128866	836	698896	28.9137	.00119617	896	802816	29.9333	.00111607
777	603729	27.8747	.00128700	837	700569	28.9310	.00119474	897	804609	29.9500	.00111483
778	605284	27.8927	.00128535	838	702244	28.9482	.00119332	898	806404	29.9666	.00111359
779	606841	27.9106	.00128370	839	703921	28.9655	.00119190	899	808201	29.9833	.00111235
780	608400	27.9285	.00128205	840	705600	28.9828	.00119048	900	810000	30.0000	.00111111

Table Q. (*concluded*)

N	N^2	$\sqrt{N}$	1/N	N	N^2	$\sqrt{N}$	1/N	N	N^2	$\sqrt{N}$	1/N
901	811801	30.0167	.00110988	936	876096	30.5941	.00106838	971	942841	31.1609	.00102987
902	813604	30.0333	.00110865	937	877969	30.6105	.00106724	972	944784	31.1769	.00102881
903	815409	30.0500	.00110742	938	879844	30.6268	.00106610	973	946729	31.1929	.00102775
904	817216	30.0666	.00110619	939	881721	30.6431	.00106496	974	948676	31.2090	.00102669
905	819025	30.0832	.00110497	940	883600	30.6594	.00106383	975	950625	31.2250	.00102564
906	820836	30.0998	.00110375	941	885481	30.6757	.00106270	976	952576	31.2410	.00102459
907	822649	30.1164	.00110254	942	887364	30.6920	.00106157	977	954529	31.2570	.00102354
908	824464	30.1330	.00110132	943	889249	30.7083	.00106045	978	956484	31.2730	.00102249
909	826281	30.1496	.00110011	944	891136	30.7246	.00105932	979	958441	31.2890	.00102145
910	828100	30.1662	.00109890	945	893025	30.7409	.00105820	980	960400	31.3050	.00102041
911	829921	30.1828	.00109769	946	894916	30.7571	.00105708	981	962361	31.3209	.00101937
912	831744	30.1993	.00109649	947	896809	30.7734	.00105597	982	964324	31.3369	.00101833
913	833569	30.2159	.00109529	948	898704	30.7896	.00105485	983	966289	31.3528	.00101729
914	835396	30.2324	.00109409	949	900601	30.8058	.00105374	984	968256	31.3688	.00101626
915	837225	30.2490	.00109290	950	902500	30.8221	.00105263	985	970225	31.3847	.00101523
916	839056	30.2655	.00109170	951	904401	30.8383	.00105152	986	972196	31.4006	.00101420
917	840889	30.2820	.00109051	952	906304	30.8545	.00105042	987	974169	31.4166	.00101317
918	842724	30.2985	.00108932	953	908209	30.8707	.00104932	988	976144	31.4325	.00101215
919	844561	30.3150	.00108814	954	910116	30.8869	.00104822	989	978121	31.4484	.00101112
920	846400	30.3315	.00108696	955	912025	30.9031	.00104712	990	980100	31.4643	.00101010
921	848241	30.3480	.00108578	956	913936	30.9192	.00104603	991	982081	31.4802	.00100908
922	850084	30.3645	.00108460	957	915849	30.9354	.00104493	992	984064	31.4960	.00100806
923	851929	30.3809	.00108342	958	917764	30.9516	.00104384	993	986049	31.5119	.00100705
924	853776	30.3974	.00108225	959	919681	30.9677	.00104275	994	988036	31.5278	.00100604
925	855625	30.4138	.00108108	960	921600	30.9839	.00104167	995	990025	31.5436	.00100503
926	857476	30.4302	.00107991	961	923521	31.0000	.00104058	996	992016	31.5595	.00100402
927	859329	30.4467	.00107875	962	925444	31.0161	.00103950	997	994009	31.5753	.00103842
928	861184	30.4631	.00107759	963	927369	31.0322	.00103842	998	996004	31.5911	.00100200
929	863041	30.4795	.00107643	964	929296	31.0483	.00103734	999	998001	31.6070	.00100100
930	864900	30.4959	.00107527	965	931225	31.0644	.00103627	1000	1000000	31.6228	.00100000
931	866761	30.5123	.00107411	966	933156	31.0805	.00103520				
932	868624	30.5287	.00107296	967	935089	31.0966	.00103413				
933	870489	30.5450	.00107181	968	937024	31.1127	.00103306				
934	872356	30.5614	.00107066	969	938961	31.1288	.00103199				
935	874225	30.5778	.00106952	970	940900	31.1448	.00103093				

Table R

Random digits

Row number										
00000	10097	32533	76520	13586	34673	54876	80959	09117	39292	74945
00001	37542	04805	64894	74296	24805	24037	20636	10402	00822	91665
00002	08422	68953	19645	09303	23209	02560	15953	34764	35080	33606
00003	99019	02529	09376	70715	38311	31165	88676	74397	04436	27659
00004	12807	99970	80157	36147	64032	36653	98951	16877	12171	76833
00005	66065	74717	34072	76850	36697	36170	65813	39885	11199	29170
00006	31060	10805	45571	82406	35303	42614	86799	07439	23403	09732
00007	85269	77602	02051	65692	68665	74818	73053	85247	18623	88579
00008	63573	32135	05325	47048	90553	57548	28468	28709	83491	25624
00009	73796	45753	03529	64778	35808	34282	60935	20344	35273	88435
00010	98520	17767	14905	68607	22109	40558	60970	93433	50500	73998
00011	11805	05431	39808	27732	50725	68248	29405	24201	52775	67851
00012	83452	99634	06288	98033	13746	70078	18475	40610	68711	77817
00013	88685	40200	86507	58401	36766	67951	90364	76493	29609	11062
00014	99594	67348	87517	64969	91826	08928	93785	61368	23478	34113
00015	65481	17674	17468	50950	58047	76974	73039	57186	40218	16544
00016	80124	35635	17727	08015	45318	22374	21115	78253	14385	53763
00017	74350	99817	77402	77214	43236	00210	45521	64237	96286	02655
00018	69916	26803	66252	29148	36936	87203	76621	13990	94400	56418
00019	09893	20505	14225	68514	46427	56788	96297	78822	54382	14598
00020	91499	14523	68479	27686	46162	83554	94750	89923	37089	20048
00021	80336	94598	26940	36858	70297	34135	53140	33340	42050	82341
00022	44104	81949	85157	47954	32979	26575	57600	40881	22222	06413
00023	12550	73742	11100	02040	12860	74697	96644	89439	28707	25815
00024	63606	49329	16505	34484	40219	52563	43651	77082	07207	31790
00025	61196	90446	26457	47774	51924	33729	65394	59593	42582	60527
00026	15474	45266	95270	79953	59367	83848	82396	10118	33211	59466
00027	94557	28573	67897	54387	54622	44431	91190	42592	92927	45973
00028	42481	16213	97344	08721	16868	48767	03071	12059	25701	46670
00029	23523	78317	73208	89837	68935	91416	26252	29663	05522	82562
00030	04493	52494	75246	33824	45862	51025	61962	79335	65337	12472
00031	00549	97654	64051	88159	96119	63896	54692	82391	23287	29529
00032	35963	15307	26898	09354	33351	35462	77974	50024	90103	39333
00033	59808	08391	45427	26842	83609	49700	13021	24892	78565	20106
00034	46058	85236	01390	92286	77281	44077	93910	83647	70617	42941
00035	32179	00597	87379	25241	05567	07007	86743	17157	85394	11838
00036	69234	61406	20117	45204	15956	60000	18743	92423	97118	96338
00037	19565	41430	01758	75379	40419	21585	66674	36806	84962	85207
00038	45155	14938	19476	07246	43667	94543	59047	90033	20826	69541
00039	94864	31994	36168	10851	34888	81553	01540	35456	05014	51176
00040	98086	24826	45240	28404	44999	08896	39094	73407	35441	31880
00041	33185	16232	41941	50949	89435	48581	88695	41994	37548	73043
00042	80951	00406	96382	70774	20151	23387	25016	25298	94624	61171
00043	79752	49140	71961	28296	69861	02591	74852	20539	00387	59579
00044	18633	32537	98145	06571	31010	24674	05455	61427	77938	91936
00045	74029	43902	77557	32270	97790	17119	52527	58021	80814	51748
00046	54178	45611	80993	37143	05335	12969	56127	19255	36040	90324
00047	11664	49883	52079	84827	59381	71539	09973	33440	88461	23356
00048	48324	77928	31249	64710	02295	36870	32307	57546	15020	09994
00049	69074	94138	87637	91976	35584	04401	10518	21615	01848	76938
00050	09188	20097	32825	39527	04220	86304	83389	87374	64278	58044
00051	90045	85497	51981	50654	94938	81997	91870	76150	68476	64659
00052	73189	50207	47677	26269	62290	64464	27124	67018	41361	82760
00053	75768	76490	20971	87749	90429	12272	95375	05871	93823	43178
00054	54016	44056	66281	31003	00682	27398	20714	53295	07706	17813
00055	08358	69910	78542	42785	13661	58873	04618	97553	31223	08420
00056	28306	03264	81333	10591	40510	07893	32604	60475	94119	01840
00057	53840	86233	81594	13628	51215	90290	28466	68795	77762	20791
00058	91757	53741	61613	62669	50263	90212	55781	76514	83483	47055
00059	89415	92694	00397	58391	12607	17646	48949	72306	94541	37408

Table R. *(continued)*

Row number										
00060	77513	03820	86864	29901	68414	82774	51908	13980	72893	55507
00061	19502	37174	69979	20288	55210	29773	74287	75251	65344	67415
00062	21818	59313	93278	81757	05686	73156	07082	85046	31853	38452
00063	51474	66499	68107	23621	94049	91345	42836	09191	08007	45449
00064	99559	68331	62535	24170	69777	12830	74819	78142	43860	72834
00065	33713	48007	93584	72869	51926	64721	58303	29822	93174	93972
00066	85274	86893	11303	22970	28834	34137	73515	90400	71148	43643
00067	84133	89640	44035	52166	73852	70091	61222	60561	62327	18423
00068	56732	16234	17395	96131	10123	91622	85496	57560	81604	18880
00069	65138	56806	87648	85261	34313	65861	45875	21069	85644	47277
00070	38001	02176	81719	11711	71602	92937	74219	64049	65584	49698
00071	37402	96397	01304	77586	56271	10086	47324	62605	40030	37438
00072	97125	40348	87083	31417	21815	39250	75237	62047	15501	29578
00073	21826	41134	47143	34072	64638	85902	49139	06441	03856	54552
00074	73135	42742	95719	09035	85794	74296	08789	88156	64691	19202
00075	07638	77929	03061	18072	96207	44156	23821	99538	04713	66994
00076	60528	83441	07954	19814	59175	20695	05533	52139	61212	06455
00077	83596	35655	06958	92983	05128	09719	77433	53783	92301	50498
00078	10850	62746	99599	10507	13499	06319	53075	71839	06410	19362
00079	39820	98952	43622	63147	64421	80814	43800	09351	31024	73167
00080	59580	06478	75569	78800	88835	54486	23768	06156	04111	08408
00081	38508	07341	23793	48763	90822	97022	17719	04207	95954	49953
00082	30692	70668	94688	16127	56196	80091	82067	63400	05462	69200
00083	65443	95659	18238	27437	49632	24041	08337	65676	96299	90836
00084	27267	50264	13192	72294	07477	44606	17985	48911	97341	30358
00085	91307	06991	19072	24210	36699	53728	28825	35793	28976	66252
00086	68434	94688	84473	13622	62126	98408	12843	82590	09815	93146
00087	48908	15877	54745	24591	35700	04754	83824	52692	54130	55160
00088	06913	45197	42672	78601	11883	09528	63011	98901	14974	40344
00089	10455	16019	14210	33712	91342	37821	88325	80851	43667	70883
00090	12883	97343	65027	61184	04285	01392	17974	15077	90712	26769
00091	21778	30976	38807	36961	31649	42096	63281	02023	08816	47449
00092	19523	59515	65122	59659	86283	68258	69572	13798	16435	91529
00093	67245	52670	35583	16563	79246	86686	76463	34222	26655	90802
00094	60584	47377	07500	37992	45134	26529	26760	83637	41326	44344
00095	53853	41377	36066	94850	58838	73859	49364	73331	96240	43642
00096	24637	38736	74384	89342	52623	07992	12369	18601	03742	83873
00097	83080	12451	38992	22815	07759	51777	97377	27585	51972	37867
00098	16444	24334	36151	99073	27493	70939	85130	32552	54846	54759
00099	60790	18157	57178	65762	11161	78576	45819	52979	65130	04860
00100	03991	10461	93716	16894	66083	24653	84609	58232	88618	19161
00101	38555	95554	32886	59780	08355	60860	29735	47762	71299	23853
00102	17546	73704	92052	46215	55121	29281	59076	07936	27954	58909
00103	32643	52861	95819	06831	00911	98936	76355	93779	80863	00514
00104	69572	68777	39510	35905	14060	40619	29549	69616	33564	60780
00105	24122	66591	27699	06494	14845	46672	61958	77100	90899	75754
00106	61196	30231	92962	61773	41839	55382	17267	70943	78038	70267
00107	30532	21704	10274	12202	39685	23309	10061	68829	55986	66485
00108	03788	97599	75867	20717	74416	53166	35208	33374	87539	08823
00109	48228	63379	85783	47619	53152	67433	35663	52972	16818	60311
00110	60365	94653	35075	33949	42614	29297	01918	28316	98953	73231
00111	83799	42402	56623	34442	34994	41374	70071	14736	09958	18065
00112	32960	07405	36409	83232	99385	41600	11133	07586	15917	06253
00113	19322	53845	57620	52606	66497	68646	78138	66559	19640	99413
00114	11220	94747	07399	37408	48509	23929	27482	45476	85244	35159
00115	31751	57260	68980	05339	15470	48355	88651	22596	03152	19121
00116	88492	99382	14454	04504	20094	98977	74843	93413	22109	78508
00117	30934	47744	07481	83828	73788	06533	28597	20405	94205	20380
00118	22888	48893	27499	98748	60530	45128	74022	84617	82037	10268
00119	78212	16993	35902	91386	44372	15486	65741	14014	87481	37220

Table R. *(continued)*

Row number					
00120	41849 84547	46850 52326	34677 58300	74910 64345	19325 81549
00121	46352 33049	69248 93460	45305 07521	61318 31855	14413 70951
00122	11087 96294	14013 31792	59747 67277	76503 34513	39663 77544
00123	52701 08337	56303 87315	16520 69676	11654 99893	02181 68161
00124	57275 36898	81304 48585	68652 27376	92852 55866	88448 03584
00125	20857 73156	70284 24326	79375 95220	01159 63267	10622 48391
00126	15633 84924	90415 93614	33521 26665	55823 47641	86225 31704
00127	92694 48297	39904 02115	59589 49067	66821 41575	49767 04037
00128	77613 19019	88152 00080	20554 91409	96277 48257	50816 97616
00129	38688 32486	45134 63545	59404 72059	43947 51680	43852 59693
00130	25163 01889	70014 15021	41290 67312	71857 15957	68971 11403
00131	65251 07629	37239 33295	05870 01119	92784 26340	18477 65622
00132	36815 43625	18637 37509	82444 99005	04921 73701	14707 93997
00133	64397 11692	05327 82162	20247 81759	45197 25332	83745 22567
00134	04515 25624	95096 67946	48460 85558	15191 18782	16930 33361
00135	83761 60873	43253 84145	60833 25983	01291 41349	20368 07126
00136	14387 06345	80854 09279	43529 06318	38384 74761	41196 37480
00137	51321 92246	80088 77074	88722 56736	66164 49431	66919 31678
00138	72472 00008	80890 18002	94813 31900	54155 83436	35352 54131
00139	05466 55306	93128 18464	74457 90561	72848 11834	79982 68416
00140	39528 72484	82474 25593	48545 35247	18619 13674	18611 19241
00141	81616 18711	53342 44276	75122 11724	74627 73707	58319 15997
00142	07586 16120	82641 22820	92904 13141	32392 19763	61199 67940
00143	90767 04235	13574 17200	69902 63742	78464 22501	18627 90872
00144	40188 28193	29593 88627	94972 11598	62095 36787	00441 58997
00145	34414 82157	86887 55087	19152 00023	12302 80783	32624 68691
00146	63439 75363	44989 16822	36024 00867	76378 41605	65961 73488
00147	67049 09070	93399 45547	94458 74284	05041 49807	20288 34060
00148	79495 04146	52162 90286	54158 34243	46978 35482	59362 95938
00149	91704 30552	04737 21031	75051 93029	47665 64382	99782 93478
00150	94015 46874	32444 48277	59820 96163	64654 25843	41145 42820
00151	74108 88222	88570 74015	25704 91035	01755 14750	48968 38603
00152	62880 87873	95160 59221	22304 90314	72877 17334	39283 04149
00153	11748 12102	80580 41867	17710 59621	06554 07850	73950 79552
00154	17944 05600	60478 03343	25852 58905	57216 39618	49856 99326
00155	66067 42792	95043 52680	46780 56487	09971 59481	37006 22186
00156	54244 91030	45547 70818	59849 96169	61459 21647	87417 17198
00157	30945 57589	31732 57260	47670 07654	46376 25366	94746 49580
00158	69170 37403	86995 90307	94304 71803	26825 05511	12459 91314
00159	08345 88975	35841 85771	08105 59987	87112 21476	14713 71181
00160	27767 43584	85301 88977	29490 69714	73035 41207	74699 09310
00161	13025 14338	54066 15243	47724 66733	47431 43905	31048 56699
00162	80217 36292	98525 24335	24432 24896	43277 58874	11466 16082
00163	10875 62004	90391 61105	57411 06368	53856 30743	08670 84741
00164	54127 57326	26629 19087	24472 88779	30540 27886	61732 75454
00165	60311 42824	37301 42678	45990 43242	17374 52003	70707 70214
00166	49739 71484	92003 98086	76668 73209	59202 11973	02902 33250
00167	78626 51594	16453 94614	39014 97066	83012 09832	25571 77628
00168	66692 13986	99837 00582	81232 44987	09504 96412	90193 79568
00169	44071 28091	07362 97703	76447 42537	98524 97831	65704 09514
00170	41468 85149	49554 17994	14924 39650	95294 00556	70481 06905
00171	94559 37559	49678 53119	70312 05682	66986 34099	74474 20740
00172	41615 70360	64114 58660	90850 64618	80620 51790	11436 38072
00173	50273 93113	41794 86861	24781 89683	55411 85667	77535 99892
00174	41396 80504	90670 08289	40902 05069	95083 06783	28102 57816
00175	25807 24260	71529 78920	72682 07385	90726 57166	98884 08583
00176	06170 97965	88302 98041	21443 41808	68984 83620	89747 98882
00177	60808 54444	74412 81105	01176 28838	36421 16489	18059 51061
00178	80940 44893	10408 36222	80582 71944	92638 40333	67054 16067
00179	19516 90120	46759 71643	13177 55292	21036 82808	77501 97427

Table R. *(continued)*

Row number										
00180	49386	54480	23604	23554	21785	41101	91178	10174	29420	90438
00181	06312	88940	15995	69321	47458	64809	98189	81851	29651	84215
00182	60942	00307	11897	92674	40405	68032	96717	54244	10701	41393
00183	92329	98932	78284	46347	71209	92061	39448	93136	25722	08564
00184	77936	63574	31384	51924	85561	29671	58137	17820	22751	36518
00185	38101	77756	11657	13897	95889	57067	47648	13885	70669	93406
00186	39641	69457	91339	22502	92613	89719	11947	56203	19324	20504
00187	84054	40455	99396	63680	67667	60631	69181	96845	38525	11600
00188	47468	03577	57649	63266	24700	71594	14004	23153	69249	05747
00189	43321	31370	28977	23896	76479	68562	62342	07589	08899	05985
00190	64281	61826	18555	64937	13173	33365	78851	16499	87064	13075
00191	66847	70495	32350	02985	86716	38746	26313	77463	55387	72681
00192	72461	33230	21529	53424	92581	02262	78438	66276	18396	73538
00193	21032	91050	13058	16218	12470	56500	15292	76139	59526	52113
00194	95362	67011	06651	16136	01016	00857	55018	56374	35824	71708
00195	49712	97380	10404	55452	34030	60726	75211	10271	36633	68424
00196	58275	61764	97586	54716	50259	46345	87195	46092	26787	60939
00197	89514	11788	68224	23417	73959	76145	30342	40277	11049	72049
00198	15472	50669	48139	36732	46874	37088	63465	09819	58869	35220
00199	12120	86124	51247	44302	60883	52109	21437	36786	49226	77837

Answers to Selected Exercises*

CHAPTER 1

1.1 a) Statistic b) Parameter c) Data
d) Data e) Inference from data f) Inference from data
g) Statistic

1.3 In accounting, numbers are used primarily to balance ledgers. In statistics, the numbers are the raw materials from which broad descriptive and inferential statements are made.

1.4 We would have established "truth" if there were no possibility of an error in measurement. However, errors in measurement do occur so, at best, we have established an approximation to "truth."

1.7 Here are a few possible examples:

All infants weighing less than five pounds at birth.
All male infants weighing less than five pounds at birth.
All female twins weighing less than five pounds at birth.
All male twins born in the month of January weighing less than five pounds at birth.

1.8 One method of selecting a random sample is to assign numerals to each member of the populations involved; then place slips of paper in a container with these numbers on them and finally select as many pieces of paper as specified. For example, in (a) you would assign *all* registered Democrats in Phoenix Arizona a numeral. The slips of paper would be placed in a container. You would then draw fifty. An alternative method would be to use the Table of Random Digits.

1.13 a) Constant b) Variable c) Variable
d) Variable e) Constant f) Variable
g) Constant h) Variable i) Variable

1.14 a) Descriptive b) Inferential c) Descriptive d) Descriptive
e) Descriptive f) Inferential g) Inferential h) Inferential
i) Inferential j) Descriptive

---

* In problems involving many steps, you may occasionally find a discrepancy between the answers you obtained and those found in the text. Where the discrepancies are small, they are probably due to rounding errors.

1.15

	Album		
Store	X	Y	Z
I	4	3	1
II	8	4	3
III	8	6	1
IV	10	7	0

The population consists of all potential purchasers, within the last month, of the three record albums in the community within which you live. The data are the numbers in the above table. The data yield neither statistics nor inferences; however, statistical procedures can be applied that will yield statistics and/or inferences.

1.16 The population is all potential purchasers of these albums during the last month throughout the United States.

1.17 The population consists of the times of completion for all 420 tasks.
The variable is the time required for completion of a task.
The data are the time scores.
The sample is the 45 time scores.

CHAPTER 2

2.1 a) $X = b - c - a$ b) $X = \dfrac{c - a}{2}$ c) $X = c - \dfrac{bY}{a}$
d) $X = ac/b$ e) $X = acY/b$ f) $X = (c^2Y/b) - a$
g) $X = \dfrac{a - cb}{b}$ h) $X = \sqrt[9]{Y}$ i) $X = \dfrac{1}{\sqrt[5]{a}}$

2.2 a) $X = 3$ b) $X = 24$ c) $X = \frac{125}{7776}$ d) $X = 27$ e) $X = 7\frac{1}{9}$

2.3 a) 100.00 b) 46.41 c) 2.96 d) 0.01
e) 16.46 f) 1.05 g) 86.21 h) 10.00

2.4 a) 0.03052 b) 0.084680 c) 0.12088
d) 0.38 e) 0.111 f) 0.85714

2.5 b) 0.2308; 23.08% c) 0.7576; 75.76% d) 0.4803; 48.03%

2.6 a) 25 b) 60 c) 37
d) 31 e) 60 f) 44

2.7 a) $\sum_{i=1}^{3} X_i$ b) $\sum_{i=1}^{N} X_i$ c) $\sum_{i=3}^{6} X_i^2$ d) $\sum_{i=4}^{N} X_i^2$

2.8 a) Ratio b) Ratio c) Nominal d) Ordinal e) Nominal

2.9 a) Ratio; continuous b) Interval; continuous
c) Nominal; discontinuous d) Ratio; continuous

2.10 a) 12.65 b) 4.00 c) 1.26 d) 0.40 e) 0.13

2.11 a) -0.5 to $+0.5$ b) 0.45 to 0.55 c) 0.95 to 1.05
d) 0.485 to 0.495 e) -4.5 to -5.5 f) -4.45 to -4.55

2.12 $\sum_{i=1}^{N} X_i^2 \neq \left(\sum_{i=1}^{N} X_i\right)^2$

$X_1 = 4$	$X_1^2 = 16$
$X_2 = 5$	$X_2^2 = 25$
$X_3 = 7$	$X_3^2 = 49$
$X_4 = 9$	$X_4^2 = 81$
$X_5 = 10$	$X_5^2 = 100$
$X_6 = 11$	$X_6^2 = 121$
$X_7 = 14$	$X_7^2 = 196$
$\sum_{i=1}^{N} X_i = 60$	$\sum_{i=1}^{N} X_i^2 = 588$

$\left(\sum_{i=1}^{N} X_i\right)^2 = 3600$

$588 \neq 3600$

2.13 b 2.14 b 2.15 a 2.16 50%

2.17 No, because we do not know the actual figures. For example, suppose Brand A increased from 100 to 140 and Brand B from 200 to 210. Brand A would show a higher percentage increase (40% as compared to 5%) but Brand B would be more popular than Brand A (210 as compared to 140).

2.18 $0.30 2.19 250

2.20 a)

Year	% Whole
1969	5.97
1970	14.43
1971	25.87
1972	27.36
1973	26.37

b)

1969 (as base year)	% Whole
1969	100.00
1970	241.67
1971	433.33
1972	458.33
1973	441.67

1971 (as base year)	% Whole
1969	23.08
1970	55.77
1971	100.00
1972	105.77
1973	101.92

c)

Year	Financial fraud	Theft	Unauthorized use	Vandalism
1969	1.49	2.99	0.00	1.49
1970	3.48	2.49	4.48	3.98
1971	10.95	8.96	2.99	2.99
1972	5.97	7.46	7.96	5.97
1973	10.45	7.46	3.98	4.48

2.21

	a)	b)	c)
Business administration	20.00%	38.10	13.33
Education	75.00%	4.76	20.00
Humanities	57.14%	14.29	26.67
Science	28.57%	23.81	13.33
Social science	50.00%	19.05	26.67

d) 58.33% Males
41.67% Females

2.23 a) Continuous b) Discrete c) Continuous d) Discrete e) Discrete

2.24 a)

Year	% Males	% Females
1969	51.29	48.71
1970	51.33	48.67
1971	51.27	48.73
1972	51.26	48.74
1973	51.26	48.74

b) i) Time Ratios, fixed base (1969)

Year	Males	Females
1969	100.00	100.00
1970	103.68	103.53
1971	98.70	98.80
1972	90.42	90.54
1973	87.06	87.17

ii) Time Ratios, moving base (preceding year)

Year	Males	Females
1969	100.0	100.0
1970	103.7	103.5
1971	95.2	95.4
1972	91.6	91.6
1973	96.3	96.3

2.25 a) 0.545 or 54.55% b) 0.909 or 90.91% c) 0.091 or 9.09%
d) 3:2 e) 10:1 f) 6:1

2.26 a) 282.27% b) 26.15%

2.27 a) i) $9.77 ii) $18.37 iii) $34.72 iv) $58.36
b) i) 13.03% ii) 15.31% iii) 17.36% iv) 19.45%

2.28 a) i) 70.9 ii) 105.5 iii) 148.7
b) i) 0.54 ii) 0.59 iii) 0.85 iv) 0.89
c) No. Individual items may not follow the general trend.
d)

Year	Price	Moving base
1969	$0.5445	100.0
1970	0.5745	105.5
1971	0.5920	103.0
1972	0.6175	104.3
1973	0.7070	114.5
1974	0.8085	114.2

2.29 a) 100% b) 50%

2.30 a) Men = 93.72% Women = 94.71%
b) Men = 2.61% Women = 4.79%
c) Men = 2.31% Women = 6.16%
d) Men = 4.18% Women = 29.78%

2.31 a) Twenty-three of the 124 cars (or 18%) averaged 20 mpg or better.
b) Of 40 cars weighing 4,000 pounds or more, none (0%) averaged 20 mph or more.
c) Of 59 manual transmissions, 23 (or 38%) averaged 20 mpg or more.
d) Of 65 automatic transmissions, none (0%) averaged 20 mpg or more.

2.32 a) 56% b) 0% c) 98% d) 18%

2.33 a)

Year	16–17	18–20	24
1968–1969	100.00	100.00	100.00
1969–1970	152.94	82.98	140.00
1970–1971	147.06	102.13	180.00
1971–1972	194.12	282.98	280.00
1972–1973	135.29	325.53	540.00

b)

	16–17	18–20	24
Before	54.84	31.91	40.70
After	45.16	68.09	59.30

c) The percentage of the 24-year old group is deceptive since the total number of collisions is relatively small and the total for the base year is extremely small. Almost any increase or decrease would appear as a large percentage of change when the base observation is only five.

2.34 0.70 made the correct decision
0.30 made the wrong decision
The results appear to contradict the common belief.

2.35

	Low dosage	High dosage
Proportion contracting cancer	0.09	0.32
Proportion not contracting cancer	0.91	0.68

CHAPTER 3

3.1

	True limits	Midpoint	Width
a)	7.5–12.5	10	5
b)	5.5–7.5	6.5	2
c)	(−0.5)–(+2.5)	1	3
d)	4.5–14.5	9.5	10
e)	(−1.5)–(−8.5)	−5	7
f)	2.45–3.55	3	1.1
g)	1.495–1.755	1.625	0.26
h)	(−3.5)–(+3.5)	0	7

3.2

	Width	Apparent limits	True limits	Midpoint
i)	7	0–6	0.5–6.5	3
ii)	1	29	28.5–29.5	29
iii)	2	18–19	17.5–19.5	18.5
iv)	4	(−30)–(−27)	(−30.5)–(−26.5)	−28.5
v)	0.01	0.30	0.295–0.305	0.30
vi)	0.006	0.206–0.211	0.2055–0.2115	0.2085

3.3

Class intervals	True limits	Midpoint	f	Cum f	Cum %
95–99	94.5–99.5	97	1	40	100.0%
90–94	89.5–94.5	92	3	39	97.5
85–89	84.5–89.5	87	4	36	90.0
80–84	79.5–84.5	82	8	32	80.0
75–79	74.5–79.5	77	11	24	60.0
70–74	69.5–74.5	72	4	13	32.5
65–69	64.5–69.5	67	3	9	22.5
60–64	59.5–64.5	62	3	6	15.0
55–59	54.5–59.5	57	0	3	7.5
50–54	49.5–54.5	52	1	3	7.5
45–49	44.5–49.5	47	1	2	5.0
40–44	39.5–44.5	42	1	1	2.5

3.4 b) $i = 3$

Class intervals	f	Class intervals	f
96–98	1	78–80	4
93–95	1	75–77	7
90–92	2	72–74	1
87–89	3	69–71	3
84–86	2	66–68	3
81–83	7	63–65	1

Class intervals	f	Class intervals	f
60–62	2	48–50	0
57–59	0	45–47	1
54–56	0	42–44	0
51–53	1	39–41	1

c) $i = 10$

Class intervals	f
90–99	4
80–89	12
70–79	15
60–69	6
50–59	1
40–49	2

d) $i = 20$

Class intervals	f
80–99	16
60–79	21
40–59	3

3.5

Class intervals	True limits	Midpoint	f
9.6–9.9	9.55–9.95	9.75	1
9.2–9.5	9.15–9.55	9.35	3
8.8–9.1	8.75–9.15	8.95	1
8.4–8.7	8.35–8.75	8.55	5
8.0–8.3	7.95–8.35	8.15	7
7.6–7.9	7.55–7.95	7.75	9
7.2–7.5	7.15–7.55	7.35	2
6.8–7.1	6.75–7.15	6.95	4
6.4–6.7	6.35–3.75	6.55	2
6.0–6.3	5.95–6.35	6.15	3
5.6–5.9	5.55–5.95	5.75	0
5.2–5.5	5.15–5.55	5.35	1
4.8–5.1	4.75–5.15	4.95	0
4.4–4.7	4.35–4.75	4.55	1
4.0–4.3	3.95–4.35	4.15	1

width = 0.4

3.6 a) $i = 7$

b) 14.5–21.5
7.5–14.5
0.5–7.5

c) 18
11
4

3.7 a)

Class intervals	f
230–244	1
215–229	0
200–214	0
185–199	1
170–184	0
155–169	1
140–154	0
125–139	2
110–124	4
95–109	9
80–94	13
65–79	20
50–64	34
35–49	47
20–34	82
5–19	36

Apparent limits	Real limits
5–19	4.5–19.5

b) $i = 25$

Class intervals	f	Class intervals	f
225–249	1	100–124	11
200–224	0	75–99	21
175–199	1	50–74	48
150–174	1	25–49	107
125–149	2	0–24	58

Apparent limits	Real limits
0–24	0.5–24.5

c) $i = 50$

Class intervals	f
200–249	1
150–199	2
100–149	13
50–99	69
0–49	165

Apparent limits	Real limits
0–49	0.5–49.5

d) $i = 1$

Class interval	f	Class interval	f	Class interval	f	Class interval	f
230	1	80	1	53	2	29	4
185	1	78	3	52	2	28	5
163	1	77	1	51	2	27	4
132	2	76	2	50	3	26	8
120	1	74	1	48	2	25	2
115	1	73	1	47	2	24	8
114	1	72	1	46	1	22	9
112	1	70	2	45	2	21	4
108	1	69	3	44	3	20	1
107	3	68	2	43	7	18	4
104	1	67	1	42	6	17	3
103	1	66	2	41	2	16	4
101	1	65	1	40	6	15	5
97	1	64	2	38	5	14	3
96	1	63	1	37	2	13	3
93	2	62	3	36	4	12	3
91	3	61	1	35	5	11	4
89	1	58	4	34	8	10	3
86	2	57	1	33	8	9	2
85	2	56	4	32	11	8	1
84	1	55	3	31	3	6	1
81	1	54	6	30	7		

Apparent limits	Real limits
6	5.5–6.5

3.8 b)

1967	% Change	1967	% Change	1967	% Change
$3\frac{1}{2}$	185.71	$26\frac{7}{8}$	16.74	35	82.86
$7\frac{3}{4}$	77.42	$27\frac{1}{4}$	110.09	$38\frac{2}{8}$	17.32
$10\frac{7}{8}$	75.86	$27\frac{3}{4}$	−4.95	$39\frac{7}{8}$	−13.48
13	25.96	$27\frac{3}{4}$	−19.82	44.78	65.72
$15\frac{1}{8}$	80.17	$28\frac{3}{8}$	1.32	46	2.17
$15\frac{7}{8}$	296.06	$29\frac{1}{8}$	−35.19	$53\frac{1}{4}$	−8.45
$18\frac{1}{2}$	129.73	$29\frac{5}{8}$	16.88	$57\frac{3}{4}$	22.51
$21\frac{3}{8}$	19.30	$30\frac{7}{8}$	10.12	$70\frac{1}{2}$	−9.04
$22\frac{1}{2}$	63.89	$32\frac{1}{4}$	28.68	102	−6.86
$25\frac{1}{2}$	80.39	33	21.21	$103\frac{1}{4}$	20.10

3.9 a) $\frac{11}{30}$ b) $\frac{10}{30}$ c) $\frac{7}{30}$ d) $\frac{6}{30}$

3.10

Class intervals	f	Class intervals	f
65–69	1	30–34	6
60–64	2	25–29	5
55–59	3	20–24	4
50–54	4	15–19	3
45–49	5	10–14	2
40–44	6	5–9	1
35–39	8		

3.11

Class intervals	f	Class intervals	f
63–67	3	28–32	0
58–62	0	23–27	9
53–57	7	18–22	0
48–52	0	13–17	5
43–47	11	8–12	0
38–42	0	3–7	1
33–37	14		

3.12

Class intervals	f	Class intervals	f
66–67	1	34–35	4
64–65	1	32–33	3
62–63	1	30–31	0
60–61	0	28–29	0
58–59	0	26–27	4
56–57	2	24–25	3
54–55	3	22–23	2
52–53	2	20–21	0
50–51	0	18–19	0
48–49	0	16–17	2
46–47	2	14–15	2
44–45	7	12–13	1
42–43	2	10–11	0
40–41	0	8–9	0
38–39	0	6–7	0
36–37	7	4–5	1

3.13

Class intervals	f
60–69	3
50–59	7
40–49	11
30–39	14
20–29	9
10–19	5
0–9	1

3.15 a) 0.200 b) 0.178 c) 0.089 d) 0.067
e) 0.044 f) 0.022 g) 0.156 h) 0.111
i) 0.067 j) 0.044 k) 0.022

3.16

Class interval	True limits	Midpoint	Frequency	Cumulative frequency	Cumulative percentage
1.5–1.6	1.45–1.65	1.55	1	45	100.00%
1.3–1.4	1.25–1.45	1.35	5	44	97.78%
1.1–1.2	1.05–1.25	1.15	12	39	86.67%
0.9–1.0	0.85–1.05	0.95	16	27	60.00%
0.7–0.8	0.65–0.85	0.75	8	11	24.44%
0.5–0.6	0.45–0.65	0.55	3	3	6.67%

3.17

Interval	Frequency	True limits	Midpoints
$88–89	2	87.5–89.5	88.5
86–87	2	85.5–87.5	86.5
84–85	2	83.5–85.5	84.5
82–83	2	81.5–83.5	82.5
80–81	2	79.5–81.5	80.5
78–79	2	77.5–79.5	78.5
76–77	4	75.5–77.5	76.5
74–75	9	73.5–75.5	74.5
72–73	4	71.5–73.5	72.5
70–71	3	69.5–71.5	70.5
68–69	2	67.5–69.5	68.5
66–67	2	65.5–67.5	66.5
64–65	2	63.5–65.5	64.5
62–63	1	61.5–63.5	62.5
60–61	1	59.5–61.5	60.5

3.18 a)

Interval	Frequency	Interval	Frequency
87–89	3	72–74	8
84–86	3	69–71	3
81–83	2	66–68	4
78–80	4	63–65	3
75–77	9	60–62	1

3.18 b)

Interval	Frequency
80–89	10
70–79	22
60–69	8

3.19 a) 1/40 b) 1/40 c) 1/8
d) 1/10 e) 1/20 f) 9/40

3.20

Class interval	Frequency	Cumulative frequency	Cumulative percentage
42–44	1	50	100.00%
39–41	0	49	98.00%
36–38	0	49	98.00%
33–35	1	49	98.00%
30–32	1	48	96.00%
27–29	1	47	94.00%
24–26	1	46	92.00%
21–23	1	45	90.00%
18–20	2	44	88.00%
15–17	3	42	84.00%
12–14	4	39	78.00%
9–11	11	35	70.00%
6–8	10	24	48.00%
3–5	8	14	28.00%
0–2	6	6	12.00%

3.21 Midpoint = 194
Width = 9
Apparent limits = 190–198
True limits = 189.5–198.5

3.22

Class interval	Frequency	Cumulative frequency	Cumulative percentage
56–59	1	50	100%
52–55	1	49	98%
48–51	1	48	96%
44–47	2	47	94%
40–43	3	45	90%
36–39	4	42	84%
32–35	7	38	76%
28–31	11	31	62%
24–27	8	20	40%
20–23	5	12	24%
16–19	2	7	14%
12–15	2	5	10%
8–11	1	3	6%
4–7	1	2	4%
0–3	1	1	2%

3.23

Class interval	Frequency	Cumulative frequency	Cumulative percentage
56–59	3	50	100%
52–55	7	47	94%
48–51	8	40	80%
44–47	6	32	64%
40–43	5	26	52%
36–39	3	21	42%
32–35	3	18	36%
28–31	5	15	30%
24–27	4	10	20%
20–23	3	6	12%
16–19	0	3	6%
12–15	1	3	6%
8–11	1	2	4%
4–7	0	1	2%
0–3	1	1	2%

3.24

Class interval	Frequency of room A	Frequency of first room B	Frequency of second room B
56–59	0	1	3
52–55	1	0	6
48–51	1	0	7
44–47	1	1	5
40–43	2	1	3
36–39	2	2	1
32–35	3	4	0
28–31	5	6	0
24–27	4	4	0
20–23	3	2	0
16–19	0	2	0
12–15	1	1	0
8–11	1	0	0
4–7	0	1	0
0–3	1	0	0

3.25

Class interval	f
28–29	1
26–27	3
24–25	5
22–23	6
20–21	8
18–19	16
16–17	17
14–15	14
12–13	27
10–11	27

3.26 a)

Class intervals	f
427–451	1
402–426	
377–401	
352–376	
327–351	1
302–326	
277–301	
252–276	
227–251	
202–226	1
177–201	2
152–176	1
127–151	1
102–126	3
77–101	2
52–76	8
27–51	13
2–26	19

b)

Class intervals	f
316–330	2
301–315	1
286–300	0
271–285	3
256–270	6
241–255	3
226–240	3
211–225	1
196–210	6
181–195	2
166–180	7
151–165	3
136–150	3
121–135	3
106–120	4
91–105	2
76–90	1
61–75	0
46–60	2

3.27

Class intervals	f 1974	f 1973
6.6–6.8		1
6.3–6.5		3
6.0–6.2		3
5.7–5.9	2	1
5.4–5.6	1	2
5.1–5.3	2	2
4.8–5.0	0	4
4.5–4.7	4	8
4.2–4.4	6	2
3.9–4.1	6	8
3.6–3.8	4	7
3.3–3.5	16	4
3.0–3.2	6	1
2.7–2.9	0	3
2.4–2.6	2	2
2.1–2.3	1	0
1.8–2.0	1	0

3.28

Class intervals	f
35.6–37.9	1
33.2–35.5	0
30.8–33.1	0
28.4–30.7	1
26.0–27.3	0
23.6–25.9	2
21.2–23.5	3
18.8–21.1	1
16.4–18.7	4
14.0–16.3	2
12.6–13.9	0
10.2–12.5	9
7.8–10.1	8
5.4–7.7	5
3.0–5.3	7
0.6–2.9	13

3.29

Class intervals	f
5070–5369	1
4770–5069	0
4470–4769	1
4170–4469	0
3870–4169	4
3570–3869	6
3270–3569	8
2970–3269	10
2670–2969	6
2370–2669	8
2070–2369	10
1770–2069	11
1470–1769	1
1170–1469	2
870–1169	5
570–869	27
270–569	28

CHAPTER 4

4.1

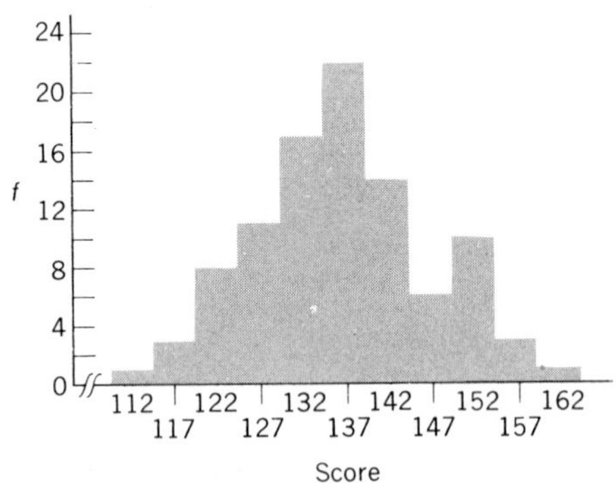

4.4 a)

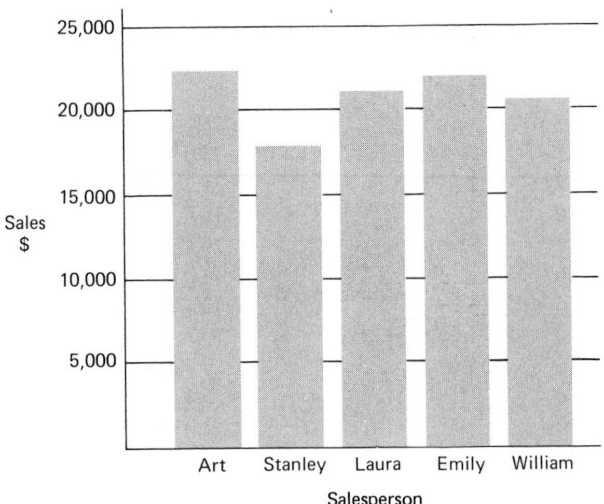

b)

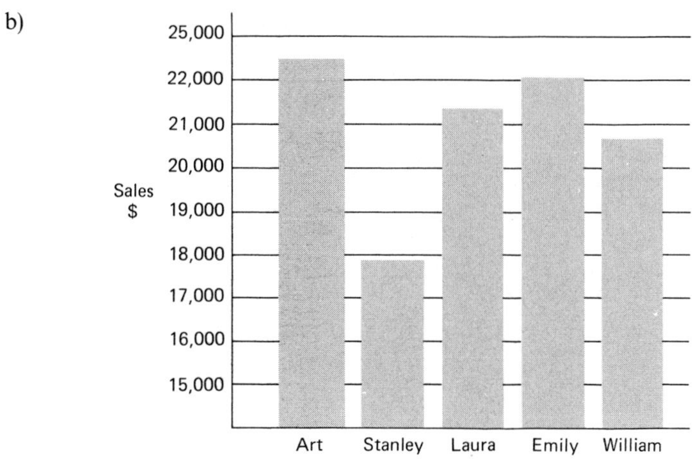
25,000
22,000
21,000
20,000
Sales
$
19,000
18,000
17,000
16,000
15,000
Art
Stanley
Laura
Emily
William
Salesperson

c)

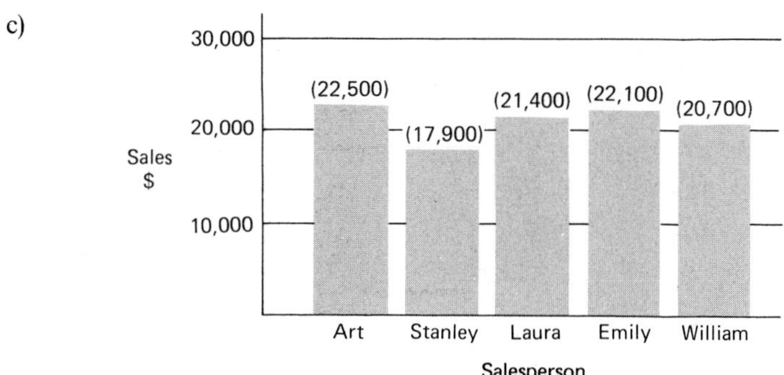
30,000
20,000
10,000
Sales
$
(22,500)
(17,900)
(21,400)
(22,100)
(20,700)
Art
Stanley
Laura
Emily
William
Salesperson

4.5

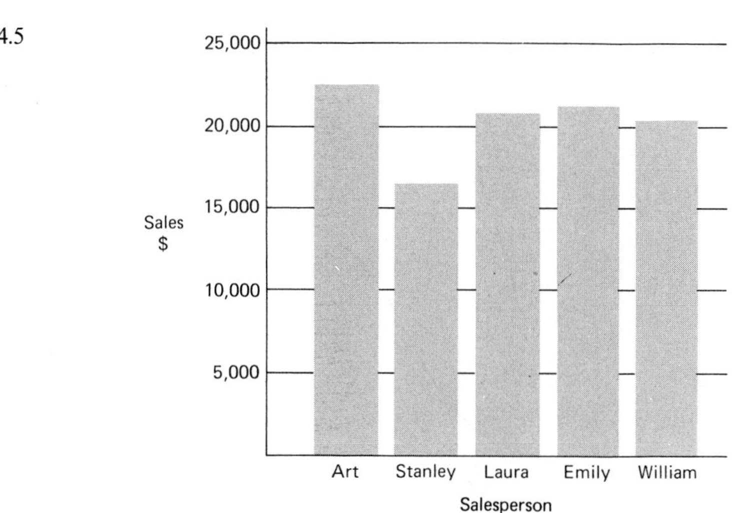
25,000
20,000
15,000
10,000
5,000
Sales
$
Art
Stanley
Laura
Emily
William
Salesperson

4.6

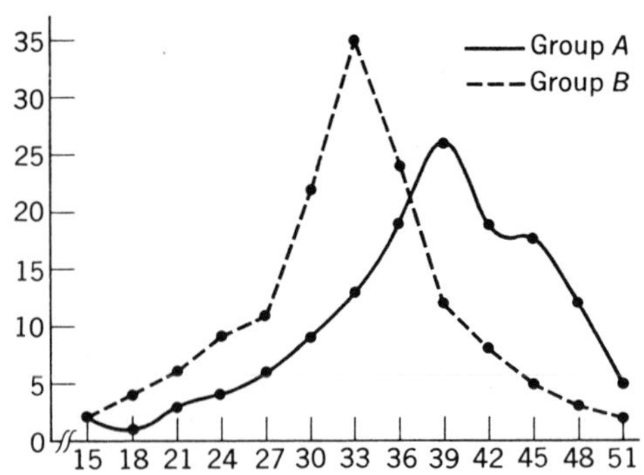

4.7 a) Positively skewed
b) Normal
c) Normal
d) Bimodal

4.8

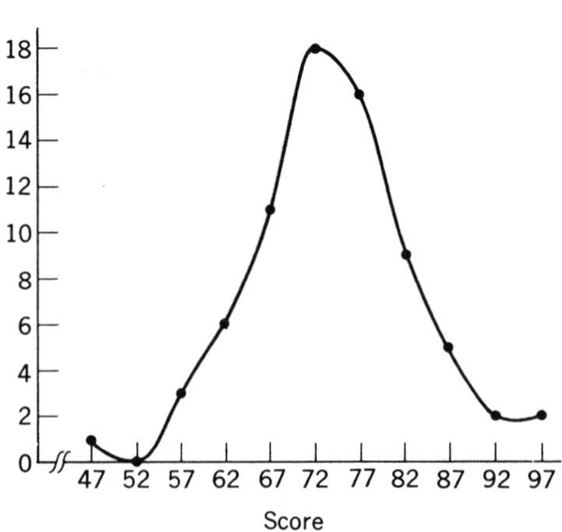

4.9

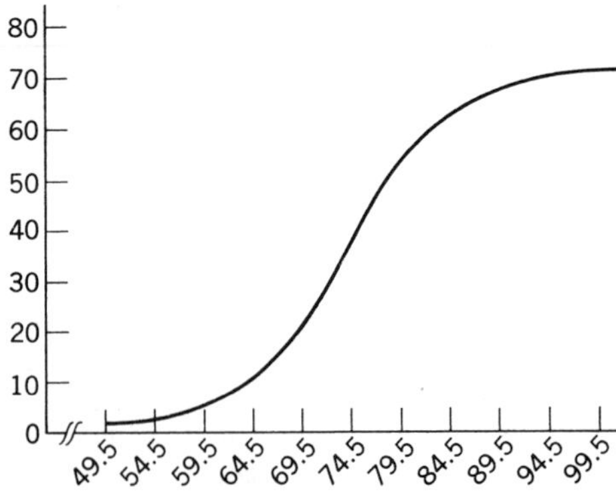

4.16 a) 34.63%

b) and c)

Class intervals	f	cf
92–94	1	52
89–91	0	51
86–88	2	51
83–85	0	49
80–82	0	49
77–79	3	49
74–76	5	46
71–73	7	41
68–70	5	34
65–67	10	29
62–64	11	19
59–61	4	8
56–58	4	4

d)

Class intervals	f	cf
90–94	1	52
85–89	2	51
80–84	0	49
75–79	7	49
70–74	9	42
65–69	14	33
60–64	15	19
55–59	4	4

4.17

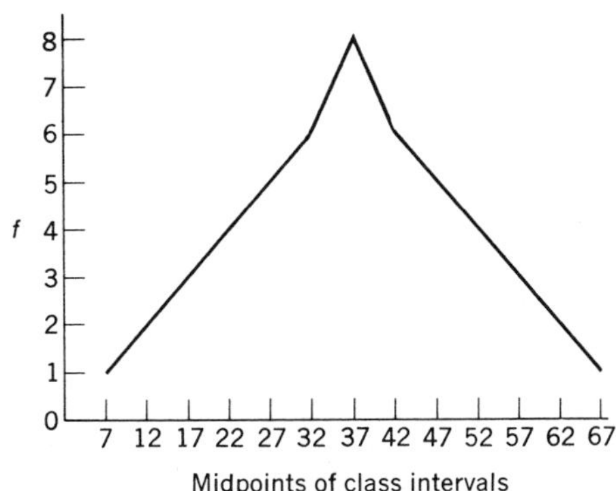

4.18

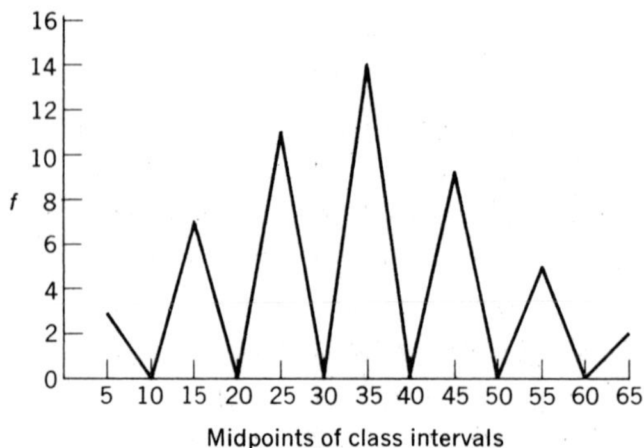
f
16
14
12
10
8
6
4
2
0
5 10 15 20 25 30 35 40 45 50 55 60 65
Midpoints of class intervals

4.19

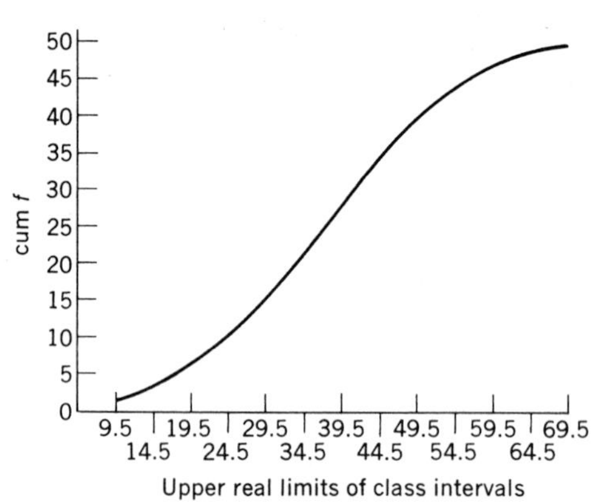
cum f
50
45
40
35
30
25
20
15
10
5
0
9.5 19.5 29.5 39.5 49.5 59.5 69.5
14.5 24.5 34.5 44.5 54.5 64.5
Upper real limits of class intervals

4.20

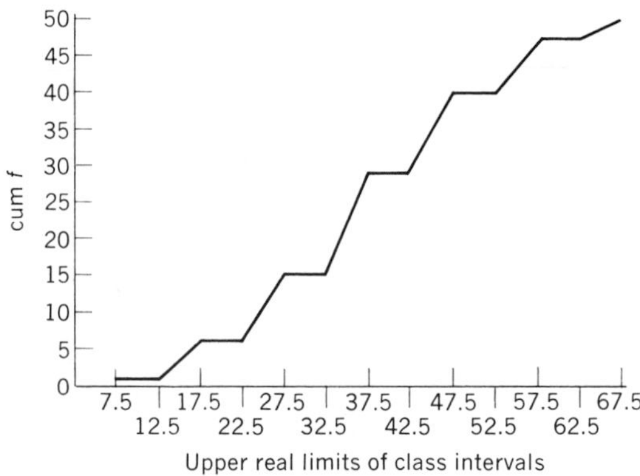
cum f
50
45
40
35
30
25
20
15
10
5
0
7.5 17.5 27.5 37.5 47.5 57.5 67.5
12.5 22.5 32.5 42.5 52.5 62.5
Upper real limits of class intervals

4.26

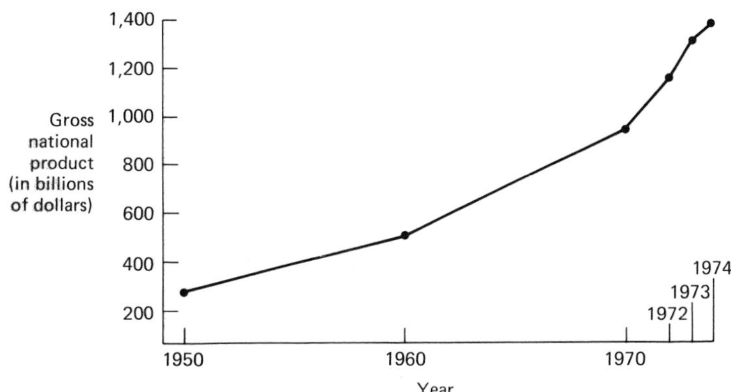
1,400
1,200
1,000
800
600
400
200
Gross national product (in billions of dollars)
1950
1960
1970
1972
1973
1974
Year

4.27

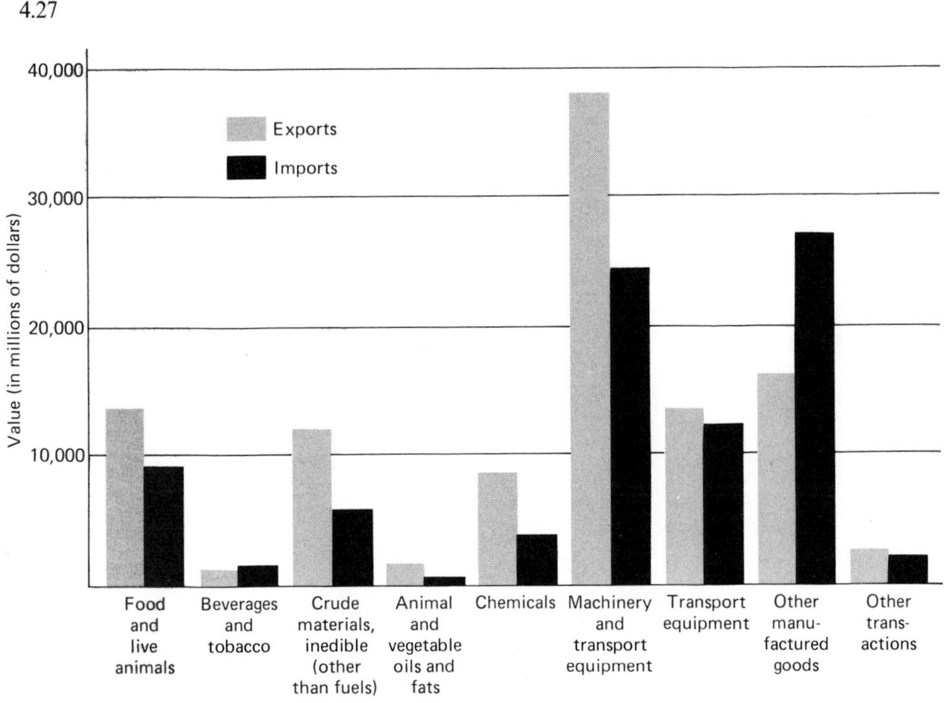
40,000
30,000
20,000
10,000
Value (in millions of dollars)
Exports
Imports
Food and live animals
Beverages and tobacco
Crude materials, inedible (other than fuels)
Animal and vegetable oils and fats
Chemicals
Machinery and transport equipment
Transport equipment
Other manu-factured goods
Other trans-actions
Commodity

4.32

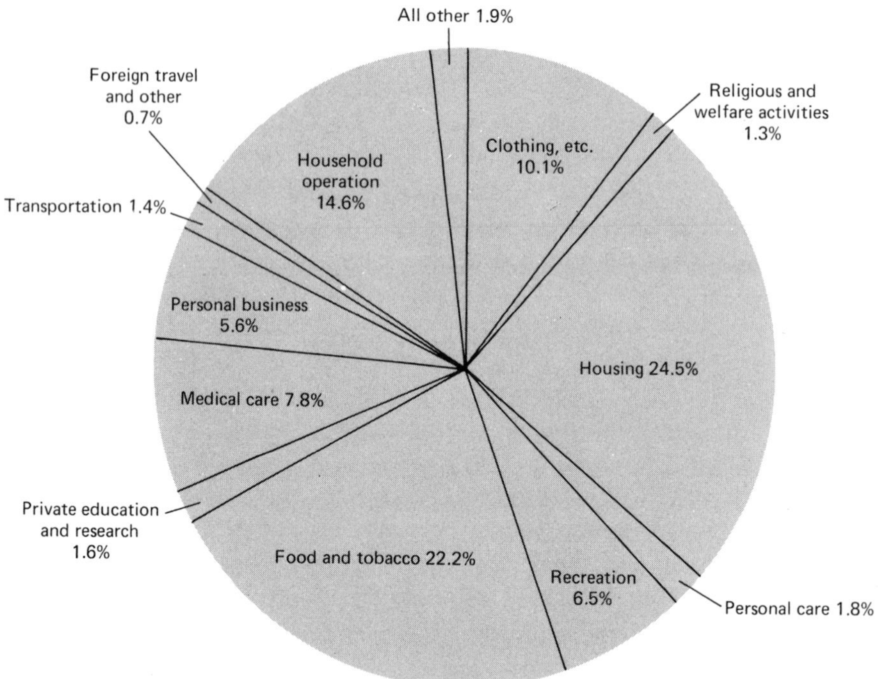
All other 1.9%
Foreign travel
and other
0.7%
Religious and
welfare activities
1.3%
Household
operation
14.6%
Clothing, etc.
10.1%
Transportation 1.4%
Personal business
5.6%
Housing 24.5%
Medical care 7.8%
Private education
and research
1.6%
Food and tobacco 22.2%
Recreation
6.5%
Personal care 1.8%

4.35 a)

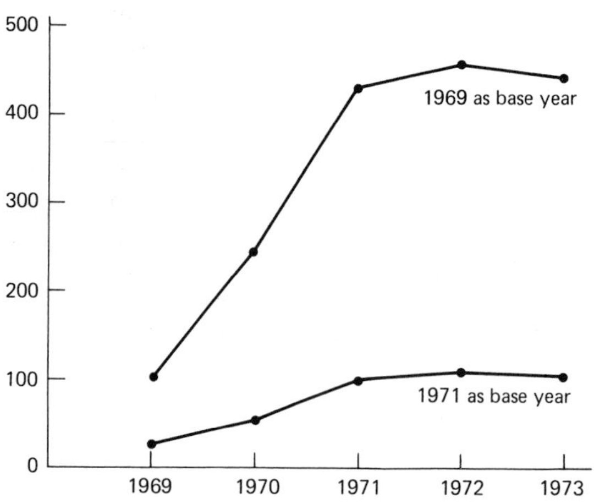
500
400
300
200
100
0
1969 as base year
1971 as base year
1969
1970
1971
1972
1973

4.35 c)

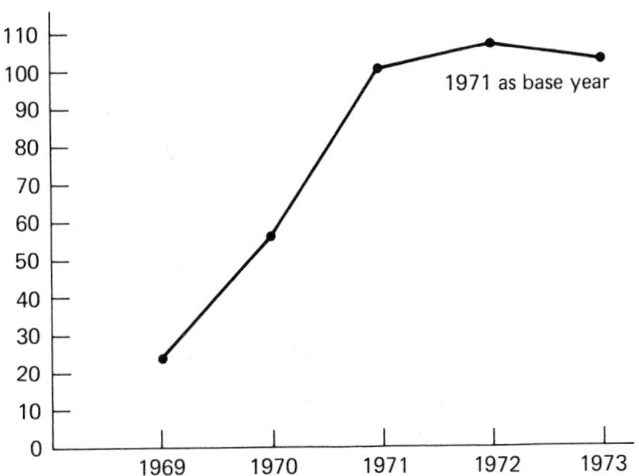

4.36 a)

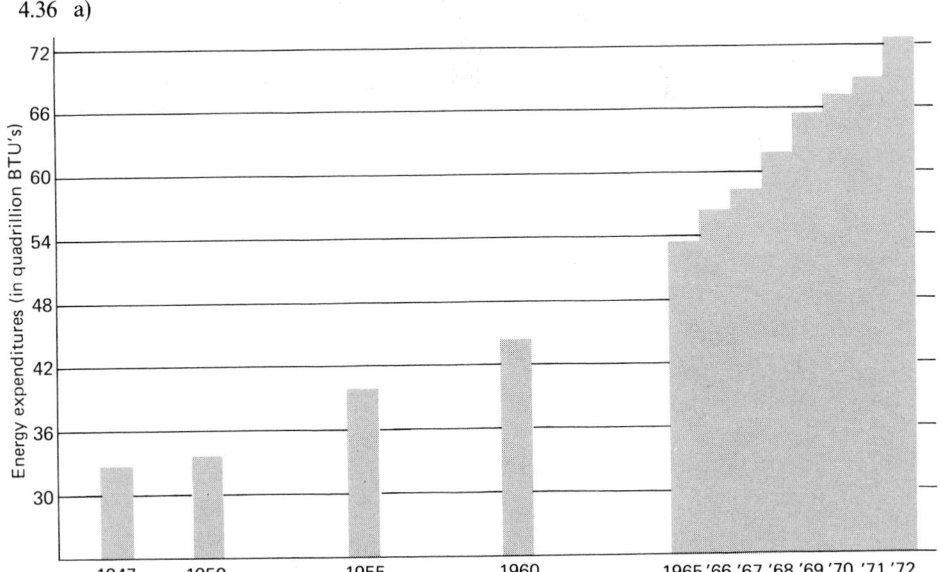

4.37 a)

Hours per week doing homework	Percent	
	Whites	Blacks
None	4	3
1–4	29	24
5–9	36	36
10–14	22	26
15–19	7	7
20 or more	2	5

4.37 b)

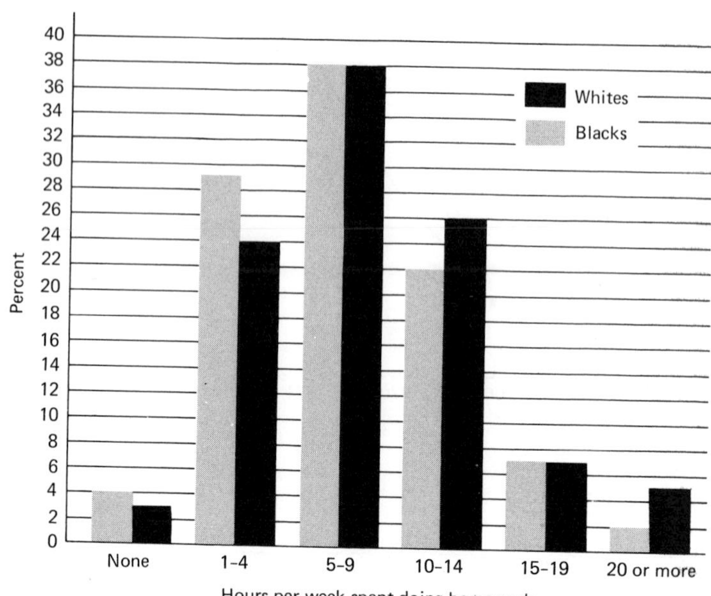
40
38
36
34
32
30
28
26
24
22
20
18
16
14
12
10
8
6
4
2
0
Percent
Whites
Blacks
None
1-4
5-9
10-14
15-19
20 or more
Hours per week spent doing homework

4.38

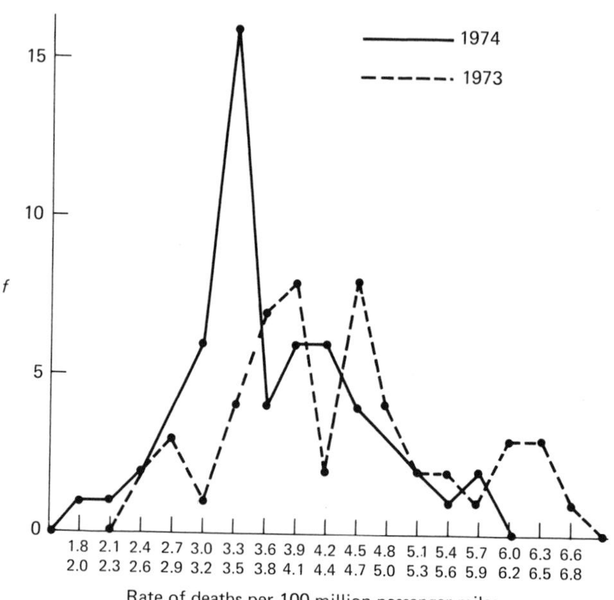
15
10
5
0
f
1974
1973
1.8 2.0
2.1 2.3
2.4 2.6
2.7 2.9
3.0 3.2
3.3 3.5
3.6 3.8
3.9 4.1
4.2 4.4
4.5 4.7
4.8 5.0
5.1 5.3
5.4 5.6
5.7 5.9
6.0 6.2
6.3 6.5
6.6 6.8
Rate of deaths per 100 million passenger miles

*4.39 a) Sum of two

scores	f	scores	f
12	1	5	6
11	2	4	5
10	3	3	4
9	4	2	3
8	5	1	2
7	6	0	1
6	7		
			$n = 49$

CHAPTER 5

5.2 a) 26.36 b) 46.82 c) 88.45

5.4 a) a) 103.88 b) 113.03 c) 124.78

5.6 a) 81.02 b) 94.76 c) 88.04

5.7 a) $\frac{16.9}{110} = 15.36\%$ b) $\frac{93.1}{110} = 84.64\%$ c) 2.45%
d) 0.18% e) 26.73% f) 3.73%

5.8 b) 57, 76 c) Law school freshmen, 60

5.9 a) 2 b) 9.6 c) 19.2 d) 36 e) 42 f) 84.8

5.10 a) 2 b) 12 c) 17.4 d) 30 e) 41.2 f) 80

5.11 a) 2 b) 12 c) 17.5 d) 31.5 e) 40 f) 82

5.12 a) 2 b) 10 c) 20.1 d) 37 e) 44 f) 84.2

5.13 a) 17.8 b) 24.5 c) 37 d) 40.3 e) 47

5.14 a) 16.5 b) 24.7 c) 36.1 d) 43.1 e) 46.4

5.15 a) 16.5 b) 24.8 c) 36.4 d) 42.5 e) 45.4

5:16 a) 17.5 b) 23.9 c) 36.6 d) 40.5 e) 47.2

5.17 a) 99.75 b) 91.0 c) 30.0
d) 12.5 e) 7.5 f) 0.25

5.18 a) 15.5 b) 23.75 c) 26.25
d) 29.318 e) 35.214 f) 43.5

5.19 a) 0.24 b) 0.76

5.20 50th percentile: a) 0.50 b) 0.50
75th percentile: a) 0.76 b) 0.24

5.21 a) 99.5; 98.5 b) 93.5; 23.5 c) 89.5; 8.5 d) 87.0; 3.5

5.22 a) 16.5; 41.5 b) 29.17; 49.5 c) 34.25; 53.33 d) 35.5; 54.17

5.23 8.77

5.24 $74.61 a) $14.49 b) $3.78 c) $8.67

5.25 and 5.26

A. 88.8	F. 42.0	K. 77.2	P. 42.0	U. 42.0
B. 77.2	G. 96.4	L. 61.6	Q. 25.6	V. 13.2
C. 25.6	H. 61.6	M. 4.8	R. 4.8	W. 61.6
D. 61.6	I. 77.2	N. 25.6	S. 0.4	X. 25.6
E. 42.0	J. 13.2	O. 88.8	T. 13.2	Y. 42.0

5.27 a) 52.82 b) 43.15 c) 78.63 d) 9.68 e) 56.85 f) 98.99 g) 62.30

5.28 a) 31.66 b) 16.92 c) 26.39 d) 14.59 e) 15.21 f) 24.64

5.29 a)

	1974		1973	
	Cum f	Cum %	Cum f	Cum %
6.6–6.8			51	100.0
6.3–6.5			50	98.0
6.0–6.2			47	92.2
5.7–5.9	51	100.0	44	86.3
5.4–5.6	49	96.1	43	84.3
5.1–5.3	48	94.1	41	80.4
4.8–5.0	46	90.2	39	76.5
4.5–4.7	46	90.2	35	68.6
4.2–4.4	42	82.4	27	52.9
3.9–4.1	36	70.6	25	49.0
3.6–3.8	30	58.8	17	33.3
3.3–3.5	26	51.0	10	19.6
3.0–3.2	10	19.6	6	12.0
2.7–2.9	4	8.0	5	9.8
2.4–2.6	4	8.0	2	4.0
2.1–2.3	2	4.0	0	0
1.8–2.0	1	2.0	0	0

b)

i)	Alabama	1973:	91.2	1974:	68.6
ii)	Arizona	1973:	87.3	1974:	88.8
iii)	California	1973:	31.0	1974:	13.7
iv)	Illinois	1973:	35.9	1974:	35.3
v)	New York	1973:	60.8	1974:	39.2

5.30

Sum	f	Cum f	Cum %	Sum	f	Cum f	Cum %
12	1	49	100	5	6	21	43
11	2	48	98	4	5	15	31
10	3	46	94	3	4	10	20
9	4	43	88	2	3	6	12
8	5	39	80	1	2	3	6
7	6	34	69	0	1	1	2
6	7	28	57				

$N = 49$

*5.31 a) 98% b) 45% c) 12% d) 12% e) 24%

CHAPTER 6

6.1 a) $\bar{X} = 4.7$ Median = 4.5 Mode = 8
b) $\bar{X} = 5.0$ Median = 5.0 Mode = 5.0
c) $\bar{X} = 17.5$ Median = 4 Mode = 4

6.2 (c)

6.3 a)

X	$(X-10)^2$	$(X-8)^2$	$(X-6)^2$	$(X-0)^2$	$(X-3)^2$	$(X-2)^2$	$(X-\bar{X})^2$
10	0	4	16	100	49	64	28.09
8	4	0	4	64	25	36	10.89
6	16	4	0	36	9	16	1.69
0	100	64	36	0	9	4	22.09
8	4	0	4	64	25	36	10.89
3	49	25	9	9	0	1	2.89
2	64	36	16	4	1	0	7.29
2	64	36	16	4	1	0	7.29
8	4	0	4	64	25	36	10.89
0	100	64	36	0	9	4	22.09
$N = 10$	405	233	141	345	153	197	124.10
$\bar{X} = 4.7$							

6.4 All measures of central tendency will be reduced by 5.

6.5 All measures of central tendency will be divided by 16.

6.6 Exercise 4.1
$\bar{X} = 136.61$ Median = 136.32 Mode = 137
Exercise 4.18
$\bar{X} = 74.05$ Median = 73.81 Mode = 72

6.7 $\bar{X} = 5.0$; median unchanged, mode unchanged

6.8 a) Negative skew b) Positive skew
c) No evidence of skew d) No evidence of skew

6.9 8c) Symmetrical 8d) Bimodal

6.10 $\bar{X} = 5.0$:
a) $X = 7.0$ b) $\bar{X} = 3.0$ c) No change d) $\bar{X} = 10.0$ e) $\bar{X} = 2.5$

6.11 Group A: median; Group B: mean

6.12 b)

	Means f	Medians f
6	1	3
5	6	5
4	11	9
3	7	6
2	4	3
1	1	4

d) Drawing from means: 1) $\frac{1}{30}$ 2) $\frac{1}{30}$ 3) $\frac{2}{30}$
Drawing from medians: 1) $\frac{3}{30}$ 2) $\frac{4}{30}$ 3) $\frac{7}{30}$

6.13 $\bar{X} = 89.9$, Median = 90.0, Mode = 90

6.14 a) \$3.03 b) \$3.13

6.15 a) 71.9 b)

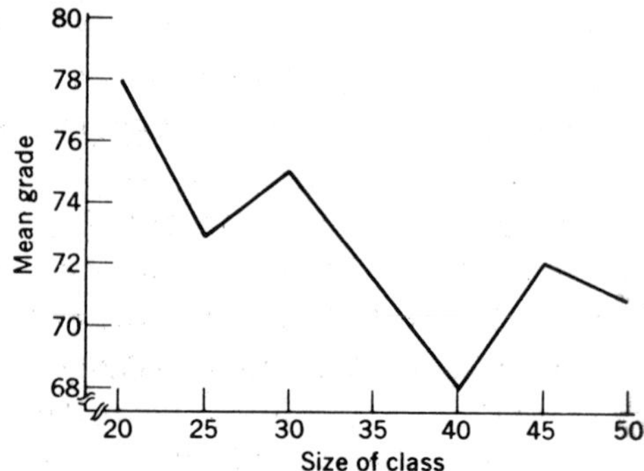

6.17 $\bar{X} = -1.44$, Median $= -0.25$, Mode $= -5.00$

6.18

	−9.0	0.0	
	−6.0	3.0	
	−5.0	4.0	
	−5.0	4.0	
	−0.5	8.5	
	0	9.0	
	0.1	9.1	
	2.0	11.0	
	4.0	13.0	
	5.0	14.0	
$\bar{X}$ =	−1.44	7.56	(−9.0) = −1.44
Median =	−0.25	8.75	(−0.0) = −0.25
Mode =	−5.0	4.0	(−9.0) = −5.0

6.22 Raise median: (c) Raise mean: (b)

6.23 Mean = 10.78 Median = 8.77 Mode = 10

6.24 Mean = \$75.02 Median = \$75.00 Mode = \$75.00

6.25 Mean = 1.00 Median = 1.0 Mode = 1.0

6.26 2.08

6.27 a) Mean = 30 Median = 30 Mode = 30
b) Mean = 30 Median = 30 Mode = 25; 30; 35
c) Mean = 25.83 Median = 27.5 Mode = 25; 30

6.28 a)

X	$(X - \bar{X})$
10	−16
25	−1
30	4
36	10
25	−1
30	4

$\sum(X - \bar{X}) = 0$

b)

X	median $(X - 27.5)$	mode $(X - 25)$	mode $(X - 30)$
10	−17.5	−15	−20
25	−2.5	0	−5
30	2.5	5	0
36	8.5	11	6
25	−2.5	0	−5
30	2.5	5	0
	$\sum$ −9.0	6	−24

6.29 a) 7; 7; 7 b) 7; 7; 7 c) 7; 7; 7

6.30 Mean = 6 Median = 6 Mode = 5; 7

6.31 $0.69

6.32 $4.90

6.33 a) either mean, median or mode b) mode

6.34 a) 820.57 b) 109,135.1

6.35 1.83%

6.36 Mean = 15.6 Median = 15 Mode = 11

6.37

Second draw	First draw 0	1	2	3	4	5	6
0	0	0.5	1.0	1.5	2.0	2.5	3.0
1	0.5	1.0	1.5	2.0	2.5	3.0	3.5
2	1.0	1.5	2.0	2.5	3.0	3.5	4.0
3	1.5	2.0	2.5	3.0	3.5	4.0	4.5
4	2.0	2.5	3.0	3.5	4.0	4.5	5.0
5	2.5	3.0	3.5	4.0	4.5	5.0	5.5
6	3.0	3.5	4.0	4.5	5.0	5.5	6.0

*6.38

	f		f
6.0	1	2.5	6
5.5	2	2.0	5
5.0	3	1.5	4
4.5	4	1.0	3
4.0	5	0.5	2
3.5	6	0	1
3.0	7		
			$n = 49$

*6.39 a) 2% b) 6% c) 43% d) 12%

*6.40 The two means are identical.

CHAPTER 7

7.1 $s^2 = 2.0$, $s = 1.41$
a) No change b) No change c) Increase d) Increase e) Decrease

	Effect on mean	Effect on standard definition
a) Add a constant to each score	Increased by constant	No change
b) Subtract a constant from each score	Decreased by constant	No change
c) Add and subtract a constant	No change	Increase
d) Multiply each score by a constant	Increased by constant factor	Increase
e) Divide each score by a constant	Decreased by constant factor	Decrease

7.3 Only the standard deviation satisfies these conditions. However, if the properties included *powers* of the constant, the variance would qualify as a measure of dispersion.

7.4 Exercise 5.4: No change
Exercise 5.5: Standard deviation will be divided by 16.

7.5 All scores have the same value.

7.6 a) 3.71 b) 2.45 c) 6.30 d) 0

7.8 a) 10 b) 8 c) 20 d) 0

7.9 The following is based on the frequency distribution presented in Exercise 3.7(a):

Midpoints	X^2	f	fX^2
237	56169	1	56169
222	49284	0	0
207	42849	0	0
192	36864	1	36864
177	31329	0	0
162	26244	1	26244
147	21609	0	0
132	17424	2	34848
117	13689	4	54756
102	10404	9	93636
87	7569	13	98397
72	5184	20	103680
57	3249	34	110466
42	1764	47	82908
27	729	82	59778
12	144	36	5184
			$\sum fX^2 = 762930$

$$\bar{X} = \frac{\sum fX}{N} = \frac{11370}{250} = 45.48$$

$$s = \sqrt{\frac{N\sum fX^2 - (\sum fX)^2}{N(N-1)}} = \sqrt{\frac{250(762930) - (11370)^2}{250(249)}}$$

$$= \sqrt{987.24} = 31.42$$

The following is based on the frequency distribution presented in Exercise 3.7(c):

Midpoints	X^2	f	fX^2
224.5	50400.25	1	50400.25
174.5	30450.25	2	60900.50
124.5	15500.25	13	201503.25
74.5	5550.25	69	382967.25
24.5	600.25	165	99041.25
			$\sum fX^2 = 794812.50$

$$\bar{X} = \frac{11375}{250} = 45.5$$

$$s = \sqrt{\frac{250(794812.50) - (11375)^2}{250(249)}} = \sqrt{1113.45} = 33.37$$

7.10 Exercise 3.10: $\bar{X} = 37.00$, $s = 14.14$
Exercise 3.11: $\bar{X} = 37.40$, $s = 14.37$

7.11 a) $\bar{X} = 63.10$ b) Range = 35, Interquartile range = 8.00, M.D. = 5.61
c) $s = 8.09$, $s^2 = 65.46$

7.12 Age: $\bar{X} = 54.29$, $s = 5.97$; years lived: $\bar{X} = 14.469$, $s = 9.330$

7.13 $\bar{X} = 68.25$, $s = 7.68$

		January, 1965	January, 1966	May, 1965	May, 1966
7.16	a) $\bar{X}$	35.58	38.65	76.90	70.42
	b) Range	39	43	42	33
	M.D.	7.78	7.40	7.47	7.81
	c) s	9.88	9.55	9.68	9.39
	s^2	97.65	91.17	93.76	88.12

7.17 a) $\bar{X}_{1974} = 3.78$; $\bar{X}_{1973} = 4.40$
b) $s_{1974} = 0.83$ $s_{1973} = 1.08$
$s^2_{1974} = 0.68$ $s^2_{1973} = 1.17$

7.18 Murder: $\bar{X} = 7.79$ Rape: $\bar{X} = 21.23$
$s = 4.36$ $s = 9.37$
$s^2 = 18.97$ $s^2 = 87.73$

7.19 a) 10.78 b) 44; 8.435 c) $s^2 = 83.11$; $s = 9.12$

7.20 a) Crude range = 4; $s^2 = 1.68$; $s = 1.30$
b) Crude range = 18; $s^2 = 29.57$; $s = 5.44$

7.22 \$0.40; \$0.40

7.23 Exercise 5.25: mean = 40; $s^2 = 100$; $s = 10$
Exercise 5.26: mean = 30; $s^2 = 100$; $s = 10$

7.24 Barrow: $s^2 = 0.08$, $s = 0.28$
Burlington: $s^2 = 0.46$, $s = 0.68$
Honolulu: $s^2 = 1.39$, $s = 1.18$
Seattle-Tacoma: $s^2 = 3.63$, $s = 1.91$

*7.25 $s = 1.43$

CHAPTER 8

8.1 a) 0.94 b) −0.40 c) 0.00
d) −1.32 e) 2.23 f) −2.53

8.2 a) 0.4798 b) 0.4713 c) 0.0987 d) 0.1554
e) 0.4505 f) 0.4750 g) 0.4901 h) 0.4951

8.3 a) i) 0.3413; 341 ii) 0.4772; 477 iii) 0.1915; 192 iv) 0.4938; 494
b) i) 0.1587; 159 ii) 0.0228; 23 iii) 0.6915; 692 iv) 0.9938; 994 v) 0.5000; 500
c) i) 0.1359; 136 ii) 0.8351; 835 iii) 0.6687; 669 iv) 0.3023; 302

8.4 a) arithmetic: $z = 1.21$; verbal comprehension: $z = 0.77$; geography: $z = -0.29$

8.5 a) 40.13; 57.93 b) 60.26

8.6 (a), (b), (c), (d)

8.7 a) $z = 09.67$
$X = 63.96$
b) $z = 0.67$
$X = 80.04$
c) $z = 1.28$
$X = 87.36$
d) $z = 0.67$
25.14% score above
e) $z = -0.5$
30.85% score below
f) $z = -0.67$ and $z = 0.67$
63.96 − 80.04
g) $z = \pm 1.64$
below 52.32, above 91.68
h) $z = \pm 2.58$
below 41.04, above 102.96

8.8 $\mu = 72, \sigma = 8$

a) 66.64 b) 77.36
c) 82.24 d) $z = 1.00$
15.87% score above
e) $z = -0.75$
22.66% score below
f) 66.64 − 77.36
g) below 58.88, above 85.12 h) below 51.36, above 92.64

$\mu = 72, \sigma = 4$

a) 69.32 b) 74.68
c) 77.12 d) $z = 2.00$
2.28% score above
e) $z = -1.5$
6.68% score below
f) 69.32 − 74.68
g) below 65.44, above 78.56 h) below 61.68, above 82.32

$\mu = 72, \sigma = 2$

a) 70.66 b) 73.34
c) 74.56 d) $z = 4.00$
0.003% score above
e) $z = -3.00$
0.13% score below
f) 70.66 – 73.34
g) below 68.72, above 75.28 h) below 66.84, above 77.16

8.9 a) 1.24% b) 1240

8.10 a) 61.70% b) 6170

8.11 a) Carrots, $V = 20.00\%$, sirloin steak, $V = 6.06\%$
b) Loin lamb, $V = 13.64\%$, sirloin steak, $V = 6.06\%$
c) Carrots, $V = 20.00\%$, corn, $V = 7.14\%$

8.12 a)

Falls		Drowning		Fires	
$\bar{X} =$	1375.67	$\bar{X} =$	727.08	$\bar{X} =$	541.92
$s =$	90.59	$s =$	533.28	$s =$	215.88
$s^2 =$	8206.55	$s^2 =$	284,383.90	$s^2 =$	46,602.27

b) Drowning: $V = 73.35\%$
Falls: $V = 6.59\%$

8.13 January, 1965, $V = 27.77\%$
May, 1965, $V = 12.59\%$

8.15 a)

	National League	American League
$\bar{X}$	44.25	41.42
s	4.41	7.73

c) American League, $V = 18.66\%$

8.16 Barrow: 77.78% Burlington: 24.46%
Honolulu: 64.84% Seattle-Tacoma: 58.77%

8.17 a) $\mu = 40; \sigma = 10.21$

b)

A.	1.47	F.	0	K.	0.98	P.	0	U.	0
B.	0.98	G.	1.96	L.	0.49	Q.	−0.49	V.	−0.98
C.	−0.49	H.	0.49	M.	−1.47	R.	−1.47	W.	0.49
D.	0.49	I.	0.98	N.	−0.49	S.	−1.96	X.	−0.49
E.	0	J.	−0.98	O.	1.47	T.	−0.98	Y.	0

c)

A.	0.9292	F.	0.50	K.	0.8365	P.	0.50	U.	0.50
B.	0.8365	G.	0.9750	L.	0.6879	Q.	0.3121	V.	0.1635
C.	0.3121	H.	0.6879	M.	0.0708	R.	0.0708	W.	0.6879
D.	0.6879	I.	0.8365	N.	0.3121	S.	0.0250	X.	0.3121
E.	0.50	J.	0.1635	O.	0.9292	T.	0.1635	Y.	0.50

d) i) 42.92% ii) 9.27% iii) 95.00%

8.18 a) $\mu = 30; \sigma = 10.21$
b) c) d) Same as Exercise 7.17

8.19 a) 25.52 b) 34.03

8.20 a) 0 b) 1.60 c) −1.07
d) −1.60 e) 0 f) 2.13

8.21 a) Males: $\bar{X} = 63.723$ Females: $\bar{X} = 68.719$
$s = 8.890$ $s = 10.784$

b) Females: $V = 15.69$

c)

		Males	Females
i)	U.S.A.:	Males, $z = 0.42$	Females, $z = 0.60$
ii)	Mexico:	Males, $z = -0.48$	Females, $z = -0.49$
iii)	Sweden:	Males, $z = 0.93$	Females, $z = 0.81$
iv)	India:	Males, $z = -2.45$	Females, $z = -2.60$

8.22 a) Birth rate: $\bar{X} = 15.87$ Death rate: $\bar{X} = 9.05$
$s = 3.10$ $s = 1.46$

b)

		Birth rate	Death rate
i)	Alaska	$z = 1.58$	$z = -3.13$
ii)	Connecticut	$z = -1.32$	$z = -0.40$
iii)	Florida	$z = -0.81$	$z = 1.27$
iv)	Pennsylvania	$z = -1.10$	$z = 0.80$
v)	Utah	$z = 3.29$	$z = -1.67$

CHAPTER 9

9.1 $r = 0.8466$

9.2 $r = 0.9107$

9.3 $r_{\text{rho}} = 0.906$

9.5 a) Nonlinear relationship   b) Truncated range   c) Truncated range

9.6 $r = 0.8429$

9.7 $r = 0$; $r_{\text{rho}} = 0.50$

9.8 $\sum xy = \sum(X - \bar{X})(Y - \bar{Y}) = \sum(XY - X\bar{Y} - Y\bar{X} + \bar{X}\bar{Y})$
$= \sum XY - \sum X\bar{Y} - \sum Y\bar{X} + N\bar{X}\bar{Y}$
Since $\sum X = N\bar{X}$ and $\sum Y - N\bar{Y}$, therefore:
$$\sum xy = \sum XY - N\bar{X}\bar{Y} - N\bar{X}\bar{Y} + N\bar{X}\bar{Y} = \sum XY - N\bar{X}\bar{Y}$$
$$= \sum XY - N\left(\frac{\sum X}{N}\right)\left(\frac{\sum Y}{N}\right) = \sum XY - \frac{(\sum X)(\sum Y)}{N}$$

9.9 $r_{\text{rho}} = -0.8571$

9.10 a) $r = -0.1659$ b) $r = 0.4327$

9.11 $r = -0.7728$

9.12 $r = 0.7614$

9.13 a) $r = -0.8924$ b) $r = -0.8720$ c) $r = 0.9729$

9.14 a) $r_{\text{rho}} = -0.8782$ b) $r_{\text{rho}} = -0.8723$ c) $r_{\text{rho}} = 0.9557$

9.15 a) $r = 0.1068$ b) $r = 0.0734$
c) Jan., 1965, $r_{\text{rho}} = -0.1700$
Jan., 1966, $r_{\text{rho}} = -0.7253$
May, 1965, $r_{\text{rho}} = 0.1321$
May, 1966, $r_{\text{rho}} = 0.5968$

9.16 a) $r_{\text{rho}} = 0.1366$ b) $r_{\text{rho}} = 0.0768$

9.17 a) $r = 0.1412$

b)

Percentage of cost per share	Cost per share
3.82	34
2.43	23
1.14	7
1.50	6
4.17	48
8.86	22
6.32	28
0.35	20
7.85	46
7.59	22
4.43	21
6.71	14
6.22	27
2.10	41
3.68	19
7.00	41
5.78	37
3.00	28
8.17	29
2.20	15
2.57	14
3.92	40
10.13	30
14.38	8
1.55	44
6.64	11
2.03	34
0.46	35
5.18	17
7.62	16
3.50	42
4.00	40
7.34	50
1.94	18
8.58	45
4.68	34
8.75	12

c) $r = -0.0356$

9.18 a) $r = 0.0785$

b) National League, $r_{\text{rho}} = -0.2074$
American League, $r_{\text{rho}} = 0.0330$

9.19 $r = 1$

9.20 b) $r = 0.8729$

9.21 $r = 0.9515$

9.22 $r = 0.9935$

9.24 $r_{\text{rho}} = -0.9705$

9.25 a) $r = -0.9536$ b) $r_{\text{rho}} = -0.9406$

9.26 $r = 0.330$

9.27 $r = -0.598$

9.28

		D	D^2
1	8	7	49
2	7	5	25
3	6	3	9
4	5	1	1
5	4	1	1
6	3	3	9
7	2	5	25
8	1	7	49
			$\sum D^2 = 504$

9.29

		D	D^2
1	1	0	0
2	2	0	0
3	3	0	0
4	4	0	0
5	5	0	0
6	6	0	0
7	7	0	0
8	8	0	0
			$\sum D^2 = 0$

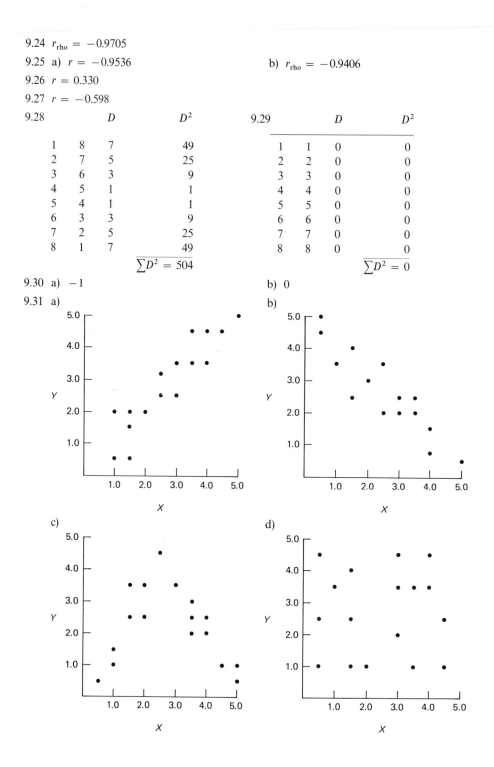

9.30 a) -1 b) 0

9.31 a)

b)

c)

d)

9.32 a) c b) a c) d d) b

9.33 $r = 0.99$

9.34 a) $r = 0.46$ b) $r = -0.03$ c) $r = -0.59$

9.35 a) $r = 0.56$ b) $r = 0.36$ c) $r = 0.52$

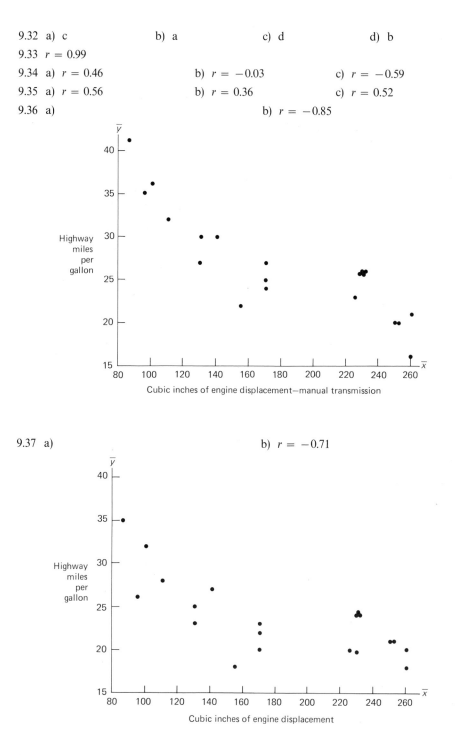

9.36 a) b) $r = -0.85$

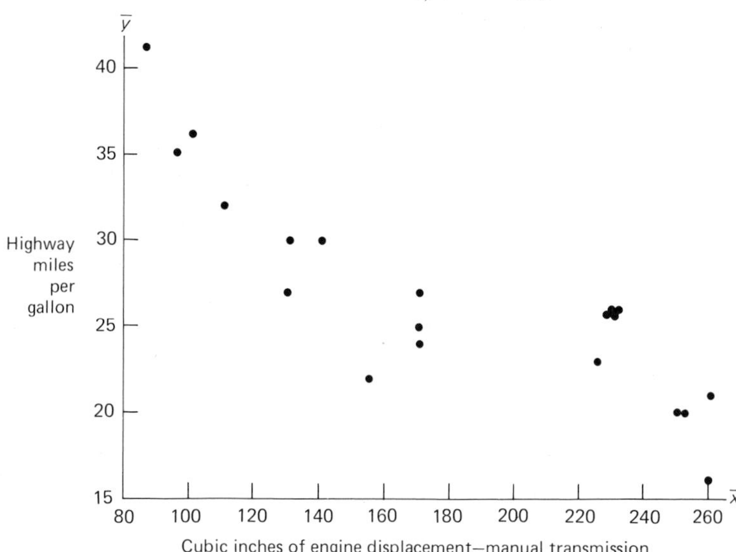

9.37 a)

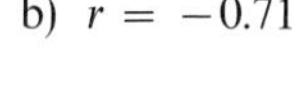
b) $r = -0.71$

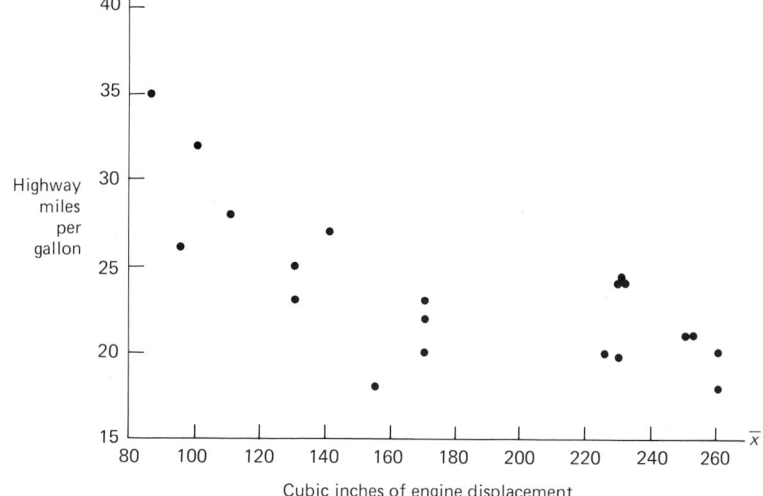

CHAPTER 10

10.1 $Y' = 3 - 0.9\,(X - 3)$

10.2 a) 1.72 b) 118.48 c) $s_{\text{est }y} = 0.3928$

10.3 a) 1.59 b) $s_{\text{est }y} = 0.47$; $s_{\text{est }x} = 11.20$ c) 0.1296

10.4 As r increases the angle formed by the regression lines decreases. In the limiting cases ($r = \pm 1.00$), the regression lines are superimposed upon each other. When $r = 0$, the regression lines are at right angles to each other.

10.5 9.0
The standard deviation for a group with the same I.Q. is the standard error of estimate.
$s_{\text{est }y} = s_y\sqrt{1 - r^2} = 9$

10.6 a) 0 b) 0.60 c) 1.20
d) 1.5 e) -0.75 f) -1.20

10.7 a) $Y' = 89.17$
b) The proportion (p) of area corresponding to a score of 110 or higher is 0.0031
c) $X = 31.54$ d) $p = 0.4129$ e) $p < 0.00003$
f) $X = 68.46$ g) 500, 93, 93, 907

10.8 a) $r = 0.9415$ b) 50, 40, 60, 70

10.10 a) 33.34 b) $p = 0.1587$ c) 48.46 d) $p = 0.8980$
e) 55.34 f) $p = 0.2709$ g) $p = 0.2981$

10.11 a) $r = -0.8469$
b) i) 4.24 ii) $p = 0.0038$

10.12 b) $r = 0.9620$
d) Chile: 432.97 Ireland: 832.25 Belgium: 846.52

10.13 29.774

10.14 $0.4348

10.16 a) 55 b) 5 c) 20

10.17 a) 34.248 b) -0.725 c) 7.047 d) 4.498

10.18 a) 86.275 b) 91.177 c) 56.863 d) 5.052

10.19 a) $r = 0.9584$
b) y for 3 = 3.222 y for 4 = 4.139
y for 5 = 5.055 y for 6 = 5.972
y for 7 = 6.889 y for 8 = 7.806
y for 9 = 8.722 y for 10 = 9.639
y for 11 = 10.556
c) 54.889 d) 4.473 e) 50.424 f) 0.919

10.20 a) 5.56 b) 4.16 c) 0.857

10.21 a) $r = -0.87$
b) Hornet, 27.39; Monaco Wagon, 9.35; Corolla, 20.31
c) $r = -0.87$
d) Hornet, 25.59; Monaco Wagon, 12.41; Corolla, 30.39

REVIEW OF SECTION I

1 a)

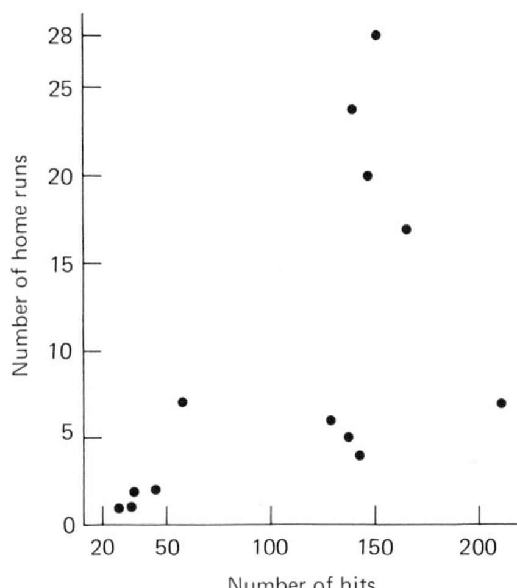

b) i) 9.46 ii) 139 iii) 53.12

c) i) $r = 0.58$ ii) $r_{\text{rho}} = 0.6827$

2 a)

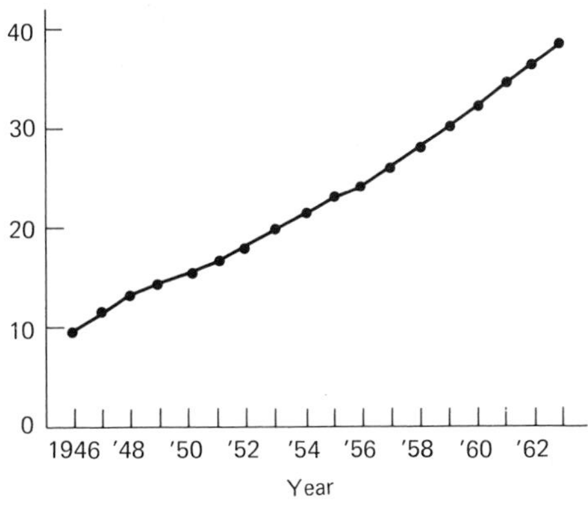

b)

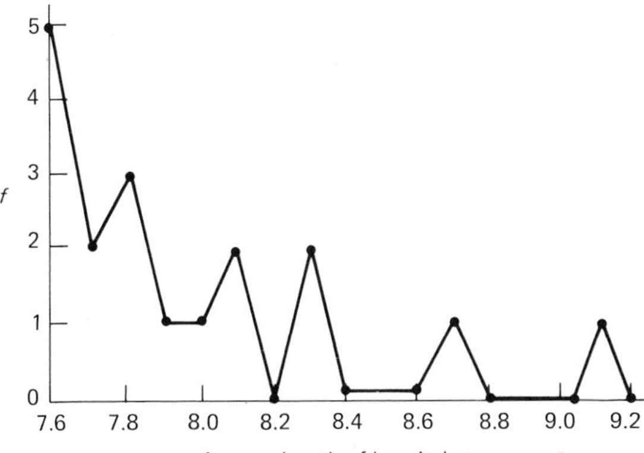

CHAPTER 11

11.1 a) 0.1250 b) 0.1250 c) 0.3750 d) 0.5000

11.2 a) 0.0192 b) 0.0769 c) 0.3077 d) 0.4231

11.3 Exercise 11.1: a) 7 to 1 against, b) 7 to 1 against
Exercise 11.2: a) 51 to 1 against, b) 12 to 1 against

11.4 a) 0.1667 b) 0.1667 c) 0.2778 d) 0.5000

11.5 a) 0.0046 b) 0.0278 c) 0.0139 d) 0.0556 e) 0.5787

11.6 a) 0.0129 b) 0.4135 c) 0.0100

11.7 a) 0.0156 b) 0.4219 c) 0.0123

11.8 a) 0.0838 b) 126.40 c) <68.64 and >131.36
d) 0.4648 e) i) 0.0070 ii) 0.2160 iii) 0.0390

11.9 (b)

11.10 $p = \frac{1}{24}$

11.11 a) $p = 0.20$ b) $p = 0.04$ c) $p = 0.80$ d) $p = 0.64$

11.12 a) 0.6561 b) 0.0001 c) 0.3438

11.13 a) $p = 0.0456$ b) $p = 0.0021$ c) $p = 0.9544$
d) $p = 0.9109$ e) $p = 0.0228$ f) $p = 0.0520$

11.14 a) $\frac{1}{3}$ b) $\frac{1}{2}$ c) $\frac{1}{6}$ d) 0

11.15 a) $\frac{1}{11}$ b) $\frac{1}{66}$ c) $\frac{14}{99}$

11.16 a) $\frac{1}{8}$ b) $\frac{1}{36}$ c) $\frac{16}{81}$

11.17 a) 40% b) 12% c) 16% d) 3.36%

11.18 a) $p = 0.3085$ b) $p = 0.2266$ c) $p = 0.4649$
d) $p = 0.2564$ e) $p = 0.3111$ f) $p = 0.0947$

11.19 a) Mean = 2; $s^2 = 1.2$ c) 0.1; 0.2; 0.4; 0.2; 0.1

11.20 a) 0.02 b) 0.02 c) 0.01 d) 0.16
e) 0.01 f) 0.04 g) 0.01 h) 0.01

11.21 a)

0.01	0.02	0.04	0.02	0.01
0.02	0.04	0.08	0.04	0.02
0.04	0.08	0.16	0.08	0.04
0.02	0.04	0.08	0.04	0.02
0.01	0.02	0.04	0.02	0.01

b) 1.00 c) Yes

11.22

4.0	3.5	3.0	2.5	2.0
3.5	3.0	2.5	2.0	1.5
3.0	2.5	2.0	1.5	1.0
2.5	2.0	1.5	1.0	0.5
2.0	1.5	1.0	0.5	0.0

11.23

Mean	f	p
4.0	1	0.04
3.5	2	0.08
3.0	3	0.12
2.5	4	0.16
2.0	5	0.20

Mean	f	p
1.5	4	0.16
1.0	3	0.12
0.5	2	0.08
0.0	1	0.04

11.26 a) 0.12 b) 0.96 c) 0.04 d) 0.72 e) 0.48

11.27 Mean = 2; $s^2 = 1.0$

11.28 c

11.29 b

11.30 a) 0.5 b) 0.1587 c) 0.3413 d) 0.1359
e) 0.5 f) 0.0002 g) 0.0228

11.31 a) 0.25 b) 0.02 c) 0.02 d) 0.25

11.32 0.0057

11.33 0.00317

11.34 a) 0.00104 b) 0 c) 0.052 d) 0.01254

*11.35 a)

0.25
0.20
0.15
0.10
0.05
0.00
Probability
0.2500
0.1875
0.1875
0.1250
0.1250
0.0625
0.0625
1.0 2.5 4.0 5.5 7.0 8.5 10.0
Sample mean when n = 2

b)

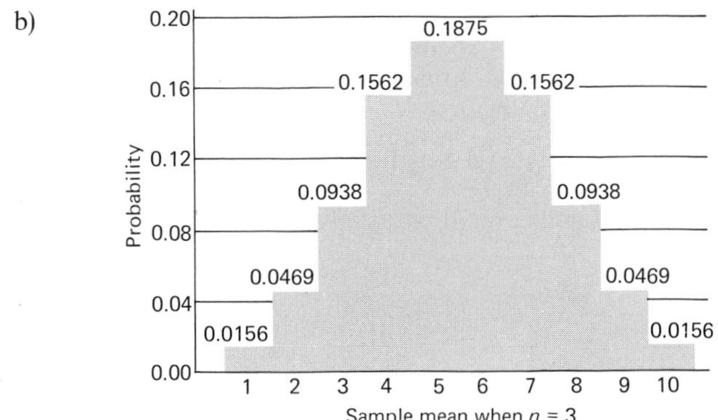

*11.36 An extreme deviation is more rare as sample size increases.

*11.37 a) 0.0625; 0.0156 b) 0.1250; 0.0312 c) 0.3750; 0.3125 d) 0.7500; 0.6250

*11.38 a) 0.6250 b) 0.1875 c) 0.2500

*11.39 a) 0.6874 b) 0.3750 c) 0.1407

*11.40 (a) and (b)

$\bar{X}$	f	$p(\bar{X})$
4	1	0.0123
3	4	0.0494
2	10	0.1235
1	16	0.1975
0	19	0.2346
−1	16	0.1975
−2	10	0.1235
−3	4	0.0494
−4	1	0.0123
	$N_{\bar{X}} = 81$	$\sum p(\bar{X}) = 1.0000$

c) Mean = 0.00; standard deviation = 1.64

*11.41 a) 0.2346 b) 0.6296 c) 0.0123 d) 0.0246
e) 0.0617 f) 0.0617 g) 0.1234 h) 0.8766

CHAPTER 12

12.2 a) In the event a critical theoretical issue is involved, the commission of a type I error might lead to false conclusion concerning the validity of the theory.
b) In many studies in which the toxic effects of new drugs are being studied, the null hypothesis is that the drug has no adverse effects. In the event of a type II error, a toxic drug might be mistakenly introduced into the market.

12.3 The null hypothesis cannot be proved. Failure to reject the null hypothesis does not constitute proof that the null hypothesis is correct.

12.4 All experiments involve two statistical hypotheses: the null hypothesis and the alternative hypothesis. One designs the experiment so that the rejection of the null hypothesis leads to affirmation of the alternative hypothesis.

12.5 H_0: There is no difference in running speed of organisms operating under different drive levels.
H_1: There is a difference in running speed of organisms operating under different drive levels.

12.6 a) $\frac{1}{1024}$ or 0.00098 b) 0.055 c) 17.28 to 1 against passing

12.7 a) H_0 b) H_1 c) H_1 d) H_0

12.8 a) Type II b) Type I c) No error d) No error

12.9 a) $p = 0.062$ b) $p = 0.773$ c) $p = 0.124$ d) $p = 0.938$

12.10 a) Type II b) Type II c) no error

12.11 It is more likely that she is accepting a false H_0 since the probability that the observed event occurred by chance is still quite low ($p = 0.02$).

12.12 By chance, one would expect five differences to be statistically significant at the 0.05 level.

12.13 By chance, one would expect five (give or take a few) statistically significant differences when $\alpha = 0.01$.

12.14 a) directional b) null c) directional d) nondirectional
e) null f) directional g) directional

12.15 b

12.16 male $= \frac{1}{2}$; female $= \frac{1}{2}$; 1:1

12.17 0.038

12.18 No

12.19 a) 1 b) 10 c) 5 d) 60
e) 50 f) 95 g) 4 h) 100

12.20 a) h b) a

12.21 a) h, Type I b) a, Type II

CHAPTER 13

13.1 a) As sample size increases, the dispersion of sample means decreases.
b) As you increase the number of samples, you are more likely to obtain extreme values of the sample mean. For example, suppose the probability of obtaining a sample mean of a given value is 0.01. If the number of samples drawn is 10, you probably will not obtain any sample means with that value. However, if you draw as many as 1,000 samples, you would expect to obtain approximately 10 sample means with values so extreme the probability of their occurrence is 0.01.

13.2 Suppose we have a population in which the mean = 100 and the standard deviation = 10. The probability of obtaining scores as extreme as 80 or less or 120 or more (± 2 standard deviations) is 0.0456 or less than five in a hundred. The probability of obtaining scores even more extreme (e.g., 50, 60, 70 or 130, 140, 150) is even lower. Thus we would expect very few (if any) of these extreme scores to occur in a given sample. Since the value of the standard deviation is a direct function of the number of extreme scores, the standard deviation of a sample will usually underestimate the standard deviation of the population.

13.3 a) 21.60–26.40 b) 20.66–27.34

13.4 a) 23.40–24.60 b) 23.15–24.85

13.5 Reject H_0; $z = 2.68$ in which $z_{0.01} = \pm 2.58$

13.6 Accept H_0; $t = 2.68$ in which $t_{0.01} = \pm 2.83$, df = 21

13.7 In Exercise 13.5 the value of the population standard deviation is known; thus we may employ the z-statistic and determine probability values in terms of areas under the normal curve. In Exercise 13.6 the value of the population standard deviation is not known and must be estimated from the sample data. Thus the test statistic is the t-ratio.

Since the t-distributions are more spread out than the normal curve, the proportion of area beyond a specific value of t is greater than the proportion of area beyond the corresponding value of z. Thus a larger value of t is required to mark off the bounds of the critical region of rejection. In Exercise 13.5, the absolute value of the obtained z must equal or exceed 2.58. In Exercise 13.6, the absolute value of the obtained t must equal or exceed 2:83.

To generalize: the probability of making a type II error is less when we know the population standard deviation.

13.8 Reject H_0; $t = 2.13$ in which $t_{0.05} = 1.83$ (one-tailed test), df $= 9$

13.9 a) $z = 2.73$, $p = 0.9968$ b) $z = -0.91$, $p = 0.1814$
c) $z = 1.82$, $p = 0.0344$ d) $p = 0.6372$

13.10 a) 36.63–43.37 b) 35.43–44.57

13.11 a) $z = 1.66$ b) $z = 7.64$

13.12 $t = 2.08$, df $= 24$

13.13 26.05–33.17

13.14 $t = 1.75$, df $= 62$

13.15 263.34–273.66

13.17 a) $p = \frac{1}{15}$ b) $p = \frac{1}{15}$ c) $p = \frac{8}{15}$ d) $p = 0$

13.18 a) $p = \frac{1}{12}$ b) $p = \frac{1}{12}$ c) $p = \frac{5}{9}$ d) $p = \frac{1}{36}$

13.19 $t = 1.83$, df $= 7$

13.20 $t = 2.38$, df $= 9$

13.21 The conclusion we make depends upon the nature of the alternative hypothesis. If H_1 is directional and we employ $\alpha = 0.05$, we reject H_0 since the obtained r_{rho} of 0.438 exceeds the value required (interpolating for $n = 17$, the critical value of r_{rho} for $\alpha = 0.05$, one-tailed test is 0.412). However, if we employ a two-tailed test, we must accept H_0.

13.22 The critical value of r_{rho} for $n = 25$ ($\alpha = 0.05$, two-tailed test) may be obtained by interpolating: 0.409 for $n = 24$, 0.392 for $n = 26$; thus the critical value for $n = 25$ is 0.400. Since the absolute value of the obtained r_{rho} is less than the critical value, we accept H_0.

13.23 a) $z = 1.11$ b) $z = 1.99$

13.24 $z = 4.23$

13.25 $z = 6$

13.26 $z = 1.4$

13.28 $t = 4.28$

13.29 $r_{\text{rho}}(0.05) = 0.450$

13.30 $t = 4$

CHAPTER 14

14.1 $t = 2.79$, reject H_0

14.2 $t = 1.07$, accept H_0

14.3 $t = 0.80$, accept H_0

14.4 a) $t = 3.52$, reject H_0 b) $F = 3.00$, accept H_0

14.5 a) $z = 1.41, p = 0.0793$ b) $z = -2.12, p = 0.9830$
c) $z = -2.12, p = 0.0170$ d) $z = -5.66, p < 0.00003$

14.6 a) $z = -0.71, p = 0.7611$ b) $z = -1.06, p = 0.8554$
c) $z = -1.41, p = 0.9207$ d) $z = -1.77, p = 0.9616$

14.7

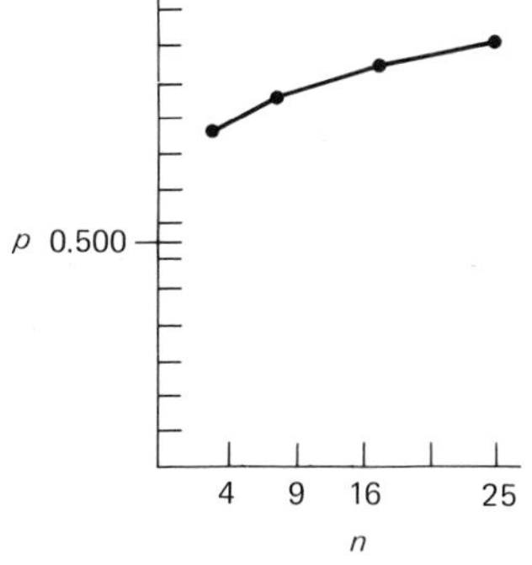

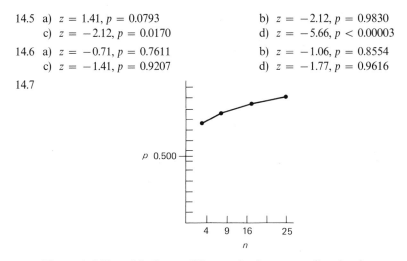

The probability of finding a difference in the correct direction between sample means increases as n increases.

14.8 $t = 2.15$, df $= 14$

14.9 $t = 2.91$, df $= 18$

14.10 $t = 2.30$, df $= 26$

14.11 $t = 1.36$, df $= 25$

14.12 $t = 1.28$, df $= 162$

14.13 $t = 2.30$, df $= 28$

14.14 $t = 0.60$, df $= 34$

14.15 $t = 0.67$, df $= 20$

14.16 a) False b) True c) False

14.17

$\bar{X}_1 - \bar{X}_2$	f
5	1
4	2
3	5
2	6
1	8
0	6
−1	5
−2	2
−3	1

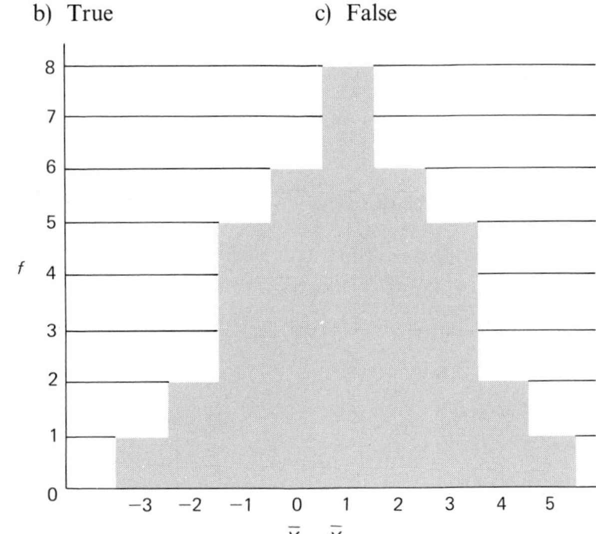

14.18 a) $p = 0.7778$ b) $p = 0.6111$ c) $p = 0.3889$ d) $p = 0.2222$
e) $p = 0.0833$ f) $p = 0$ g) $p = 0.4722$

14.19 $t = -5.40$

14.20 $z = 3.33$

14.21 $t = 24.50$

14.22 $t = 5.02$

14.23 $F = 1.66$

14.24 $t = 1.76$

14.25 $z = 2.29$

14.26 $t = -6.88$

14.27 $F = 1.27$

14.28 $t = 1.63$, df = 8

14.29 $t = 0.77$, df = 8

14.30 a) $t = 4.60$, df = 18 b) $t = 8.00$, df = 18
c) $t = 2.02$, df = 18 d) $t = 6.78$, df = 18

CHAPTER 15

15.1 $t = 3.027$, df = 15; reject H_0.

15.2 $t = 2.08$, df = 32; accept H_0.

15.3 $t = 1.571$, $A = 0.465$, df = 9; accept H_0.

15.4 r_{rho} between final standing and homeruns in American League is 0.04; for National League it is 0.36. Both correlations are too low to justify use of correlated samples.

15.5 $t = 1.590$, df = 18, accept H_0.
The obtained t-ratio is closer to the rejection region because the number of degrees of freedom is larger (18 as compared to 9) when we employ independent samples.

15.6 a) Greater risk of a type II error because of the loss of degrees of freedom.
b) Less risk of a type II error because of the greater sensitivity of the standard error of the difference.

15.7 a) $A = 0.048$ b) $A = 0.048$ c) $A = 0.778$

15.8 $A = 0.199$

15.9 Manufacturer A: $t = \dfrac{104.82 - 100}{2.23} = 2.16$, df = 10

Manufacturer B: $t = \dfrac{101.91 - 100}{1.52} = 1.26$, df = 10

Employing $\alpha = 0.05$ (two-tailed test), we accept H_0 in both cases.

15.10 Manufacturer A: $A = 0.195$
Manufacturer B: $A = 0.573$

15.11 $A = 0.216$

15.12 $A = 5.480$

15.13 $A = 0.226$

15.14 $A = 0.230$

15.15 $A = 0.422$

15.16 $A = 1.625$

15.17 $t = 8.22$

15.18 $A = 0.3056$

15.19 $A = 0.6406$

15.20 $A = 12.781$

15.21 $A = 0.0457$

15.22 $A = 0.3696$

15.23 $A = 19$

15.24 $A = 0.295$

15.25 $A = 5.5$

CHAPTER 16

16.1 $F = 47.02$; reject H_0.

16.2 All comparisons are significant at 0.01 level. It is clear that death rates are lowest in summer, next to lowest in spring, next to highest in fall, and highest in winter.

16.3 t (winter vs. spring) = 7.540, t (spring vs. summer) = 3.889,
t (winter vs. summer) = 11.429, t (spring vs. fall) = 3.095,
t (winter vs. fall) = 4.44, t (summer vs. fall) = 6.984
All comparisons are significant at 0.01 level. It is clear that death rates are lowest in summer, next to lowest in spring, next to highest in fall, and highest in winter.

16.4 $F = 0.643$

16.5 When we employ the 0.05 level, one out of every twenty comparisons will be significant *by chance.* Thus a significant difference at the 0.05 level in one of ten studies can be attributed to chance.

16.6 $F = 0.30$, df = 3/12

16.7 $F = 10.43$, df = 2/12

16.8 $F = 12.99$, df = 5/18

16.9

Source of variation	Sum squares	Degrees of freedom	Variance estimate	F
Between groups	1.083	2	0.5415	0.6228
Within groups	13.042	15	0.8695	
Total	14.125	17		

16.10

Source of variation	Sum squares	Degrees of freedom	Variance estimate	F
Between groups	12.067	1	12.067	3.102
Within groups	108.934	28	3.8905	
Total	121.0	29		

16.11

Source of variation	Sum squares	Degrees of freedom	Variance estimate	F
Between groups	919.793	3	306.598	6.43
Within groups	954.167	20	47.708	
Total	1873.96	23		

16.12

Source of variation	Sum squares	Degrees of freedom	Variance estimate	F
Between groups	26.80	2	13.4	9.348
Within groups	17.20	12	1.43	
Total	44.00	14		

16.13 All comparisons are significant at 0.01 level. It is clear that Group C is superior, next Group B, followed by Group A.

16.14

Source of variation	Sum squares	Degrees of freedom	Variance estimate	F
Between groups	285.431	2	142.716	1.606
Within groups	2843.312	32	88.854	
Total	3128.743	34		

16.15

Source of variation	Sum squares	Degrees of freedom	Variance estimate	F
Between groups	15.8645	3	5.288	15.57
Within groups	6.7917	20	0.3396	
Total	22.6562	23		

Comparison of A and B: HSD $= 1.19$ for $\alpha = 0.01$.

CHAPTER 17

17.1 a) $p = 0.070$ b) $p = 0.020$ c) $p = 0.040$

17.2 a) $p < 0.0003$ b) $p = 0.0003$ c) $p = 0.1335$

17.3 $P = 0.05, Q = 0.95, N = 10$

$$p(1) = \frac{10!}{1!9!}(0.05)^1(0.95)^9$$
$$= 0.3151$$

17.4 Inflated N

17.5 $\chi^2 = 6.00$, df $= 3$

17.6 Binomial Test $N = 8, \chi = 6, P = \frac{1}{3}, Q = \frac{2}{3}$; reject H_0 at $\alpha = 0.05$

17.7 $\chi^2 = 1.66$, df $= 1$

17.8 $P = 0.20(\frac{1}{5}), Q = 0.80(\frac{4}{5}), N = 6$

	6	5	4	3	2	1	0
$(P + Q)^6$	$\frac{1}{15625}$	$\frac{24}{15625}$	$\frac{240}{15625}$	$\frac{1280}{15625}$	$\frac{3840}{15625}$	$\frac{6144}{15625}$	$\frac{4096}{15625}$
	0.000	0.002	0.015	0.082	0.246	0.393	0.262

17.9 $\chi^2 = 1.789$, df = 1

17.10 $f_o = 5 \quad 15$
$f_e = 10 \quad 10$

$$\lambda^2 = \frac{(5-10)^2}{10} + \frac{(15-10)^2}{10} = 5.00$$

The value of z^2 (without correcting for continuity) is:

$$z^2 = \left(\frac{(5-10)^2}{\sqrt{5.00}}\right) = 5.00 = \lambda^2.$$

17.11 $\chi^2 = 6.00$, df = 3

17.12 $\chi^2 = 3.60$, df = 3

17.13 $z = 1.327$, $\chi^2 = 1.761$

17.14 $\chi^2 = 19.3$, df = 1

17.15 $\chi^2 = 12.5$, df = 2

17.16 $\chi^2 = 3.5$, df = 2

17.17 $\chi^2 = 7.86$, df = 5

17.18 $z = 2.108$

17.19 $\chi^2 = 4.44$, df = 1

17.20 $\chi^2 = 12.81$, df = 2

17.21 $\chi^2 = 9.67$, df = 1

CHAPTER 18

18.1 a) Sign test $N = 10$, $x = 3$
b) $N = 10$, $T = 68$
c) $U = 23$, $U' = 77$

18.2 $U = 9.5$, $U' = 62.5$

18.3 $U = 18$, $U' = 103$

18.4 $T = 11.5$

18.5 $T = 57$

18.6 $U = 25$, $U' = 75$

18.7 Sign test = 0.754
Wilcoxon, $T = -12$

18.8 $T = -7.5$

18.9 $U = 86.5$, $U' = 13.5$

REVIEW OF SECTION II—INFERENTIAL STATISTICS

A. Parametric Tests

1. a) $\mu = 4$, $\sigma = 1.15$

b)

$\bar{X}$	f
2	1
2.5	4
3	10
3.5	16
4	19
4.5	16
5	10
5.5	4
6	1

c)

$\bar{X}$	p
2	0.0123
2.5	0.0494
3	0.1235
3.5	0.1975
4	0.2346
4.5	0.1975
5	0.1235
5.5	0.0494
6	0.0123

d) 0.82

e) 0.82

f) i) 0.0123; ii) 0.1852; iii) 0.3704

2. a) $t = 1.856$, df $= 8$; accept H_0. b) $t = 2.727$, df $= 18$; reject H_0.
3. a) $A = 0.210$, df $= 4$; reject H_0. b) $A = 0.111$, df $= 9$; reject H_0.
4. $F = 1.64$, df $= 2/27$

B. Nonparametric Tests

1. $U = 54.5$, $U' = 89.5$
2. Sign test $N = 12$, $x = 2$
3. $T = 11$

Index